Lecture Notes in Computer Science 10553

Commenced Publication in 1973
Founding and Former Series Editors:
Gerhard Goos, Juris Hartmanis, and Jan van Leeuwen

More information about this series at http://www.springer.com/series/7412

M. Jorge Cardoso · Tal Arbel et al. (Eds.)

Deep Learning in Medical Image Analysis and Multimodal Learning for Clinical Decision Support

Third International Workshop, DLMIA 2017
and 7th International Workshop, ML-CDS 2017
Held in Conjunction with MICCAI 2017
Québec City, QC, Canada, September 14, 2017
Proceedings

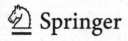

Springer

Editors
M. Jorge Cardoso
University College London
London
UK

Tal Arbel
McGill University
Montreal, QC
Canada

Workshop Editors *see next page*

ISSN 0302-9743   ISSN 1611-3349   (electronic)
Lecture Notes in Computer Science
ISBN 978-3-319-67557-2   ISBN 978-3-319-67558-9   (eBook)
DOI 10.1007/978-3-319-67558-9

Library of Congress Control Number: 2017953400

LNCS Sublibrary: SL6 – Image Processing, Computer Vision, Pattern Recognition, and Graphics

Printed on acid-free paper

This Springer imprint is published by Springer Nature
The registered company is Springer International Publishing AG
The registered company address is: Gewerbestrasse 11, 6330 Cham, Switzerland

Workshop Editors

Third International Workshop on Deep Learning in Medical Image Analysis, DLMIA 2017

Gustavo Carneiro ⓘ
University of Adelaide
Adelaide, SA
Australia

Jacinto C. Nascimento
Instituto Superior Técnico
Lisboa
Portugal

João Manuel R.S. Tavares ⓘ
Universidade do Porto
Porto
Portugal

Jaime S. Cardoso
Universidade do Porto
Porto
Portugal

Andrew Bradley
University of Queensland
Brisbane, QLD
Australia

Vasileios Belagiannis
University of Oxford
Oxford
UK

João Paulo Papa
Universidade Estadual Paulista
Bauru
Brazil

Zhi Lu
University of South Australia
Adelaide, SA
Australia

7th International Workshop on Multimodal Learning for Clinical Decision Support, ML-CDS 2017

Tanveer Syeda-Mahmood
IBM Research – Almaden
San Jose, CA
USA

Anant Madabhushi
Case Western Reserve University
Cleveland, OH
USA

Hayit Greenspan
Tel Aviv University
Tel Aviv
Israel

Mehdi Moradi
IBM Research – Almaden
San Jose, CA
USA

Preface DLMIA 2017

Welcome to the third MICCAI workshop on Deep Learning in Medical Image Analysis (DLMIA). Deep learning methods have experienced an immense growth in interest from the medical image analysis community because of their ability to process very large training sets, to transfer learned features between different databases, and to analyse multimodal data. Deep Learning in Medical Image Analysis (DLMIA) is a workshop dedicated to the presentation of work focused on the design and use of deep learning methods in medical image analysis applications. We believe that this workshop is setting the trend and identifying the challenges facing the use of deep learning methods in medical image analysis. For the keynote talks, we invited Associate Prof. Christopher Pal from the École Polytechnique de Montréal, Dr. Kevin Zhou from Siemens Healthineers, and Dr. Ronald M. Summers from the National Institutes of Health (NIH), who are prominent researchers in the field of deep learning in medical image analysis.

The first call for papers for the 3rd DLMIA was released on 25 March 2017 and the last call was done on 7 June 2017, with the paper deadline set to 16 June 2017. The submission site of DLMIA received 82 paper registrations, from which 73 papers turned into full-paper submissions, where each submission was reviewed by between two and five reviewers. The chairs decided to select 38 out of the 73 submissions, based on the scores and comments made by the reviewers (i.e., a 52% acceptance rate). The top 14 papers with the best reviews were selected for oral presentation and the remaining 24 accepted papers for poster presentation. We would like to acknowledge the financial support provided by Nvidia and the Butterfly Network for the realization of the workshop.

Finally, we would like to acknowledge the support from the Australian Research Council for the realization of this workshop (discovery project DP140102794). We would also like to thank the reviewers of the papers.

September 2017

Gustavo Carneiro
Joao Manuel R.S. Tavares
Andrew Bradley
Joao Paulo Papa
Jacinto C. Nascimento
Jaime S. Cardoso
Vasileios Belagiannis
Zhi Lu

Organization

Organizing Committee

Gustavo Carneiro	University of Adelaide, Australia
João Manuel R.S. Tavares	Universidade do Porto, Portugal
Andrew Bradley	University of Queensland, Australia
João Paulo Papa	Universidade Estadual Paulista, Brazil
Jacinto C. Nascimento	Instituto Superior Tecnico, Portugal
Jaime S. Cardoso	Universidade do Porto, Portugal
Vasileios Belagiannis	University of Oxford, UK
Zhi Lu	University of South Australia, Australia

Program Committee

Aaron Carass	Johns Hopkins University, USA
Adrian Barbu	Florida State University, USA
Adrien Depeursinge	HES-SO and EPFL, Switzerland
Albert Montillo	University of Texas Southwestern Medical Center, USA
Amarjot Singh	University of Cambridge, UK
Amin Khatami	Deakin University, Australia
Amir Jamaludin	University of Oxford, UK
Angshul Majumdar	Indian Institute of Technology Delhi, India
Ankush Gupta	University of Oxford, UK
Anne Martel	Sunnybrook Research Institute, Canada
Ariel Benou	Ben-Gurion University of the Negev, Israel
Ayelet Akselrod-Ballin	IBM-Research, Israel
Carlos Arteta	University of Oxford, UK
Carole Sudre	University College London, UK
Daguang Xu	Siemens Healthineers, USA
Dana Cobzas	University of Alberta, Canada
Daniel Worrall	University College London, UK
Daniela Iacoviello	Sapienza University of Rome, Italy
Dario Oliveira	IBM Research, Brazil
Deepak Mishra	Indian Institute of Technology Delhi, India
Diogo Pernes	INESC TEC and University of Porto, Portugal
Erik Smista	Norwegian University of Science and Technology, Norway
Evangelia Zacharaki	CentraleSupélec, France
Felix Achilles	Technical University of Munich, Germany
Gabriel Maicas	University of Adelaide, Australia
Ghassan Hamarneh	Simon Fraser University, Canada

Gregory Slabaugh	City University London, UK
Grzegorz Chlebus	Fraunhofer MEVIS, Bremen, Germany
Guilherme Aresta	INESC TEC and University of Porto, Portugal
Gustavo Souza	Federal University of Sao Carlos, Brazil
Haifang Qin	Peking University, China
Hariharan Ravishankar	GE Global Research, USA
Hayit Greenspan	Tel Aviv University, Israel
Helder Oliveira	Universidade do Porto, Portugal
Holger Roth	Nagoya University, Japan
Hyun Jun Kim	VUNO Inc., South Korea
Islem Rekik	University of Dundee, UK
Itzik Avital	Tel Aviv University, Israel
Jianming Liang	Arizona State University, USA
John (Zhaoyang) Xu	Queen Mary University of London, UK
Jose Costa Pereira	INESC TEC and University of Porto, Portugal
Kelwin Fernandes	University of Porto, Portugal
Kyu-Hwan Jung	VUNO Inc., South Korea
Le Lu	National Institutes of Health, USA
Maria Gabrani	IBM Research Zurich, Switzerland
Marleen de Bruijne	Erasmus MC Rotterdam, The Netherlands/University of Copenhagen, Denmark
Mehmet Aygün	Istanbul Technical University, Turkey
Michal Drozdzal	Polytechnique Montréal, Canada
Mohammad Arafat Hussain	University of British Columbia, Canada
Narayanan Babu	GE Global Research, USA.
Neeraj Dhungel	University of British Columbia, Canada
Nico Hoffmann	Technical University of Dresden, Germany
Nishikant Deshmukh	Johns Hopkins University, USA
Pew-Thian Yap	University of North Carolina-Chapel Hill, USA
Pheng-Ann Heng	The Chinese University of Hong Kong, China
Philippe C. Cattin	University of Basel, Switzerland
Prasad Sudhakar	GE Global Research, USA
Rafeef Abugharbieh	University of British Columbia, Canada
Rami Ben-Ari	IBM-Research, Israel
Roger Tam	University of British Columbia, Canada
S. Chakra Chennubhotla	University of Pittsburgh, USA
Saad Ullah Akram	University of Oulu, Finland
Shadi Albarqouni	Technical University of Munich, Germany
Simon Pezold	MIAC, University of Basel, Switzerland
Siqi Bao	Hong Kong University of Science and Technology, China
Song Wang	University of South Carolina, USA
Steffen Schneider	RWTH-Aachen University, Germany
Takayuki Kitasaka	Aichi Institute of Technology, Japan
Tammy Riklin Raviv	Ben-Gurion University of the Negev, Israel
Teresa Araujo	INESC TEC and University of Porto, Portugal

Thomas Schultz	University of Bonn, Germany
Tom Brosch	University of British Columbia, Canada
Vijay Kumar	University of Adelaide, Australia
Vivek Vaidya	GE Global Research, USA
Weidong Cai	University of Sydney, Australia
Xiang Xiang	Johns Hopkins University, USA
Xiao Yang	University of North Carolina Chapel Hill, USA
Xiaodong Wu	University of Iowa, USA
Xiaoguang Lu	Siemens Healthineers, USA
Xiaohui Xie	University of California, Irvine, USA
Yefeng Zheng	Siemens, USA
Yong Xia	Northwestern Polytechnical University, China
Youngjin Yoo	University of British Columbia, Canada
Zhi Huang	Purdue University, USA
Zhibin Liao	University of Adelaide, Australia
Zita Marinho	Instituto Superior Tecnico, Portugal
Ziyue Xu	National Institutes of Health, USA

Preface ML-CDS 2017

On behalf of the organizing committee, we welcome you to the 7th Workshop on Multimodal Learning for Clinical Decision Support. The goal of this series of workshops is to bring together researchers in medical imaging, medical image retrieval, data mining, text retrieval, and machine learning/AI to discuss the latest techniques in multimodal mining/retrieval and their use in clinical decision support. Although the title of the workshop has changed slightly over the years, the common theme preserved is the notion of clinical decision support and the need for multimodal analysis. The previous six workshops on this topic, held in Athens (2016), Munich (2015), Nagoya (2013), Nice (2012), Toronto (2011), and London (2009), were well-received at MICCAI.

Continuing the momentum built up by these workshops, we have expanded the scope this year to include decision support focusing on multimodal learning. As has been the norm with these workshops, the papers were submitted in an eight-page double-blind format and were accepted after review. As in previous years, the program features an invited lecture by a practicing radiologist to bridge the gap between medical image interpretation and clinical informatics. This year we chose to stay with an oral format for all the presentations. The day will end with a lively panel composed of more doctors, medical imaging researchers, and industry experts.

With less than 5% of medical image analysis techniques translating to clinical practice, workshops on this topic have helped raise the awareness of our field to clinical practitioners. The approach taken in the workshop is to scale it to large collections of patient data, exposing interesting issues of multimodal learning and its specific use in clinical decision support by practicing physicians. With the introduction of intelligent browsing and summarization methods, we hope to also address the ease-of-use in conveying derived information to clinicians to aid their adoption. Finally, the ultimate impact of these methods can be judged when they begin to affect treatment planning in clinical practice. We hope you will enjoy the program we have assembled and actively participate in the discussion on the topics of the papers and the panel.

September 2017

Tanveer Syeda-Mahmood
Hayit Greenspan
Anant Madabhushi
Mehdi Moradi

Organization

Organizing Committee

Tanveer Syeda-Mahmood   IBM Research - Almaden, San Jose, CA, USA
Mehdi Moradi   IBM Research - Almaden, San Jose, CA, USA
Hayit Greenspan   Tel-Aviv University, Tel-Aviv, Israel
Anant Madabhushi   Case Western Reserve University, Cleveland, Ohio, USA

Program Committee

Amir Amini   University of Louisville, USA
Sameer Antani   National Library of Medicine, USA
Rivka Colen   MD Andersen Research Center, USA
Keyvan Farahani   National Cancer Institute, USA
Alejandro Frangi   University of Sheffield, UK
Guido Gerig   University of Utah, USA
David Gutman   Emory University, USA
Allan Halpern   Memorial Sloan-Kettering Research Center, USA
Ghassan Hamarneh   Simon Fraser University, Canada
Jayshree Kalpathy-Cramer   Massachusetts General Hospital, USA
Ron Kikinis   Harvard University, USA
Georg Langs   Medical University of Vienna, Austria
Robert Lundstrom   Kaiser Permanente, USA
B. Manjunath   University of California, Santa Barbara, USA
Dimitris Metaxas   Rutgers, USA
Nikos Paragios   Ecole centrale de Paris, France
Daniel Racoceanu   National University of Singapore, Singapore
Eduardo Romero   Universidad National de Colombia, Colombia
Daniel Rubin   Stanford University, USA
Russ Taylor   Johns Hopkins University, USA
Agma Traina   Sao Paulo University, Brazil
Max Viergewer   Utrecht University, The Netherlands
Sean Zhou   Siemens Corporate Research, USA

Contents

**7th International Workshop on Multimodal Learning
for Clinical Decision Support, ML-CDS 2017**

Third International Workshop on Deep Learning in Medical Image Analysis, DLMIA 2017

Simultaneous Multiple Surface Segmentation Using Deep Learning

Abhay Shah[1(✉)], Michael D. Abramoff[1,3], and Xiaodong Wu[1,2]

[1] Department of Electrical and Computer Engineering,
University of Iowa, Iowa City, USA
abhay-shah-1@uiowa.edu

[2] Department of Radiation Oncology, University of Iowa, Iowa City, USA

[3] Department of Ophthalmology and Visual Sciences,
University of Iowa, Iowa City, USA

Abstract. The task of automatically segmenting 3-D surfaces representing boundaries of objects is important for quantitative analysis of volumetric images, and plays a vital role in biomedical image analysis. Recently, graph-based methods with a global optimization property have been developed and optimized for various medical imaging applications. Despite their widespread use, these require human experts to design transformations, image features, surface smoothness priors, and re-design for a different tissue, organ or imaging modality. Here, we propose a Deep Learning based approach for segmentation of the surfaces in volumetric medical images, by learning the essential features and transformations from training data, without any human expert intervention. We employ a regional approach to learn the local surface profiles. The proposed approach was evaluated on simultaneous intraretinal layer segmentation of optical coherence tomography (OCT) images of normal retinas and retinas affected by age related macular degeneration (AMD). The proposed approach was validated on 40 retina OCT volumes including 20 normal and 20 AMD subjects. The experiments showed statistically significant improvement in accuracy for our approach compared to state-of-the-art graph based optimal surface segmentation with convex priors (G-OSC). A single Convolutional Neural Network (CNN) was used to learn the surfaces for both normal and diseased images. The mean unsigned surface positioning errors obtained by G-OSC method 2.31 voxels (95% CI 2.02-2.60 voxels) was improved to 1.27 voxels (95% CI 1.14-1.40 voxels) using our new approach. On average, our approach takes 94.34 s, requiring 95.35 MB memory, which is much faster than the 2837.46 s and 6.87 GB memory required by the G-OSC method on the same computer system.

Keywords: Optical Coherence Tomography (OCT) · Deep learning · Convolution Neural Networks (CNNs) · Multiple surface segmentation

1 Introduction

For the diagnosis and management of disease, segmentation of images of organs and tissues is a crucial step for the quantification of medical images. Segmen-

© Springer International Publishing AG 2017
M.J. Cardoso et al. (Eds.): DLMIA/ML-CDS 2017, LNCS 10553, pp. 3–11, 2017.
DOI: 10.1007/978-3-319-67558-9_1

tation finds the boundaries or, limited to the 3-D case, the surfaces, that separate organs, tissues or regions of interest in an image. Current state-of-the-art methods for automated 3-D surface segmentation use expert designed graph search/graph cut approaches [5] or active shape/contour modelling [8], all based on classical expert designed image properties, using carefully designed transformations including mathematical morphology and wavelet transformations for segmentation of retinal surfaces in OCT volumes. For instance, OCT is a 3-D imaging technique that is widely used in the diagnosis and management of patients with retinal diseases. The tissue boundaries in OCT images vary by presence and severity of disease. An example is shown in Fig. 1 to illustrate the difference in profile for the Internal Limiting Membrane (ILM) and Inner Retinal Pigment Epithelium (IRPE) in a normal eye and in an eye with AMD. In order to overcome these different manifestations, graph based methods [5–7] with transformations, smoothness constraints, region of interest extraction and multiple resolution approaches designed by experts specifically for the target surface profile have been used. The contour modelling approach [8] requires surface specific and expert designed region based and shape prior terms. The current methods for surface segmentation in OCT images are highly dependent on expert designed target surface specific transformations (cost function design) and therefore, there is a desire for approaches which do not require human expert intervention. Deep learning, where all transformation levels are determined from training data, instead of being designed by experts, has been highly successful in a large number of computer vision [4] and medical image analysis tasks [1,3], substantially outperforming all classical image analysis techniques, and given the spatial coherence that is characteristic of images are typically implemented as Convolutional Neural Networks (CNN) [4]. All these examples are where CNNs are used to identify pixels or voxels as belonging to a certain class, called classification and not to identify the exact location of the surface boundaries in the images, i.e. surface segmentation.

In this study, we propose a CNN based deep learning approach for boundary surface segmentation of a target object, where both features are learnt from training data without intervention by a human expert. We are particularly interested in terrain-like surfaces. An image volume is generally represented as a 3-D volumetric cube consisting of voxel columns, wherein a terrain-like surface intersects each column at exactly one single voxel location. The smoothness of a given surface may be interpreted as the piecewise change in the surface positions of two neighboring columns. The graph based optimal surface segmentation methods [5,7] use convex functions to impose the piecewise smoothness while globally minimizing the objective function for segmentation. In order to employ CNNs for surface segmentation, two key questions need to be answered. First, since most of the CNN based methods have been used for classification, how can a boundary be segmented using a CNN? Second, how can the CNN learn the surface smoothness and surface distance between two interacting surfaces implicitly? We answer these questions by representing consecutive target surface positions for a given input image as a vector. For example, m_1 consecutive target surface positions for a single surface are represented as a m_1-D vector, which may be interpreted as a point in the m_1-D space, while maintaining a strict order with respect to the consecutiveness of the target surface positions. The ordering of the target surface positions partially encapsulates the smoothness of the surface. Thereafter, the error (loss) function utilized in the CNN to back propagate the error is chosen as a Euclidean loss function shown in Eq. (1), wherein the network adjusts the weights of the various nonlinear transformations within the network to minimize the Euclidean distance between the CNN output and the target surface positions in the m_1-D space. Similarly, for detecting λ surfaces, the surface positions are represented as a m_2-D vector, where $\lambda = \{1, 2, \ldots \lambda\}$, $m_2 = \lambda \times m_1$ and m_1 consecutive surface positions for a surface index i $(i \in \lambda)$ are given by $\{((i-1) \times m_1) + 1, ((i-1) \times m_1) + 2, \ldots ((i-1) \times m_1) + m_1\}$ index elements in the m_2-D vector.

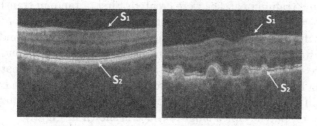

Fig. 1. Illustration of difference in surface profiles on a single B-scan. (left) Normal Eye (right) Eye with AMD. S_1 = ILM and S_2 = IRPE, are shown in red. (Color figure online)

In the currently used methods for segmentation of surfaces, the surface smoothness is piecewise in nature. The surface smoothness penalty (cost) enforced in these methods is the sum of the surface smoothness penalty ascertained using the difference of two consecutive surface positions. Thus, such methods require an expert designed, and application specific, smoothness term to attain accurate segmentations. On the contrary, segmentation using CNN should be expected to also learn the different smoothness profiles of the target surface. Because the smoothness is piecewise, it should be sufficient for the CNN to learn the different local surface profiles for individual segments of the surface with high accuracy because the resultant surface is a combination of these segments. Hence, the CNN is trained on individual patches of the image with segments of the target surface.

2   Method

Consider a volumetric image $I(x, y, z)$ of size $X \times Y \times Z$. A surface is defined as $S(x, y)$, where $x \in \mathbf{x} = \{0, 1, \ldots X - 1\}$, $y \in \mathbf{y} = \{0, 1, \ldots Y - 1\}$ and $S(x, y) \in \mathbf{z} = \{0, 1, \ldots Z - 1\}$. Each (x, y) pair forms a voxel column parallel to the z-axis, wherein the surface $S(x,y)$ intersects each column at a single voxel location. For simultaneously segmenting λ ($\lambda \geq 2$) surfaces, the goal of the CNN is to learn the surface positions $S_i(x,y)$ ($i \in \lambda$) for columns formed by each (x,y) pair. In this work, we present a slice by slice segmentation of a 3-D volumetric image applied on OCT volumes. Patches are extracted from B-scans with the target Reference Standard (RS). A patch $P(x_1, z)$ is of size $N \times Z$, where $x_1 \in \mathbf{x_1} = \{0, 1, ...N-1\}$, $z \in \mathbf{z} = \{0, 1, ...Z-1\}$ and N is a multiple of 4. The target surfaces S_i's to be learnt simultaneously from P is $\overline{S_i}(x_2) \in \mathbf{z} = \{0, 1, ...Z-1\}$, where $x_2 \in \mathbf{x_2} = \{\frac{N}{4}, \frac{N}{4}+1, ...\frac{3N}{4}-1\}$. Essentially, the target surfaces to be learnt is the surface locations for the middle $\frac{N}{2}$ consecutive columns in P. The overlap between consecutive patches ensures no abrupt changes occur at patch boundaries. By segmenting the middle $N/2$ columns in a patch size with N columns, the boundary of patches overlap with the consecutive surface segment patch. Then, data augmentation is performed, where for each training patch, three additional training patches were created. First, a random translation value was chosen between -250 and 250 voxels such that the translation was within the range of the patch size. The training patch and the corresponding RS for surfaces $\overline{S_i}$'s were translated in the z dimension accordingly. Second, a random rotation value was chosen between $-45°$ and $45°$. The training patch and the corresponding RS for surfaces $\overline{S_i}$'s were rotated accordingly. Last, a combination of rotation and translation was used to generate another patch. Examples of data augmentation on patches for a single surface is shown in Fig. 2.

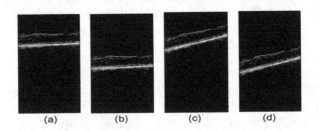

(a)   (b)   (c)   (d)

Fig. 2. Illustration of data augmentation applied to an input patch. The target surface is shown in red. (a) Extracted patch from a B-scan, (b) Translation (c) Rotation (d) Translation and rotation, as applied to (a). (Color figure online)

For segmenting λ surfaces simultaneously, the CNN learns λ surfaces for each patch. The CNN architecture used in our work is shown in Fig. 3, employed for $\lambda = 2$ and patches with $N = 32$. The CNN contains three convolution layers [4], each of which is followed by a max-pooling layer [4] with stride length of two. Thereafter, it is followed by two fully connected layers [4], where the last fully connected layer represents the final output of the middle $\frac{N}{2}$ surface positions for 2 target surfaces in P. Lastly, a Euclidean loss function (used for regressing to real-valued labels) as shown in Eq. (1) is utilized to compute the error between CNN outputs and RS of S_i's ($i \in \lambda$) within P for back propagation during the

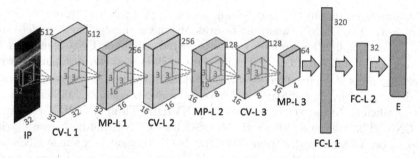

Fig. 3. The architecture of the CNN learned in our work for $N = 32$ and $\lambda = 2$. The numbers along each side of the cuboid indicate the dimensions of the feature maps. The inside cuboid (*green*) represents the convolution kernel and the inside square (*green*) represents the pooling region size. The number of hidden neurons in the fully connected layers are marked aside. IP = Input Patch, CV-L = Convolution Layer, MP-L = Max-Pooling Layer, FC-L = Fully Connected Layer, E = Euclidean Loss Layer. (Color figure online)

training phase. Unsigned mean surface positioning error (UMSPE) [5] is one of the commonly used error metric for evaluation of surface segmentation accuracy. The Euclidean loss function (E), essentially computes sum of the squared unsigned surface positioning error over the $\frac{N}{2}$ consecutive surface position for S_i's of the CNN output and the RS for P.

$$E = \sum_{i=1}^{i=\lambda} \sum_{k_1=0}^{k_1=\frac{N}{2}-1} (\overline{a}_{k_1}^i - a_{k_2}^i)^2 \tag{1}$$

where, $k_2 = ((i - 1) \times N/2) + k_1$, $\overline{a}_{k_1}^i$ and $a_{k_2}^i$ is the k_1-th surface position of reference standard and CNN output respectively for surface S_i in a given P.

3   Experiments

The experiments compare segmentation accuracy of the proposed CNN based method (CNN-S) and G-OSC method [7]. The two surfaces simultaneously segmented in this study are S_1-ILM and S_2-IRPE as shown in Fig. 1. 115 OCT scans of normal eyes, 269 OCT scans of eyes with AMD and their respective reference standards (RS) (created by experts with aid of the DOCTRAP software [2]) were obtained from the publicly available repository [2]. The 3-D volume size was $1000 \times 100 \times 512$ voxels. The data volumes were divided into a training set (79 normal and 187 AMD), a testing set (16 normal and 62 AMD) and a validation set (20 normal and 20 AMD). The volumes were denoised by applying a median filter of size $5 \times 5 \times 5$ and normalized with the resultant voxel intensity varying from -1 to 1. Thereafter, patches of size $N \times 512$ with their respective RS for the middle $\frac{N}{2}$ consecutive surface positions for S_1 and S_2 is extracted using data augmentation, for training and testing volumes, resulting in a training set of 340,000 and testing set of 70,000 patches. In our work, we use $N = 32$. The UMSPE was used to evaluate the accuracy. The complete surfaces for each validation volume were segmented using the CNN-S method by creating $\frac{1016}{N/2}$ patches from each B-scan where each B-scan was zero padded with 8 voxel columns at each extremity. Statistical significance of observed differences was determined by paired Student t-tests for which p value of 0.05 was considered significant. In our study we used one NVIDIA Titan X GPU for training the CNN. The validation using the G-OSC and CNN-S method were carried out on a on a Linux workstation (3.4 GHz, 16 GB memory). A single CNN was trained to infer on both the normal and AMD OCT scans. For a comprehensive comparison, three experiments were performed with the G-OSC method. The first experiment (G-OSC 1) involved segmenting the surfaces in both normal and AMD OCT scans using a same set of optimized parameters. The second (G-OSC 2) and third (G-OSC 3) experiment involved segmenting the normal and AMD OCT scans with different set of optimized parameters, respectively.

4 Results

The quantitative comparisons between the proposed CNN-S method and the G-OSC method on the validation volumes is summarized in Table 1. For the entire validation data, the proposed method produced significantly lower UMSPE for surfaces S_1 ($p < 0.01$) and S_2 ($p < 0.01$), compared to the segmentation results of G-OSC 1, G-OSC 2 and G-OSC 3. Illustrative results of segmentations from the CNN-S, G-OSC 2 and G-OSC 3 methods on validation volumes are shown in Fig. 4. It can be observed that CNN-S method yeilds consistent and qualitatively superior segmentations with respect to the G-OSC method. On closer analysis of some B-scans in the validation data, the CNN-S method produced high quality segmentation for a few cases where the RS was not accurate enough as verified by an expert (4th row in Fig. 4). The CNN required 17 days to train on the GPU. The CNN-S method with average computation time of 94.34 s (95.35 MB memory) is much faster than G-OSC with average computation time of 2837.46 s (6.87 GB memory).

Table 1. UMSPE expressed as (mean ± 95% CI) in voxels. RS - Reference Standard. $N = 32$ was used as the patch size (32×512).

Surface	Normal and AMD		Normal		AMD	
	G-OSC 1 vs. RS	CNN-S vs. RS	G-OSC 2 vs. RS	CNN-S vs. RS	G-OSC 3 vs. RS	CNN-S vs.RS
S_1	1.45 ± 0.19	0.98 ± 0.08	1.19 ± 0.05	0.89 ± 0.07	1.37 ± 0.22	1.06 ± 0.11
S_2	3.17 ± 0.43	1.56 ± 0.15	1.41 ± 0.11	1.28 ± 0.10	2.88 ± 0.54	1.83 ± 0.26
Overall	2.31 ± 0.29	1.27 ± 0.13	1.31 ± 0.07	1.08 ± 0.08	2.13 ± 0.39	1.44 ± 0.19

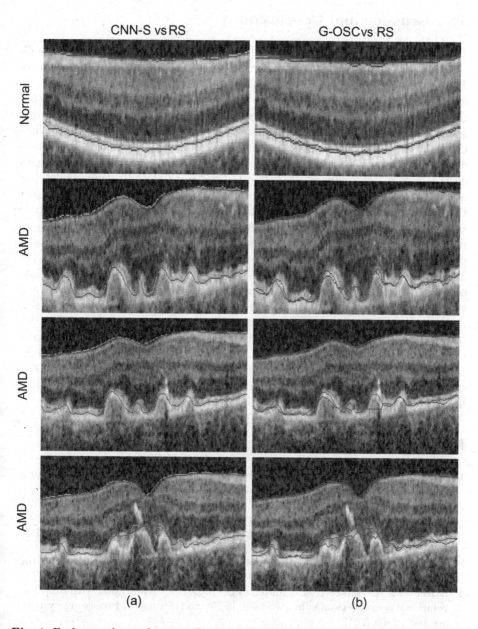

Fig. 4. Each row shows the same B-scan from a Normal or AMD OCT volume. (a) CNN-S vs. RS (b) G-OSC vs. RS, for surfaces S_1 = ILM and S_2 = IRPE. RS = Reference Standard, Red = reference standard, Green = Segmentation using proposed method and Blue = Segmentation using G-OSC method. In the 4th row, we had the reference standard reviewed by a fellow-ship trained retinal specialist, who stated that the CNN-S method is closer to the real surface than the reference standard. (Color figure online)

5 Discussion and Conclusion

The results demonstrate superior quality of segmentations compared to the G-OSC method, while eliminating the requirement of expert designed transforms. The proposed method used a single CNN to learn various local surface profiles for both normal and AMD data. Our results compared to G-OSC 1 show that the CNN-S methods outperforms the G-OSC method. If the parameters are tuned specifically for each type of data by using expert prior knowledge while using the G-OSC method, as in the cases of G-OSC 2 and G-OSC 3, the results depict that the CNN-S method still results in superior performance. The inference using CNN-S is much faster than G-OSC method and requires much less memory. Consequently, the inference can be parallelized for multiple patches, thereby further reducing the computation time, thus making it potentially more suitable for clinical applications. However, a drawback of any such learning approach in medical imaging is the limited amount of available training data. The proposed method was trained on images from one type of scanner and hence it is possible that the trained CNN may not produce consistent segmentations on images obtained from a different scanner due to difference in textural and spatial information. The approach can readily be extended to perform 3-D segmentations by employing 3-D convolutions. In this paper, we proposed a CNN based method for segmentation of surfaces in volumetric images with implicitly learned surface smoothness and surface separation models. We demonstrated the performance and potential of the proposed method through application on OCT volumes to segment the ILM and IRPE surface. The experiment results show higher segmentation accuracy as compared to the G-OSC method.

References

1. Challenge, M.B.G.: Multimodal brain tumor segmentation benchmark: change detection. http://braintumorsegmentation.org/. Accessed 5 Nov 2016
2. Farsiu, S., Chiu, S.J., O'Connell, R.V., Folgar, F.A., Yuan, E., Izatt, J.A., Toth, C.A.: Quantitative classification of eyes with and without intermediate age-related macular degeneration using optical coherence tomography. Ophthalmology **121**(1), 162–172 (2014)
3. Kaggle: diabetic retinopathy detection. http://www.kaggle.com/c/diabetic-retinopathy-detection/. Accessed 15 July 2016
4. Krizhevsky, A., Sutskever, I., Hinton, G.E.: Imagenet classification with deep convolutional neural networks. In: Advances in Neural Information Processing Systems, pp. 1097–1105 (2012)
5. Lee, K., Garvin, M., Russell, S., Sonka, M., Abràmoff, M.: Automated intraretinal layer segmentation of 3-d macular oct scans using a multiscale graph search. Invest. Ophthalmol. Vis. Sci. **51**(13), 1767 (2010)
6. Shah, A., Bai, J., Hu, Z., Sadda, S., Wu, X.: Multiple surface segmentation using truncated convex priors. In: Navab, N., Hornegger, J., Wells, W.M., Frangi, A.F. (eds.) MICCAI 2015. LNCS, vol. 9351, pp. 97–104. Springer, Cham (2015). doi:10.1007/978-3-319-24574-4_12

7. Song, Q., Bai, J., Garvin, M.K., Sonka, M., Buatti, J.M., Wu, X.: Optimal multiple surface segmentation with shape and context priors. IEEE Trans. Med. Imag. **32**(2), 376–386 (2013)
8. Yazdanpanah, A., Hamarneh, G., Smith, B., Sarunic, M.: Intra-retinal layer segmentation in optical coherence tomography using an active contour approach. In: Yang, G.-Z., Hawkes, D., Rueckert, D., Noble, A., Taylor, C. (eds.) MICCAI 2009. LNCS, vol. 5762, pp. 649–656. Springer, Heidelberg (2009). doi:10.1007/978-3-642-04271-3_79

A Deep Residual Inception Network for HEp-2 Cell Classification

Yuexiang Li and Linlin Shen[✉]

Computer Vision Institute, Shenzhen University, Shenzhen, China
llshen@szu.edu.cn

Abstract. Indirect-immunofluorescence (IIF) of Human Epithelial-2 (HEp-2) cells is a commonly-used method for the diagnosis of autoimmune diseases. Traditional approach relies on specialists to observe HEp-2 slides via the fluorescence microscope, which suffers from a number of shortcomings like being subjective and labor intensive. In this paper, we proposed a hybrid deep learning network combining the latest high-performance network architectures, i.e. ResNet and Inception, to automatically classify HEp-2 cell images. The proposed Deep Residual Inception (DRI) net replaces the plain convolutional layers in Inception with residual modules for better network optimization and fuses the features extracted from shallow, medium and deep layers for performance improvement. The proposed model is evaluated on publicly available I3A (Indirect Immunofluorescence Image Analysis) dataset. The experiment results demonstrate that our proposed DRI remarkably outperforms the benchmarking approaches.

Keywords: HEp-2 cells · Image classification · Deep Learning Network

1 Introduction

Detecting antinuclear antibodies (ANAs) by Human Epithelial-2 (HEp-2) cell patterns is a recommended approach for autoimmune disease diagnosis. However, in current clinical practice, inspection of the indirect-immunofluorescence (IIF) slides requires highly-skilled pathologists, resulting in a time-consuming analysis subject to inter-observer variations. To address the problem, computer-aided diagnostic (CAD) systems are proposed to assist pathologists. As one of the most challenging tasks, HEp-2 staining pattern recognition has received increasing attention from the research community. Recent contests [1–3] of HEp-2 cell image processing further promoted the developments of CAD systems for automatic classification of HEp-2 cells.

The performances of previous works on HEp-2 cell pattern classification highly rely on the choices of hand-crafted features. Nosaka et al. [4] proposed a rotation invariant Local Binary Pattern descriptor (CoALBP) for HEp-2 cell classification, which won the first prize of ICPR 2012 contest. In ICIP 2013 contest, the framework combining LBP and bag of words proposed by Shen et al. [5] achieved the highest accuracy, i.e. 83.65%. Manivannan et al. [6], the winner of ICPR 2014 contest, extracted Root-SIFT features and multi-resolution local patterns from HEp-2 cell

© Springer International Publishing AG 2017
M.J. Cardoso et al. (Eds.): DLMIA/ML-CDS 2017, LNCS 10553, pp. 12–20, 2017.
DOI: 10.1007/978-3-319-67558-9_2

images and employed ensembles of SVMs for classification. In more recent studies, an increasing number of methods using hand-crafted features were proposed to automatically classify HEp-2 cell images. For examples, Xu et al. [7] trained a linear SVM with a Co-occurrence Differential Texton (CoDT) feature to identify the staining patterns of HEp-2 cells. Taalimi et al. [8] employed joint sparse representation of HEp-2 cell images for classification. Kastaniotis et al. [9] developed a method, named VHAR, hierarchically aggregating the residual of the sparse representation of feature vectors derived from SIFT descriptor for HEp-2 images classification.

Different from traditional approaches, convolutional neural networks (CNN) do not require the design of hand-crafted features. They automatically construct feature representations from input images through multi-layer processing. The feature learned from CNN is verified to provide better classification performance than hand-crafted features [10]. Researchers have made their efforts to implement CNN for HEp-2 classification. Gao et al. [11] presented the first work using CNN to classify HEp-2 cell images. Their model is a VGG-based network [12], which consists of 8 layers. Bayramoglu et al. reported their new progress on using CNN for HEp-2 classification in [13]. Phan et al. [14] finetuned a model pre-trained on ImageNet to HEp-2 dataset. The published CNN frameworks for HEp-2 classification share the same characteristics: the network architectures are straightforward with a single softmax classifier. As the deep learning model developed, multi-branches networks, e.g. ResNet [15], Inception [16, 17], have gradually surpassed straightforward networks and become the main-stream model for researchers.

In this paper, we proposed a hybrid multi-branches model, Deep Residual Inception (DRI), instead of straightforward CNN to improve the accuracy of classifying HEp-2 cell images. The proposed model combines the architectures of two advanced high-performance deep learning networks, i.e. ResNet and Inception, by replacing the plain convolutional layers in Inception with residual modules. Therefore, the proposed DRI can exploit the strengths of both ResNet and Inception for network training and image classification, i.e. easy network optimization from ResNet and multi-scale feature extractors from Inception. Furthermore, to better utilize the features learned by DRI, we proposed a novel scheme to fuse the features extracted by the shallow, medium and deep layers. Performance evaluation has been conducted on I3A dataset. Experiment results demonstrate that our DRI can provide outstanding HEp-2 cell classification performance.

2 Deep Residual Inception

2.1 Network Architecture

The architecture of proposed Deep Residual Inception network is shown in Fig. 1. The input cell images are reshaped to a size of 76×76. The parameters of 1/2 and 1/3 represent the stride size of 2 and 3, respectively. The green layers are convolutional layers. The numbers represent the size and amount of convolutional kernels. Our DRI uses two kinds of pooling, i.e. max pooling and average pooling. The numbers in pooling layers represent the kernel size. Layers named with 'FC' are fully-connected

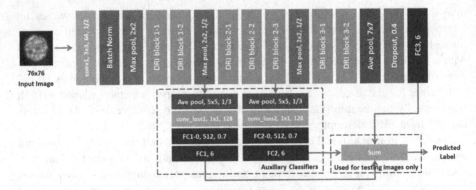

Fig. 1. Network architecture of Deep Residual Inception. (Color figure online)

layers. The second parameter of fully-connected layers is the number of containing neurons. The third parameter of FC1-0 and FC2-0 indicates that the layers are followed by a dropout layer [18] of 0.7. The non-linear activation function used in DRI is PReLU [19]. Two auxiliary classifiers, i.e. FC1 and FC2, and one primary classifier, i.e. FC3, are involved in the proposed DRI. The three classifier branches are trained with separate softmax loss (L), as defined in (1), respectively.

$$L = \frac{1}{N}\sum_i L_i = \frac{1}{N}\sum_i -\log(\frac{e^{f_{y_i}}}{\sum_j e^{f_j}}) \qquad (1)$$

where f_j denotes the j-th element ($j \in [1, K]$, K is the number of classes) of vector of class scores f, y_i is the label of i-th input feature and N is the number of training data.

Different weights (w) are assigned to the softmax loss of different classifiers. Assume L_n represents the softmax loss of classifier FCn, the total joint loss (L_{total}) for DRI can be defined as:

$$L_{total} = \sum_{n=1}^3 w_n \times L_n \qquad (2)$$

where $w = 0.3$, 0.3 and 1 for L_1, L_2 and L_3, respectively.

According to the calculated total loss, the optimization of DRI is performed using the stochastic gradient descent (SGD) with back-propagation [20].

The original Inception net uses auxiliary classifiers to improve the convergence of very deep network during training. In our experiments, we found the uses of auxiliary classifiers not only assist network convergence, but also have positive effect to the testing classification performance. Thus, the proposed DRI integrates the features from shallow (FC1), medium (FC2) and deep (FC3) classifier branches with a sum layer and accordingly predicts the staining cell pattern for testing HEp-2 image.

2.2   DRI Module

The DRI has 7 multi-branches modules, named DRI block, whose architecture is presented in Fig. 2. Branches with different sizes of convolutional kernels, i.e. 1×1, 3×3, 5×5, enable the network to extract multi-scale features. Inspired by [21], the Batch

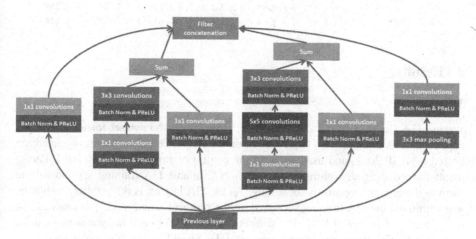

Fig. 2. Architecture of DRI module. Two primary changes of module structure: (1) Two identity shortcuts are added to 3×3 and 5×5 branches. (2) The depth of 5×5 branch is increased.

Normalization [22] and PReLU are placed in front of convolutional layers. Compared to the original Inception module [16], two identity shortcut connections [15] are added to 3×3 and 5×5 branches to alleviate the gradient vanishing problem as network goes deeper. Another difference between the proposed DRI module and the original inception module is that we increased the depth of 5×5 branch by adding a 3×3 convolutional layer to the end of 5×5 convolutions. This change of network structure is inspired by the recent research [23], which illustrates the different-depth branches can benefit the overall performance of multi-branches model.

A detailed configuration of DRI is shown in Table 1. The first number of each entry in Table 1 is the kernel size and the second number is the amount of kernels. Similar to original Inception, the amount of convolutional kernels in our DRI smoothly increases for extracting features with better representation capacity.

2.3   Network Training

The proposed DRI is implemented utilizing Caffe toolbox [24]. The network is initialized with the 'xavier' scheme and trained with a mini-batch size of 24 on one GPU (GeForce GTX TITANX, 12 GB RAM). A momentum of 0.9 is used to assist network optimization. The initial learning rate is set to 0.001 and decayed with a gamma of 0.79. The network converges after 290,000 iterations.

Table 1. The configuration of the proposed DRI architecture

	DRI 1-1	DRI 1-2	DRI 2-1	DRI 2-2	DRI 2-3	DRI 3-1	DRI 3-2
1 × 1 branch	1 × 1, 64	1 × 1, 128	1 × 1, 192	1 × 1, 160	1 × 1, 244	1 × 1, 256	1 × 1, 384
3 × 3 branch	1 × 1, 96	1 × 1, 128	1 × 1, 112	1 × 1, 128	1 × 1, 144	1 × 1, 160	1 × 1, 192
	3 × 3, 128	3 × 3, 192	3 × 3, 224	3 × 3, 256	3 × 3, 288	3 × 3, 320	3 × 3, 384
5 × 5 branch	1 × 1, 16	1 × 1, 32	1 × 1, 16	1 × 1, 24	1 × 1, 24	1 × 1, 32	1 × 1, 48
	5 × 5, 32	5 × 5, 96	5 × 5, 48	5 × 5, 64	5 × 5, 64	5 × 5, 128	5 × 5, 128
	3 × 3, 64	3 × 3, 128	3 × 3, 96	3 × 3, 128	3 × 3, 128	3 × 3, 256	3 × 3, 256
3 × 3 max	1 × 1, 32	1 × 1, 64	1 × 1, 64	1 × 1, 64	1 × 1, 64	1 × 1, 128	1 × 1, 128

3   Results

3.1   Dataset

We use I3A (Indirect Immunofluorescence Image Analysis) dataset for testing in this paper. The **I3A dataset**[1] was first released on the HEp-2 cell classification competition hosted by ICIP 2013, and used again in the contest organized by ICPR 2014. Participants trained their algorithms on the publicly available I3A training set and submit them to the contest organizers for testing. As the I3A test set is not publicly available, we partitioned the I3A training set based on the protocol stated in [11] to evaluate the performance of proposed DRI. The dataset contains 13,596 cell images which can be separated to six categories: Homogeneous (**H**), Speckled (**S**), Nucleolar (**N**), Centromere (**C**), Nuclear membrane (**Nm**) and Golgi (**G**) (Fig. 3).

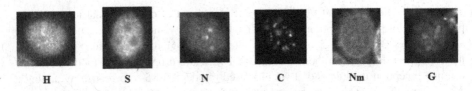

| H | S | N | C | Nm | G |

Fig. 3. Example images of I3A dataset

3.2   Data Augmentation

The I3A dataset contains 13,596 cell images. For comparison convenience, we partition the dataset according to the same percentages, i.e. 64% for training, 16% for validation and 20% for testing, as reported in [11]. A class-balanced augmentation scheme [25] is employed to the training set. Each Golgi (**G**) image was rotated for 59°, i.e. 6° for each rotation, which is three times more than that for the images of other cell patterns, i.e. 18° for each rotation. The class-balanced augmentation approach addressed the problem of unbalanced volumes of cell images contained in I3A dataset. Detailed information of

---

[1] http://nerone.diem.unisa.it/hep2-benchmarking/dbtools/.

Table 2. Details of augmented training datasets

	H	S	N	C	Nm	G	Total
I3A	2,494	2,831	2,598	2,741	2,208	724	13,596
Training set	1,596	1,812	1,663	1,754	1,413	463	8,701
Augmented set	31,920	36,240	33,260	35,080	28,260	27,780	192,540

Table 3. Performance evaluations of the proposed DRI. (a) MCA of using DRI Module (%) (b) MCA of using auxiliary classifiers (%)

	MCA
Shallow Inception	96.54
SRI	97.85
Inception [16]	97.93
DRI	98.37

(a)

	MCA
DRI_3	96.23
DRI_{1+3}	97.41
DRI_{2+3}	96.23
DRI_{1+2+3}	98.37

(b)

augmented training set is listed in Table 2. The original training set is augmented from 8,701 to 192,540 images.

3.3 Performance Analysis

We use the mean class accuracy (MCA) adopted by ICPR 2014 competition [3] as the criterion for performance evaluation. It measures the average of per-class accuracies, as defined in (3).

$$MCA = \frac{1}{K}\sum_{k=1}^{K} CCR_k \tag{3}$$

where CCR_k is the correct classification rate for class k and K is the number of classes.

DRI Module. To investigate the effect of DRI model and network depths, we developed a Shallow Residual Inception (SRI) from the DRI by removing the *DRI 1-2*, *DRI 2-3*, and *DRI 3-2* modules. Shallow Inception was also developed by replacing the DRI modules in SRI to original Inception modules. Table 3(a) listed the MCA of Shallow Inception, SRI, and the deep frameworks using original Inception/DRI modules evaluated on testing set. It can be observed that with similar network depth, the frameworks using the proposed DRI modules, i.e. SRI and DRI, outperform the ones with original Inception modules, i.e. Shallow and Deep Inceptions, illustrating the DRI is a more advanced network architecture compared to the original Inception. Furthermore, the deeper networks, i.e. Inception and DRI, yield better results, i.e. 97.93% and 98.37%, than the shallow ones, i.e. Shallow Inception (96.54%) and SRI (97.85%), which demonstrates that the network depth enables the models to learn better feature representations for HEp-2 cell classification.

Auxiliary Classifiers. Auxiliary classifiers were only used as regularizers to improve network convergence during training in original Inception. In our experiments, combinations of the features extracted by auxiliary classifier branches, i.e. FC-1, FC-2, are found to produce improvements of classification performance. Table 3(b) presents the accuracies of combinations of different classifier branches. The MCA increases from 96.23% to 97.41% by adding FC1 branch. As the FC2 branch extracts similar features compared to FC3, there is no improvement of classification accuracy by only using FC2 and FC3. However, FC2 branch can be used as a corrector to amend errors occurred in the model using shallow classifier branch, i.e. DRI_{1+3}. The model using all branches, i.e. DRI_{1+2+3}, achieves the highest MCA of 98.37%.

3.4 Comparisons

Table 4 compares our DRI with the benchmarking algorithms, i.e. the VHAR approach [9], the Bag-of-features (BoF) model [11], the Fisher Vector (FV) model [11] and the CNN model proposed by Gao et al. [11]. The average classification accuracy (ACA) [11] is also measured for evaluation. All deep-learning approaches outperform the models using hand-crafted features, i.e. *VHAR*, *BoF* and *FV*. Our DRI achieves the highest MCA, i.e. 98.37%, and ACA, i.e. 98.49%, which are 1.61% and 1.25% higher than that of the straight-forward CNN model proposed by Gao et al.

Table 4. Comparsion with benchmarking algorithms (%)

	VHAR [9]	BoF	FV	Gao et al. [11]	SRI (ours)	DRI (ours)
MCA	93.60	94.23	95.73	96.76	97.85	**98.37**
ACA	-	94.38	96.07	97.24	97.98	**98.49**

4 Conclusion

In this paper, we presented a hybrid deep-learning network, named Deep Residual Inception, combining the architectures of two most advanced frameworks, i.e. ResNet and Inception, for HEp-2 cell image classification. The DRI model adds short-cut connections to the original Inception model for better network convergence and fuses the features extracted from shallow, medium and deep classifier branches to improve classification performance. The experimental results show that our DRI provides a significant improvement compared to existing HEp-2 classification methods.

References

1. Foggia, P., Percannella, G., Soda, P., Vento, M.: Benchmarking HEp-2 cells classification methods. IEEE Trans. Med. Imag. **32**, 1878–1889 (2013)
2. Hobson, P., Lovell, B.C., Percannella, G., Vento, M., Wiliem, A.: Benchmarking human epithelial type 2 interphase cells classification methods on a very large dataset. Artif. Intell. Med. **65**, 239–250 (2015)

3. Hobson, P., Lovell, B.C., Percannella, G., Saggese, A., Vento, M., Wiliem, A.: HEp-2 staining pattern recognition at cell and specimen levels: datasets, algorithms and results. Pattern Recognit. Lett. **82**, 12–22 (2016)
4. Nosaka, R., Fukui, K.: HEp-2 cell classification using rotation invariant co-occurrence among local binary patterns. Pattern Recognit. **47**, 2428–2436 (2014)
5. Shen, L., Lin, J., Wu, S., Yu, S.: HEp-2 image classification using intensity order pooling based features and bag of words. Pattern Recognit. **47**, 2419–2427 (2014)
6. Manivannan, S., Li, W., Akbar, S., Wang, R., Zhang, J., Mckenna, S.J.: An automated pattern recognition system for classifying indirect immunofluorescence images of HEp-2 cells and specimens. Pattern Recognit. **51**, 12–26 (2016)
7. Xu, X., Lin, F., Ng, C., Leong, K.P.: Adaptive co-occurrence differential texton space for HEp-2 cells classification. In: Navab, N., Hornegger, J., Wells, W.M., Frangi, A.F. (eds.) MICCAI 2015. LNCS, vol. 9351, pp. 260–267. Springer, Cham (2015). doi:10.1007/978-3-319-24574-4_31
8. Taalimi, A., Ensafi, S., Qi, H., Lu, S., Kassim, A.A., Tan, C.L.: Multimodal dictionary learning and joint sparse representation for HEp-2 cell classification. In: Navab, N., Hornegger, J., Wells, W.M., Frangi, A.F. (eds.) MICCAI 2015. LNCS, vol. 9351, pp. 308–315. Springer, Cham (2015). doi:10.1007/978-3-319-24574-4_37
9. Kastaniotis, D., Fotopoulou, F., Theodorakopoulos, I., Economou, G., Fotopoulos, S.: HEp-2 cell classification with vector of hierarchically aggregated residuals. Pattern Recognit. **65**, 47–57 (2017)
10. Lecun, Y., Bengio, Y., Hinton, G.: Deep learning. Nature **521**, 436–444 (2015)
11. Gao, Z., Wang, L., Zhou, L., Zhang, J.: HEp-2 cell image classification with deep convolutional neural networks. IEEE J. Biomed. Health Inform. **21**(2), 416–428 (2016)
12. Simonyan, K., Zisserman, A.: Very deep convolutional networks for large-scale image recognition. arXiv e-print arXiv:1409.1556 (2015)
13. Bayramoglu, N., Kannala, J., Heikkila, J.: Human epithelial type 2 cell classification with convolutional neural networks. In: BIBE, pp. 1–6 (2015)
14. Phan, H.T.H., Kumar, A., Kim, J., Feng, D.: Transfer learning of a convolutional neural network for HEp-2 cell image classification. In: ISBI, pp. 1208–1211 (2016)
15. He, K., Zhang, X., Ren, S., Sun, J.: Deep residual learning for image recognition. In: CVPR, pp. 770–778 (2016)
16. Szegedy, C., Liu, W., Jia, Y., Sermanet, P.: Going deeper with convolutions. In: CVPR, pp. 1–9 (2015)
17. Szegedy, C., Ioffe, S., Vanhoucke, V., Alemi, A.: Inception-v4, inception-ResNet and the impact of residual connections on learning. arXiv e-print arXiv:1602.07261 (2016)
18. Hinton, G.E., Srivastava, N., Krizhevsky, A., Sutskever, I., Salakhutdinov, R.R.: Improving neural networks by preventing co-adaptation of feature detectors. arXiv e-print arXiv:1207.0580 (2012)
19. He, K., Zhang, X., Ren, S., Sun, J.: Delving deep into rectifiers: surpassing human-level performance on imagenet classification. In: ICCV, pp. 1026–1034 (2015)
20. Lecun, Y., Boser, B., Denker, J.S., Henderson, D., Howard, R.E., Hubbard, W., Jackel, L.D.: Backpropagation applied to handwritten zip code recognition. Neural Comput. **1**, 541–551 (1989)
21. He, K., Zhang, X., Ren, S., Sun, J.: Identity mappings in deep residual networks. In: ECCV, pp. 630–645 (2016)
22. Ioffe, S., Szegedy, C.: Batch normalization: accelerating deep network training by reducing internal covariate shift. In: ICML, pp. 448–456 (2015)
23. Zhao, L., Wang, J., Li, X., Tu, Z., Zeng, W.: On the connection of deep fusion to ensembling. arXiv e-print arXiv:1611.07718 (2016)

24. Jia, Y., Shelhamer, E., Donahue, J., Karayev, S., Long, J., Girshick, R., Guadarrama, S., Darrell, T.: Caffe: convolutional architecture for fast feature embedding. arXiv e-print arXiv:1408.5093 (2014)
25. Jia, X., Shen, L., Zhou, X., Yu, S.: Deep convolutional neural network based HEp-2 cell classification. In: ICPR Contest and Workshop: Pattern Recognition Techniques for Indirect Immunofluorescence Images Analysis (2016)

Joint Segmentation of Multiple Thoracic Organs in CT Images with Two Collaborative Deep Architectures

Roger Trullo[1,2]([✉]), Caroline Petitjean[1], Dong Nie[2], Dinggang Shen[2], and Su Ruan[1]

[1] Normandie Univ., UNIROUEN, UNIHAVRE, INSA Rouen, LITIS, 76000 Rouen, France
rogertrullo@gmail.com

[2] Department of Radiology and BRIC, UNC-Chapel Hill, Chapel Hill, USA

Abstract. Computed Tomography (CT) is the standard imaging technique for radiotherapy planning. The delineation of Organs at Risk (OAR) in thoracic CT images is a necessary step before radiotherapy, for preventing irradiation of healthy organs. However, due to low contrast, multi-organ segmentation is a challenge. In this paper, we focus on developing a novel framework for automatic delineation of OARs. Different from previous works in OAR segmentation where each organ is segmented separately, we propose two collaborative deep architectures to jointly segment all organs, including esophagus, heart, aorta and trachea. Since most of the organ borders are ill-defined, we believe spatial relationships must be taken into account to overcome the lack of contrast. The aim of combining two networks is to learn anatomical constraints with the first network, which will be used in the second network, when each OAR is segmented in turn. Specifically, we use the first deep architecture, a deep SharpMask architecture, for providing an effective combination of low-level representations with deep high-level features, and then take into account the spatial relationships between organs by the use of Conditional Random Fields (CRF). Next, the second deep architecture is employed to refine the segmentation of each organ by using the maps obtained on the first deep architecture to learn anatomical constraints for guiding and refining the segmentations. Experimental results show superior performance on 30 CT scans, comparing with other state-of-the-art methods.

Keywords: Anatomical constraints · CT segmentation · Fully Convolutional Networks (FCN) · CRF · CRFasRNN · Auto-context model

1 Introduction

In medical image segmentation, many clinical settings include the delineation of multiple objects or organs, e.g., the cardiac ventricles, and thoracic or abdominal organs. From a methodological point of view, the ways of performing multi-organ

© Springer International Publishing AG 2017
M.J. Cardoso et al. (Eds.): DLMIA/ML-CDS 2017, LNCS 10553, pp. 21–29, 2017.
DOI: 10.1007/978-3-319-67558-9_3

segmentation are diverse. For example, multi-atlas approaches in a patch based setting have been shown effective for segmenting abdominal organs [11]. Many other approaches combine several techniques, such as in [4] where thresholding, generalized hough transform and an atlas-registration based method are used. The performance of these approaches is bound to the use of separate methods that can also be computationally expensive. Usually, organs are segmented individually ignoring their spatial relationships, although this information could be valuable to the segmentation process.

In this paper, we focus on the segmentation of OAR, namely the aorta, esophagus, trachea and heart, in thoracic CT (Fig. 1), an important prerequisite for radiotherapy planning in order to prevent irradiation of healthy organs. Routinely, the delineation is largely manual with poor intra- or inter-practitioners agreement. Note that the automated segmentation of the esophagus has hardly been addressed in research works as it is exceptionally challenging: the boundaries in CT images are almost invisible (Fig. 2). Radiotherapists manually segment it based on not only the intensity information, but also the anatomical knowledge, i.e., the esophagus is located behind the trachea in the upper part, behind the heart in the lower part, and also next to the aorta in several parts. More generally, this observation can be made for the other organs as well. Our aim is to design a framework that would learn this kind of constraints automatically to improve the segmentation of all OAR and the esophagus in particular.

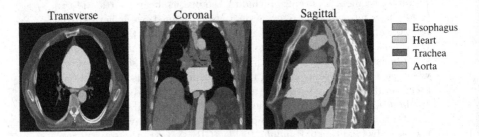

Fig. 1. Typical CT scan with manual segmentations of the esophagus, heart, trachea and aorta.

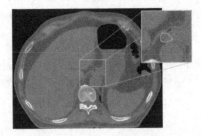

Fig. 2. CT scan with manual delineation of the esophagus. Note how the esophagus is hardly distinguishable.

We propose to tackle the problem of segmenting OAR in a joint manner through the application of two collaborative deep architectures, which will implicitly learn anatomical constraints in each of the organs to mitigate the difficulty caused by lack of image contrast. In particular, we perform an initial segmentation by using a first deep Sharpmask network, inspired by the refinement framework presented in [8] which allows an effective combination of low-level features and deep high-level representations. In order to enforce the spatial and intensity relationships between the organs, the initial segmentation result is further refined by Conditional Random Fields (CRF) with the CRFasRNN architecture. We propose to use a second deep architecture which is designed to be able to make use of the segmentation maps obtained by the first deep architecture of all organs, to learn the anatomical constraints for the one organ that is currently under refinement of its segmentation. We show experimentally that our framework outperforms other state-of-the-art methods. Note that our framework is also generic enough to be applied to other multi-label joint segmentation problems.

2   Method

2.1   SharpMask Feature Fusion Architecture and CRF Refinement

The first deep architecture performs initial segmentation, with its output as a probability map of each voxel belonging to background, esophagus, heart, aorta, or trachea. In order to alleviate the loss of image resolution due to the use of pooling operations in regular Convolutional Neural Networks (CNN); Fully Convolutional Networks (FCN) [5] and some other recent works such as the U-Net [9] and Facebooks SharpMask (SM) [8] have used skip connections, outperforming many traditional architectures. The main idea is to add connections from early to deep layers, which can be viewed as a form of multiscale feature fusion, where low-level features are combined with highly-semantic representations from deep layers.

  In this work, we use an SM architecture that has been shown superior to the regular FCNs for thoracic CT segmentation [12]. The CRF refinement is done subsequently with the CRFasRNN architecture, which formulates the mean field approximation using backpropagable operations [15], allowing the operation to be part of the network (instead of a separated postprocessing step) and even to learn some of its parameters. Thus, a new training is performed for fine-tuning the learned weights from the first step, and also for learning some parameters of the CRF [12]. In the second deep architecture as described below, the segmentation initial results of the surrounding organs by the first deep architecture will be used to refine the segmentation of each target organ separately.

2.2   Learning Anatomical Constraints

The second deep architecture, using SharpMask, is trained to distinguish between background and each target organ under separate refinement. This architecture has two sets of inputs, i.e., (1) the original CT image and (2) the initial segmentation results of the neighbouring organs around the target organ under refinement of segmentation. The main difference of this second deep architecture, compared to the first deep architecture with multiple output channels representing different organs and background, is that it only has two output channels in the last layer, i.e., a probability map representing each voxel belonging to background or a target organ under refinement of segmentation. The basic assumption is that the second deep architecture will learn the anatomical constraints around the target organ under refinement of segmentation and thus help to produce better segmentation for the target organ. In Fig. 3 we show the full framework with both the first deep architecture (top) and the second deep architecture.

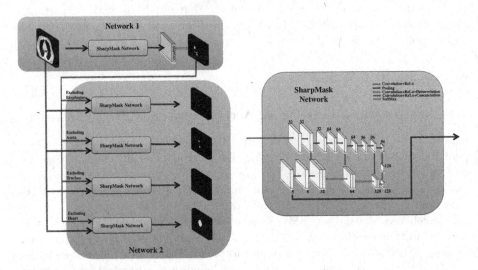

Fig. 3. Proposed architecture for multi-organ segmentation. The core sharpmask network is detailed on the right. Numbers indicate the number of channels at each layer.

Note that our framework (using two deep architectures) shares some similarity with a refinement step, called AutoContext Model (ACM) [13], which has been successfully applied to brain image segmentation [14], by using traditional classifiers such as Random Forests. The main idea of ACM is to iteratively refine the posterior probability maps over the labels, given not only the input features, but also the previous probabilities of a large number of context locations which provide information about neighboring organs, forcing the deep network to learn the spatial constrains for each target organ [14]. In practice, this translates to train several classifiers iteratively, where each classifier is trained not only with the original image data, but also with the probability maps obtained from the previous classifier, which gives additional context information to the new classifier. Comparing our proposed framework with the ACM, we use a deep architecture. Overall, our method has three advantages: (1) it can avoid the design of hand-crafted features, (2) our network can automatically learn the selection of context features, and (3) our method uses less redundant information. Note that, in the classical ACM, the selection of these features must be hard-coded; that is, the algorithm designer has to select a specific number of sparse locations (i.e., using sparse points from rays at different angles from a center point [13]), which makes the context information limited within a certain range by the algorithm designer. On the other hand, in our method, the context information can be automatically learned by the deep network, and limited only by the receptive field of the network (which can even be the whole range of the image in deep networks). Regarding the redundancy, ACM uses the probability maps of all organs as input, which is often very similar to the ground-truth label maps. In this way, the ACM is not able to further refine the results. In our method, we use only the complementary information, the label probability maps of the neighboring organs around the target organ under refinement of segmentation.

3   Experiments

In our implementation, the full slices of the CT scan are the inputs to our proposed framework. Both the first and second architectures use large filters (i.e., 7×7, or $7 \times 7 \times 7$), as large filters have been shown beneficial for CT segmentation [2]. Both 2D and 3D settings are implemented in our study. We have found that, different from MRI segmentation [7], small patches are not able to produce good results for CT segmentation. Thus, we use patches of size $160 \times 160 \times 48$ as the training samples. Specifically, we first build a 3D mesh model for each organ in all the training CT images, and then define each vertex as the mean of a certain Gaussian distribution with diagonal covariance, from which we can sample points as the centers of the respective patches. In this way, the training samples will contain important boundary information and also background information. In particular, the elements in the diagonal are chosen to be 5, in such a way, the kernel size used would include them when centered in the boundary. In addition, it is also important to sample inside the organs and thus, we also sample in an uniform grid.

3.1   Dataset and Pre-processing

The dataset used in this paper contains 30 CT scans, each with lung cancer or Hodgkin lymphoma and 6-fold cross validation is performed. Manual segmentations of the four OAR are available for each CT scan, along with the body contour (which can used to remove background voxels during the training). The scans have $512 \times 512 \times (150 \sim 284)$ voxels with a resolution of $0.98 \times 0.98 \times 2.5 \, \mathrm{mm}^3$. For each CT scan, its intensities are normalized to have zero mean and unit variance, and it is also augmented to generate more CT samples (for improving the robustness of training) through a set of random affine transformations and random deformation fields (generated with B-spline interpolation [6]). In particular, an angle between -5 to $5°$ and a scale factor between 0.9 and 1.1 were randomly selected for each CT scan to produce the random affine transformation. These values were selected empirically trying to produce realistic CT scans similar to those of the available dataset.

3.2 Training

For organ segmentation in CT images, the data samples are often highly imbalanced, i.e., with more background voxels than the target organ voxels. This needs to be considered when computing the loss function in the training. We utilize a weighted cross-entropy loss function, where each weight is calculated as the complement of the probability of each class. In this way, more importance will be given to small organs, and also each class will contribute to the loss function in a more equally way. We have found that this loss function leads to better performance than using a regular (equally-weighted) loss function. However, the results are still not reaching our expected level. For further improvement, we use our above-obtained weights as initialization for the network, and then fine-tune them by using the regular cross-entropy loss. This new integrated strategy always outperforms the weighted or the regular cross-entropy loss function. In our optimization, stochastic gradient descent is used as optimizer, with an initial learning rate of 0.1 that is divided by 10 every 20 epochs, and the network weights are initialized with the Xavier algorithm [3].

3.3 Results

In Fig. 4, we illustrate the improvement on the esophagus segmentation by using our proposed framework with learned anatomical constraints. The last column shows the results using the output of the first network as anatomical constraints. We can see how the anatomical constraints can help produce a more accurate result on the segmentation of the esophagus, even when having air inside (black voxels inside the esophagus). Interestingly, the results obtained by using the output of the first network or the ground-truth manual labels as anatomical constraints are very similar, almost with negligible differences. Similar conclusions can also be drawn for segmentations of other organs. In Fig. 5, we show the segmentation results for the aorta, trachea, and heart, with and without anatomical constraints. In the cases of segmenting the aorta and trachea, the use of anatomical constraints improves the segmentation accuracy. For the trachea, our network is able to generalize to segment the whole part on the right lung (i.e., left side of the image) even when it was segmented partially in the manual ground-truth. On the other hand, for the heart, there are some false positives when using anatomical constraints, as can be seen in the third column. However, accurate contours are obtained, which are even better than those obtained without anatomical constraints, as can be seen in the fourth column.

In Table 1, we report the Dice ratio (DR) obtained using each of the comparison methods, including a state-of-the-art 3D multi-atlas patch-based method (called OPAL (Optimized PatchMatch for Near Real Time and Accurate Label Fusion [10])) and different deep architecture variants using 2D or 3D as well

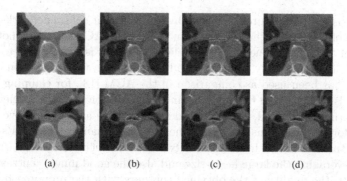

(a)   (b)   (c)   (d)

Fig. 4. Segmentation results for the esophagus. (a) Input data to the second network, with the anatomical constraints overlapped; results using (b) only the first network without anatomical constraints, (c) manual labels on the neighboring organs as anatomical constraints, (d) the output of the first network as anatomical constraints.

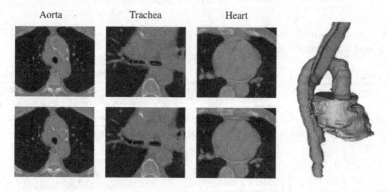

Aorta   Trachea   Heart

Fig. 5. Segmentation without (1st row) and with (2nd row) anatomical constraints. Green contours denote manual ground-truths, and red contours denote our automatic segmentation results. Right panel shows the 3D rendering for our segmented four organs, i.e., aorta (blue), heart (beige), trachea (brown), and esophagus (green). (Color figure online)

as different combinations of strategies. Specifically, SM2D and SM3D refer to the use of the Network 1 in Fig. 3 using 2D or 3D respectively. We also tested their refinement with ACM and CRF, and finally, the proposed framework is denoted as SM2D + Constraints. As OPAL mainly compares patches for guiding the segmentation, OPAL should be effective in segmenting the clearly observable organs, such as the trachea (an identifiable black area), which is true as indicated by the table. But, for the organs with either low contrast or large intensity variation across slices and subjects, which is the case for the esophagus, the respective performance is seriously affected, as the table shows. The highest performance for each organ is obtained by the SM2D-based architectures, while all 3D-based architectures do not improve the segmentation performance. This is possibly due to large slice thickness in the CT scans, as noticed also in [1], where the authors preferred to handle the third dimension by the recurrent neural networks, instead of 3D convolutions. Another observation is that the ACM model is not able to outperform the CRF refinement. We believe that this is mainly due to the fact that the CRF used is fully connected and not based on the neighboring regions. The latter has been used as comparison in the ACM [13], for claiming that the advantage is coming from the context range information that the framework can reach. On the other hand, our proposed framework is able to improve the performance for all the organs, except the heart whose quantitative results are very similar to those obtained by the first network, and which can be well-segmented by it, by leveraging the large heart size and also the good image contrast around it. However, the quality of the obtained contours with the proposed framework is better as shown in Fig. 5. Although room for improvement is still left for the esophagus (with mean DR value of 0.69), the experimental results show that our proposed framework does bring an improvement, compared to the other methods.

Table 1. Comparison of mean DR ± stdev by different methods. Last column indicates our proposed framework.

	OPAL	SM3D	SM3D + ACM	SM2D	SM2D + CRF	SM2D + ACM	SM + Constraints
Esoph.	0.39 ± 0.05	0.55 ± 0.08	0.56 ± 0.05	0.66 ± 0.08	0.67 ± 0.04	0.67 ± 0.04	**0.69 ± 0.05**
Heart	0.62 ± 0.07	0.77 ± 0.05	0.83 ± 0.02	0.89 ± 0.02	0.90 ± 0.01	**0.91 ± 0.01**	0.90 ± 0.03
Trach.	0.80 ± 0.03	0.71 ± 0.06	0.82 ± 0.03	0.83 ± 0.06	0.82 ± 0.06	0.79 ± 0.06	**0.87 ± 0.02**
Aorta	0.49 ± 0.10	0.79 ± 0.06	0.77 ± 0.04	0.85 ± 0.06	0.86 ± 0.05	0.85 ± 0.06	**0.89 ± 0.04**

4 Conclusions

We have proposed a novel framework for joint segmentation of OAR in CT images. It provides a way to learn the relationship between organs which can give anatomical contextual constraints in the segmentation refinement procedure to improve the performance. Our proposed framework includes two collaborative architectures, both based on the SharpMask network, which allows for effective combination of low-level features and deep highly-semantic representations. The main idea is to implicitly learn the spatial anatomical constraints in the second deep architecture, by using the initial segmentations of all organs (but a target organ under refinement of segmentation) from the first deep architecture. Our experiments have shown that the initial segmentations of the surrounding organs can effectively guide the refinement of segmentation of the target organ. An interesting observation is that our network is able to automatically learn spatial constraints, without specific manual guidance.

Acknowledgment. This work is co-financed by the European Union with the European regional development fund (ERDF, HN0002137) and by the Normandie Regional Council via the M2NUM project.

References

1. Chen, J., Yang, L., Zhang, Y., Alber, M., Chen, D.Z.: Combining fully convolutional and recurrent neural networks for 3d biomedical image segmentation. In: NIPS, pp. 3036–3044 (2016)
2. Dou, Q., Chen, H., Jin, Y., Yu, L., Qin, J., Heng, P.: 3d deeply supervised network for automatic liver segmentation from CT volumes. CoRR abs/1607.00582 (2016)
3. Glorot, X., Bengio, Y.: Understanding the difficulty of training deep feedforward neural networks. In: AISTATS (2010)
4. Han, M., Ma, J., Li, Y., Li, M., Song, Y., Li, Q.: Segmentation of organs at risk in CT volumes of head, thorax, abdomen, and pelvis. In: Proceedings of SPIE, vol. 9413 (2015). Id: 94133J-6
5. Long, J., Shelhamer, E., Darrell, T.: Fully convolutional networks for semantic segmentation. In: CVPR (2015)
6. Milletari, F., Navab, N., Ahmadi, S.: V-net: fully convolutional neural networks for volumetric medical image segmentation. CoRR abs/1606.04797 (2016)
7. Nie, D., Wang, L., Gao, Y., Shen, D.: Fully convolutional networks for multi-modality isointense infant brain image segmentation. In: ISBI, pp. 1342–1345 (2016)
8. Pinheiro, P.H.O., Lin, T., Collobert, R., Dollár, P.: Learning to refine object segments. CoRR abs/1603.08695 (2016)
9. Ronneberger, O., Fischer, P., Brox, T.: U-Net: convolutional networks for biomedical image segmentation. In: Navab, N., Hornegger, J., Wells, W.M., Frangi, A.F. (eds.) MICCAI 2015. LNCS, vol. 9351, pp. 234–241. Springer, Cham (2015). doi:10.1007/978-3-319-24574-4_28
10. Ta, V.-T., Giraud, R., Collins, D.L., Coupé, P.: Optimized PatchMatch for near real time and accurate label fusion. In: Golland, P., Hata, N., Barillot, C., Hornegger, J., Howe, R. (eds.) MICCAI 2014. LNCS, vol. 8675, pp. 105–112. Springer, Cham (2014). doi:10.1007/978-3-319-10443-0_14
11. Tong, T., et al.: Discriminative dictionary learning for abdominal multi-organ segmentation. Med. Image Anal. **23**, 92–104 (2015)
12. Trullo, R., Petitjean, C., Ruan, S., Dubray, B., Nie, D., Shen, D.: Segmentation of organs at risk in thoracic CT images using a sharpmask architecture and conditional random fields. In: ISBI (2017)
13. Tu, Z., Bai, X.: Auto-context and its application to high-level vision tasks and 3d brain image segmentation. IEEE Trans. Pattern Anal. Mach. Intell. **32**(10), 1744–1757 (2010)
14. Wang, L., et al.: Links: learning-based multi-source integration framework for segmentation of infant brain images. NeuroImage **108**, 160–172 (2015)
15. Zheng, S., et al.: Conditional random fields as recurrent neural networks. In: ICCV (2015)

Accelerated Magnetic Resonance Imaging by Adversarial Neural Network

Ohad Shitrit[✉] and Tammy Riklin Raviv

Department of Electrical Engineering, The Zlotowski Center for Neuroscience,
Ben-Gurion University of the Negev, Beersheba, Israel
shohad25@gmail.com, rrtammy@bgu.ac.il

Abstract. A main challenge in Magnetic Resonance Imaging (MRI) for clinical applications is speeding up scan time. Beyond the improvement of patient experience and the reduction of operational costs, faster scans are essential for time-sensitive imaging, where target movement is unavoidable, yet must be significantly lessened, e.g., fetal MRI, cardiac cine, and lungs imaging. Moreover, short scan time can enhance temporal resolution in dynamic scans, such as functional MRI or dynamic contrast enhanced MRI. Current imaging methods facilitate MRI acquisition at the price of lower spatial resolution and costly hardware solutions.

We introduce a practical, software-only framework, based on deep learning, for accelerating MRI scan time allows maintaining good quality imaging. This is accomplished by partial MRI sampling, while using an adversarial neural network to estimate the missing samples. The interplay between the generator and the discriminator networks enables the introduction of an adversarial cost in addition to a fidelity loss used for optimizing the peak signal-to-noise ratio (PSNR). Promising image reconstruction results are obtained for 1.5T MRI where only 52% of the original data are used.

1 Introduction

Magnetic Resonance Imaging (MRI) is a non-ionizing imaging modality, and is therefore widely used in diagnostic medicine and biomedical research. The physical principles of MRI are based on a strong magnetic field and pulses of radio frequency (RF) electromagnetic radiation. Images are produced when hydrogen atoms, which are prevalent in living organisms, emit the absorbed RF energy that is then received by antennas in close proximity to the anatomy being examined. Spatial localization of the detected MRI signals is obtained by varying the magnetic field gradients. The discretized RF output is presented in a Fourier space (called K-space), where the x-axis is refers to the frequency and the y-axis to the phase. An inverse fast Fourier transform (IFFT) of the K-space is then

© Springer International Publishing AG 2017
M.J. Cardoso et al. (Eds.): DLMIA/ML-CDS 2017, LNCS 10553, pp. 30–38, 2017.
DOI: 10.1007/978-3-319-67558-9_4

used for generating anatomically meaningful MRI scans. Figure 1 presents K-space traversal patterns used in conventional imaging. Each row of the k-space is acquired after one RF excitation pulse. The number of rows multiplied by the number of slices (z-axis) determines the total scan time.

The duration of standard single structural MRI acquisition is approximately 5 min. Usually, several scans of different modalities or a sequence of scans are acquired such that the overall scan time is much longer. Lengthy imaging process reduces patient comfort and is more vulnerable to motion artifacts. In cases where motion is inevitable, e.g., fetal MRI, cardiac cine, and lungs imaging, scan time must be significantly shortened, otherwise the produced images might be useless. Moreover, in dynamic MRI sequences, acquisition must be brief such that the temporal resolution of the sequence would allow capturing significant temporal changes, e.g., instantaneous increment of the contrast-enhanced material concentration in DCE-MRI or differences in hemodynamic response expressed in fMRI [10].

A straight forward reduction of the scan time can be obtained by sampling fewer slices, thus reducing the spatial resolution in the z-axis. Spatial distances between adjacent slices of fetal MRI or fMRI, for example, are often as high as 0.5 cm. Therefore, a significant portion of the potential input is not conveyed through imaging. On the other hand, under-sampling in the x-y domain leads to aliasing, as predicted by the Nyquist sampling theorem.

Numerous research groups as well as leading MRI scanner manufacturers make significant efforts to accelerate the MRI acquisition process. Hardware solutions allow parallel imaging by using multiple coils [17] to sample k-space data. There exist two major approaches [3] that are currently implemented in commercial MRI machines. Both reconstruct an image from the under-sampled k-space data provided by each of the coils. The sensitivity encoder (SENSE) transforms the partial k-spaces into images, then merges the resulting aliased images into one coherent image [14]. The GeneRalized Autocalibrating Partial Parallel Acquisition (GRAPPA) techniques [7] operate on signal data within the complex frequency domain before the IFFT.

The compressed sensing (CS) technique [4] allows efficient acquisition and reconstruction of a signal with fewer samples than the Nyquist-Shannon sampling theorem requires, if the signal has sparse representation in a known transform domain. Using CS for MRI reconstruction by sampling a small subset of the k-space grid had been proposed in [9]. The underlying assumption is that the undersampling is random, such that the zero-filled Fourier reconstruction exhibits incoherent artifacts that behave similarly to additive random noise. This, however, would require specified pulse programming.

Recently machine learning techniques based on manifold learning [1,18] and dictionary learning [2,16] were suggested for MRI reconstruction. MRI reconstruction using convolutional neural networks (CNN) was introduced in [19]. The network learns the mapping between zero-filled and fully-sampled MR images. In [12], residual network was proposed for MRI super-resolution. Their model is able to receive multiple inputs acquired from different viewing planes for better image reconstruction. Both works address the reconstruction problems in the image domain rather than the k-space domain. The proposed framework utilizes recent advances in deep learning, while similarly to the CS methods addresses MRI reconstruction directly from the k-space. Specifically, we use generative adversarial networks (GAN) [6,13,15]. GANs are based on the interplay between two networks: a generator and a discriminator. The generator is capable of learning the distribution over a data-base, and sample realizations of this distribution. The discriminator is trained to distinguish between 'generated' samples and real ones. This powerful combination has been used for generating Computed Tomography (CT)-like images from MRIs [11]. Here, the generator is used for reconstruction of the entire k-space grid from under-sampled data. Its loss is a combination of an adversarial loss, based on the discriminator output and a fidelity loss with respect to the fully sampled MRI. Promising results are obtained for brain MRI reconstruction using only 52% of the data.

The paper is organized as follows. Section 2 presents some theoretical foundation and our method. Section 3 describes the experimental results. Conclusions and future directions are describes in Sect. 4.

2   Method

2.1   K-space

Let u denote the desired signal, a 2D MR image, obtained by the IFFT of the complex k-space signal s_0. Let M_F denote a full sampling mask such that the reconstructed MR image is:

$$u = F^H M_F \odot s_0 \tag{1}$$

where H is the Hermitian transpose operation, $\odot$ denotes element-wise multiplication, and F^H is an orthonormal 2D IFFT operator, such that $F^H F = I$. While sampling part of the k-space, using M_p as a sampling mask, the reconstructed MR image suffers from artifacts and aliasing. An example of the under-sampling (52%) artifacts is shown in Fig. 1.

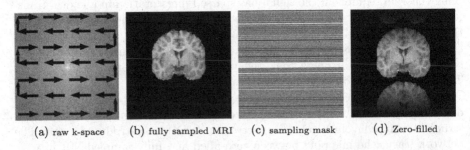

(a) raw k-space   (b) fully sampled MRI   (c) sampling mask   (d) Zero-filled

Fig. 1. Under-sampling artifacts: the arrows illustrate the sampling methodology

2.2 Objective

Let $s_p = M_p \odot s_0$ denote the under-sampled k-space. Given a sampling mask and a model f, defined by the set of parameters Θ, our goal is to estimate the missing k-space samples such that:

$$\Theta = \arg \min_{\Theta} L(F^H f (s_p; \Theta), \mathbf{u}) \tag{2}$$

where $L(\cdot)$ is the loss function. While choosing the loss to be L2 norm is reasonable for natural images, for the k-space, which has different spatial features, this may not be enough. As mentioned in [13], L2 minimization provides a blurry solution. Averaging the high frequency details in the k-space domain results in very poor reconstruction. In order to address this problem, we used the adversarial loss, based on GAN.

We trained our model using the adversarial strategy, as described in [6,15]. This method is based on a generator G, which takes noise z with uniform distribution $p_u(z)$ as input and generates samples from the data distribution. A discriminator D is trained to distinguish between "real" examples from the data and generated ("fake") examples from G. During the training process, we optimize G to maximize the discriminator's probability of error. Simultaneously, D is getting better and provides more accurate predictions.

Let s_0 denote a "real" k-space sample from the distribution $p_r(s_0)$. The following optimization process can be described by two-players min-max game:

$$\min_{G} \max_{D} \mathbb{E}_{s_0 \sim p_r(s_0)} \log [D(\mathbf{x})] + \mathbb{E}_{z \sim p_u(z)} \log [1 - D(G(z))] \tag{3}$$

In equilibrium, the generator G is able to generate samples that look like the real data. In our case, G estimates the missing k-space samples from a linear combination of the sampled data and a uniform noise with distribution $p_u(z)$. An L2 fidelity constraint is added to the adverbial loss of the generator, as follows:

$$L_G = \alpha \cdot E_{z \sim p_u(z)} \log \left[1 - D\left(F^{-1}(\hat{s}_0)\right)\right]$$
$$+ \beta \cdot \| (1 - M_p) \odot (\hat{s}_0 - s_0) \|_2^2 \tag{4}$$

where $\hat{s}_0$ is the estimated k-space and $\alpha = 1$, $\beta = 0.8$ are hyperparameters tuned by a cross-validation process. The discriminator's input is the reconstructed MR image, i.e., after IFFT. By that, we are integrating the reconstruction phase in our optimization.

2.3 Network Architecture

The generator input is a two-channel image representing the real and the imaginary parts of the partially sampled k-space image, s_p. Each missing sample is initialized by uniform i.i.d. noise. The pixel (i, j) in the generator input image is:

$$G_{in}(i, j) = s_{p_{i,j}} + (1 - M_p)_{i,j} z_{i,j} \tag{5}$$

Due to the combination of the adversarial and the fidelity loss, G produces reasonable k-space samples from a given samples and noise distribution $p_u(z)$. In order to use the sampled data, s_p, and estimate only the missing samples we used a residual network [8] as used in [12], such that:

$$\hat{s}_0 = s_p + (1 - M_p) \odot G_{out} \qquad (6)$$

where G_{out} is the generator output. Figure 2 describes our framework:

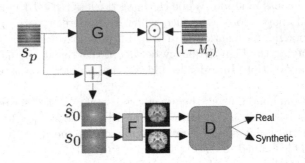

Fig. 2. Framework architecture: G and D are the generator and discriminator networks, respectively. F is a $2D$ IFFT operator.

A common architecture is used for the discriminator, composed of convolutional layers, batch normalization, and leaky-ReLU as suggested in [15]. For the generator, we compose a dedicated architecture based on multi-channel input for representing the real and imaginary components. Both architectures are shown in Fig. 3. The training methodology is doing k_g generator update steps for each discriminator single step.

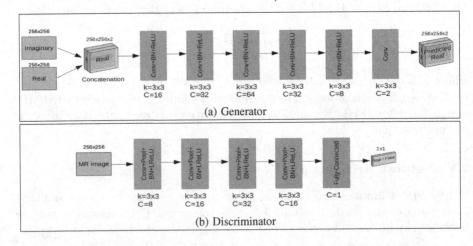

Fig. 3. Networks architecture. The generator input is a two-channel signal, real and imaginary. For each layer, k is the kernel size and C is the number of output channels.

3   Experimental Results

The training data consists of 55 3D brain MRI (T1) scans of different patients, acquired by 1.5T MR machine with resolution of 256×256 pixels. We used $3k$ 2D slices for training and $1.2k$ for testing. In order to create k-space images for training, inverse orthonormal 2D FFT is applied to the fully-sampled MR images. We sample the k-space images along the phase axis using Cartesian

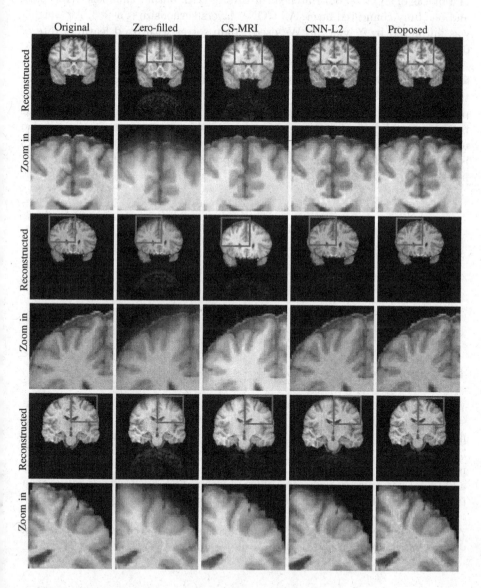

Fig. 4. Examples of reconstructed MR images from under-sampled k-space.

binary mask, while keeping 5% from the centered phases, which is the area that contains 95% of the energy (see Fig. 1). Data augmentation is created by random offsets of the proposed mask and image flipping. This leads us to reconstruction of the MR image from only 52% of the original k-space data.

The generator is composed of 5 blocks of CONV-BatchNorm-ReLU, with output channels $16, 32, 64, 32, 8$, respectively. The last layer is CONV with two outputs channels (for real and imaginary parts). The discriminator is composed of 4 blocks of CONV-Pool-BatchNorm-LReLU with output channels $8, 16, 32, 16$ and one fully-connected layer. All CONV layers kernel size is 3×3. All weights was initialized by Xavier [5]. We used Adam solver with fixed learning rate of 0.0005 and set k_g to 20.

We compare the proposed method to reconstruction results obtained by using a conventional compressed sensing method CS-MRI [9] and Zero-filling. In addition, we trained a generator (G) using only L2 loss (CNN-L2). The same sampling mask was used for all cases. Visual comparison is shown in Fig. 4. MRIs reconstructed by using the suggested adversarial loss have stronger contrast and no significant aliasing or artifacts. Quantitative evaluation using PSNR values is presented in Table 1. Note that the proposed method outperforms the other methods.

Table 1. Error in PSNR

PSNR	Mean	std
Zero-filled	30.48	0.13
CS-MRI	36.78	0.4
CNN-L2	37.12	0.82
Proposed	**37.95**	0.6

4   Conclusions

We proposed a software-only framework, using GANs for accelerating MRI acquisition. Specifically, high-quality MRI reconstruction using only 52% of the original k-space data is demonstrated. The key idea is based on utilizing an adversarial loss in addition to L2 loss. It is worth mentioning that the proposed sampling mask is currently implemented in commercial MRI machines with no need for additional hardware or dedicated pulse programming. Future work will concentrate on generation of MRI in the presence of pathologies.

Acknowledgment. This research is partially supported by the Israel Science Foundation (T.R.R. 1638/16) and the IDF Medical Corps (T.R.R.).

References

1. Bhatia, K.K., Caballero, J., Price, A.N., Sun, Y., Hajnal, J.V., Rueckert, D.: Fast reconstruction of accelerated dynamic MRI using manifold kernel regression. In: Navab, N., Hornegger, J., Wells, W.M., Frangi, A.F. (eds.) MICCAI 2015. LNCS, vol. 9351, pp. 510–518. Springer, Cham (2015). doi:10.1007/978-3-319-24574-4_61
2. Caballero, J., Price, A.N., Rueckert, D., Hajnal, J.V.: Dictionary learning and time sparsity for dynamic MR data reconstruction. IEEE Trans. Med. Imaging **33**(4), 979–994 (2014)
3. Deshmane, A., Gulani, V., Griswold, M.A., Seiberlich, N.: Parallel MR imaging. J. Magn. Reson. Imaging **36**(1), 55–72 (2012)
4. Donoho, D.L.: Compressed sensing. IEEE Trans. Inf. Theory **52**(4), 1289–1306 (2006)
5. Glorot, X., Bengio, Y.: Understanding the difficulty of training deep feedforward neural networks. In: AISTATS, vol. 9, pp. 249–256 (2010)
6. Goodfellow, I., Pouget-Abadie, J., Mirza, M., Xu, B., Warde-Farley, D., Ozair, S., Courville, A., Bengio, Y.: Generative adversarial nets. In: Advances in Neural Information Processing Systems, pp. 2672–2680 (2014)
7. Griswold, M.A., Jakob, P.M., Heidemann, R.M., Nittka, M., Jellus, V., Wang, J., Kiefer, B., Haase, A.: Generalized Autocalibrating Partially Parallel Acquisitions (GRAPPA). Magn. Reson. Med. **47**(6), 1202–1210 (2002)
8. He, K., Zhang, X., Ren, S., Sun, J.: Deep residual learning for image recognition. In: Proceedings of the IEEE Conference on Computer Vision and Pattern Recognition, pp. 770–778 (2016)
9. Lustig, M., Donoho, D., Pauly, J.M.: Sparse MRI: the application of compressed sensing for rapid MR imaging. Magn. Reson. Med. **58**(6), 1182–1195 (2007)
10. Moeller, S., Yacoub, E., Olman, C.A., Auerbach, E., Strupp, J., Harel, N., Uğurbil, K.: Multiband multislice GE-EPI at 7 tesla, with 16-fold acceleration using partial parallel imaging with application to high spatial and temporal whole-brain fMRI. Magn. Reson. Med. **63**(5), 1144–1153 (2010)
11. Nie, D., Trullo, R., Petitjean, C., Ruan, S., Shen, D.: Medical image synthesis with context-aware generative adversarial networks. arXiv preprint arXiv:1612.05362 (2016)
12. Oktay, O., et al.: Multi-input cardiac image super-resolution using convolutional neural networks. In: Ourselin, S., Joskowicz, L., Sabuncu, M.R., Unal, G., Wells, W. (eds.) MICCAI 2016. LNCS, vol. 9902, pp. 246–254. Springer, Cham (2016). doi:10.1007/978-3-319-46726-9_29
13. Pathak, D., Krahenbuhl, P., Donahue, J., Darrell, T., Efros, A.A.: Context encoders: feature learning by inpainting. In: Proceedings of the IEEE Conference on Computer Vision and Pattern Recognition, pp. 2536–2544 (2016)
14. Pruessmann, K.P., Weiger, M., Scheidegger, M.B., Boesiger, P., et al.: Sense: sensitivity encoding for fast MRI. Magn. Reson. Med. **42**(5), 952–962 (1999)
15. Radford, A., Metz, L., Chintala, S.: Unsupervised representation learning with deep convolutional generative adversarial networks. arXiv preprint arXiv:1511.06434 (2015)
16. Ravishankar, S., Bresler, Y.: MR image reconstruction from highly undersampled k-space data by dictionary learning. IEEE Trans. Med. Imaging **30**(5), 1028–1041 (2011)
17. Roemer, P.B., Edelstein, W.A., Hayes, C.E., Souza, S.P., Mueller, O.: The NMR phased array. Magn. Reson. Med. **16**(2), 192–225 (1990)

18. Usman, M., Vaillant, G., Atkinson, D., Schaeffter, T., Prieto, C.: Compressive manifold learning: estimating one-dimensional respiratory motion directly from undersampled k-space data. Magn. Reson. Med. **72**(4), 1130–1140 (2014)
19. Wang, S., Su, Z., Ying, L., Peng, X., Zhu, S., Liang, F., Feng, D., Liang, D.: Accelerating magnetic resonance imaging via deep learning. In: 2016 IEEE 13th International Symposium on Biomedical Imaging (ISBI), pp. 514–517. IEEE (2016)

Left Atrium Segmentation in CT Volumes with Fully Convolutional Networks

Honghui Liu, Jianjiang Feng$^{(\boxtimes)}$, Zishun Feng, Jiwen Lu, and Jie Zhou

Department of Automation, Tsinghua University, Beijing, China
liuhh15@mails.tsinghua.edu.cn, jfeng@tsinghua.edu.cn

Abstract. Automatic segmentation of the left atrium (LA) is a fundamental task for atrial fibrillation diagnosis and computer-aided ablation operation support systems. This paper presents an approach to automatically segmenting left atrium in 3D CT volumes using fully convolutional neural networks (FCNs). We train FCN for automatic segmentation of the left atrium, and then refine the segmentation results of the FCN using the knowledge of the left ventricle segmented using ASM based method. The proposed FCN models were trained on the STACOM'13 CT dataset. The results show that FCN-based left atrium segmentation achieves Dice coefficient scores over 93% with computation time below 35s per volume, despite of the high variation of LA.

1 Introduction

Among the most common and hazardous cardiovascular diseases, atrial fibrillation (AF) is usually characterized by abnormally rapid and irregular heart rhythm. In recent years, surgical treatment for AF, typically ablation procedure (AP), has gradually become the mainstream [9]. AP is minimally invasive, which is the main consideration for some patients. Computer aided atrial detection and precise segmentation can help doctors gain valuable preoperative information. Computed tomography (CT) has been widely used for diagnosis and treatment for cardiovascular disease. However, automated LA segmentation from CT data is still a non trivial task. Due to larger shape variations of LA than other organs, especially four pulmonary veins (PV) and left atrial appendage (LAA). Some existing methods have achieved good results, but the training cost is high [13]. In this paper, we propose a fully automatic LA segmentation system on conventional CT. Based on state-of-the-art fully convolution network (FCN), and statistical shape models, the proposed method is efficient, robust, and is able to obtain good results with small training dataset.

2 Related Work

Image segmentation is a fundamental task in medical image analysis. Producing accurate segmentation is difficult due to many influencing factors: noise, pathology, occlusion, and object shape complexity. Some semi-automated or automated

© Springer International Publishing AG 2017
M.J. Cardoso et al. (Eds.): DLMIA/ML-CDS 2017, LNCS 10553, pp. 39–46, 2017.
DOI: 10.1007/978-3-319-67558-9_5

algorithms have been applied to LA segmentation problem. Daoudi et al. [3] proposed an algorithm based on active contour, with region growing and snakes. This type of classical methods are simple and fast, but sensitive to image quality. Sandoval et al. [6] proposed an algorithm based on multi-atlas. Multi-atlas has remarkable advantages of robustness and making good use of a priori anatomical information. The main shortcoming of multi-atlas is computing cost. Image registration for 3D volume is quite time consuming even though GPU parallel acceleration strategy has been applied. Zuluaga et al. [15] proposed another multi-atlas propagation based segmentation. Nowadays, statistical shape models are widely used in image segmentation. Zheng et al. [14] proposed an algorithm based on a multi-part shape model and marginal space learning, which divides LA into six-parts: LA body, LAA and four PVs. This algorithm is efficient, and robust to image quality. However, a large manually labelled data set is required. For medical image application, this limitation cannot be ignored till now. Moreover, compared to LA body, the LAA has larger anatomical variations, thus method using strong shape priors is not suitable [5].

Our work draws on recent progress of deep neural nets [12]. Image segmentation can be viewed as a pixel-wise classification task. In recent years, convolutional neural networks (CNN) achieve success on image classification problem. Therefore, taking advantage of coarse but highly abstract hidden layer output,

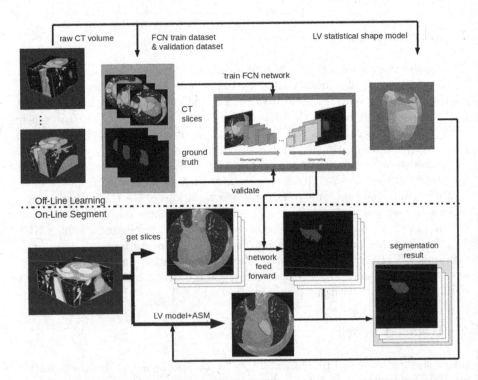

Fig. 1. Pipeline of the proposed approach: off-line training and on-line testing

pixel-wise classification can also be solved with CNN [10]. For segmentation task, the target is a dense label map, which is of the same size as the input. Specifically, to achieve this coarse-to-fine processing, we can add upsampling layer and deconvolution layer upon CNN, and then obtain the so-called fully convolutional network (FCN). Owing to its powerful ability to learn both local and global information, FCN has made great progress in image segmentation.

In this work, we present an automatic LA segmentation method for 3D CT. We transferred deep neural network architectures for natural scenes to medical task, thus good generalization ability can be expected. Furthermore, to achieve both strong deformable ability and shape constraints, we combined FCN and statistical shape models, and thus improve the accuracy while preserving the efficiency.

In the following sections, we will demonstrate our proposed pipeline, including preprocessing (Sect. 3.1), FCN (Sect. 3.2) and shape model based postprocessing (Sect. 3.3), report experiments (Sect. 4), and summarize the paper (Sect. 5).

3   Method

We trained a hourglass-shaped architecture with a per-pixel logistic loss, and validated with the standard pixel intersection over union metric. Our network includes 12 stacked convolution layers, each followed by max pooling layer and ReLU activation layer. Two deconvolution layers were added on the top of the CNN architecture.

Training FCN with small training dataset has been an awkward problem. To resolve this problem, we adopted two measures. First, we pre-train our network on a big dataset [4] with supervision. Next, the network was fine-tuned on our medical dataset. We selected slices from raw volume data along different axes. Thereby an abundant training dataset with more than 1000 slices can meet the demand for network training. This treatment may inevitably lose some 3D structure informatinon. However, with postprocessing from the 3D perspective, this 3D information lost will be minimized. Figure 1 shows the steps of our proposed approach.

3.1   Preprocessing

Our preprocessing consists of two aspects. First, the image contrast is increased with histogram equalization. Next, we convert the gray scale image to pseudo color image. Taking advantage of three color channels, network training can be improved.

3.2   Fully Convolutional Network

We denote the 3D raw CT volume as I. For our two class problem, the set of possible labels is $L = \{0, 1\}$. Foreground LA voxels are marked with label 1 while other background voxels are marked with 0. For each voxel v, we define a variable $y_v \in L$ that denotes the assigned label. Given the image I, the FCN calculates the probability of assigning label k to v, described by $P(y_v = k|I)$.

The first step in the training process, we trained FCN-32s network. FCN-32s networks give coarse segmentation results, much local information of the input is dropped while passing convolution and pooling layers. Hence we added links form lower layers to the final layer, and the whole network turns into a directed acyclic graph (DAG). Compared with simple linear networks, DAG-style networks combine the global structure with local information, thus the results became more sophisticated. Combining the second-to-last and then the third-to-last pooling layers, FCN-16s and FCN-8s structure were generated in succession.

Our dataset included 1200 slices, which were extracted from 10 CT volumes. These slices were divided into two parts: 1000 slices as training set, and the other 200 as validation set.

Due to the fact that FCN was trained on slices, some 3D structure information was lost unavoidably, continuity and smoothness cannot be ensured for the final 3D model. Some tiny tissues were misclassified as foreground. Therefore, we extract the maximum connected component from the FCN segmentation result, and apply hole filling algorithm to the maximum connected component. By doing this, most false positives can be excluded, then segmentation result become a smooth and solid model.

3.3   Shape Constraints

The knotty problem for LA segmentation task is determination for the fuzzy boundary of LA and LV. These two chambers are connected in structure, and may have similar gray-scale for some images (see Fig. 2(a)). Therefore, many machine learning methods [8] have shortcoming of deciding the boundary of LA and LV, as well as FCN method. These methods mainly focus on LA segmentation, while LV is ignored.

Some previous work [1] adopted 3D CRF as postprocessing to improve the segmentation results. Dense CRF mainly focuses on prior and posterior distrib-

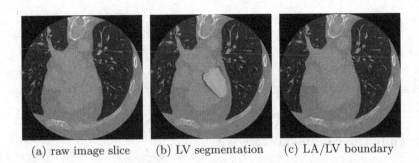

(a) raw image slice   (b) LV segmentation   (c) LA/LV boundary

Fig. 2. Segment LV with ASM based method, to get the LA/LV boundary

ution of pixel classes, rather than global shape constraints. Furthermore, CRF method is time consuming, for both offline learning and online testing.

For fuzzy boundary problem, shape models with constraints are shown to be more robust [13]. Compared with LA wall, LV epicardium is much thicker, and the clearly visible posterior borders of the LV and LA are flat. We establish a statistic shape model to segment LV epicardium with active shape model (ASM) method [2]. The top of LV is regarded as the boundary of these two chambers. These two chambers can therefore be clearly distinguished (see Fig. 2).

Table 1. LA segmentation performance comparison by three evaluation metrics. Acronyms of methods are from [13].

Method	Evaluation criterion		
	DC	Time(s)	Train dataset size
BECHAR [3]	0.66	900	10
INRIA [8]	0.82	1500	10
LTSI_VRG [6]	0.88	4700	10
SIE_PMB [14]	0.94	3	457
UCL_1C [15]	0.93	4200	30
Proposed	**0.93**	**32**	**30**

4 Experiments

We trained networks on the benchmark CT dataset for Left Atrial Segmentation Challenge (LASC) carried out at the STACOM'13 workshop [13]. Ten CT volumes were provided with expert manual segmentations, and the other twenty volumes were used for algorithm evaluation. For statistical shape model method, we randomly selected other twenty volumes from patients who underwent a CTA examination using a Philips Brilliance iCT256 scanner. The volumes in the whole dataset set contain 210 to 455 slices while the size of all slices is of 512×512 pixels. The resolution inside each slice is isotropic but varies between $0.314\,mm$ and $0.508\,mm$ for different volumes, and the slice thickness varies between $0.450\,mm$ and $0.510\,mm$.

For the training process, the mean validation pixel-wise accuracy of FCN-32s, 16s, and 8s is 86%, 92%, and 95%, respectively. In general, 32s networks give coarse results, with very low false positive rate. Furthermore, 16s networks give higher accuracy, but high false positive rate as well. Moreover, 8s networks give obviously better results than the others.

The test data set includes 20 volumes. We choose Dice coefficient as evaluation criterion, which is defined by:

$$DC = \frac{2|V_{GT} \bigcap V_{SEG}|}{|V_{GT}| + |V_{SEG}|}, \tag{1}$$

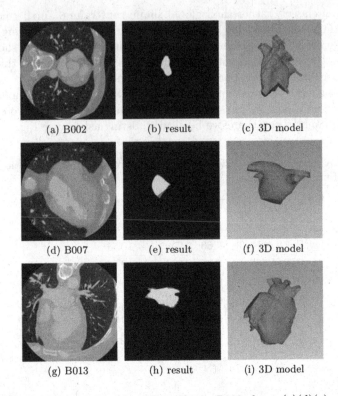

(a) B002   (b) result   (c) 3D model

(d) B007   (e) result   (f) 3D model

(g) B013   (h) result   (i) 3D model

Fig. 3. Segmentation examples for B002, B007, B013 data. (a)(d)(g): raw image, (b)(e)(h): segmentation result, and (c)(f)(i): segmented 3D model. Yellow region indicates true positive, red indicates false negative, and magenta indicates false positive. (Color figure online)

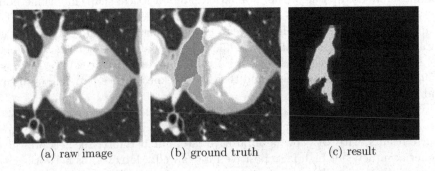

(a) raw image   (b) ground truth   (c) result

Fig. 4. LAA segmentation example. Note that LA body ground truth is from the original STACOM' 13 dataset, and LAA ground truth was labelled by ourselves.: (a) raw image, (b) ground truth, LAA is labelled with blue, and (c) segmentation result, the meaning for each color is the same as Fig. 3. (Color figure online)

where we denote the ground truth volume mask by V_{GT}, and the segment result mask by V_{SEG}.

Table 1 shows the results of some different methods. Segmentation accuracy is the main consideration for algorithm evaluation, while the other two aspects affect the applicability.

As a comparison experiment, we choose dense 3D CRF [7] as postprocessing method, the result shows that ASM based LV segmentation gives simpler yet robust LA/LV boundary. For the other comparison experiment, we establish LA statistical shape model, and apply ASM method for both LA and LV, to compare FCN with ASM method for LA segmentation. Table 2 shows the results of these comparison experiments.

As an example, Fig. 3 shows segmentation results for CT volume B002, B007, and B013. The result and ground truth are compared in Fig. 3(b), (e), (h).

STACOM' 13 challenge ignored LAA while evaluating each method. However, atrial fibrillation usually leads to LAAs emptying obstruction, blood stasis and induces thrombosis [11]. An additional advantage of our method, also not reflected in Table 1, is that it can also segments LAA accurately (see Fig. 4), which is hard for ASM based methods.

Table 2. Segmentation performance comparison by different LA/LV processing

Method	Evaluation criterion		
	DC	Time(s)	Train dataset size
FCN + CRF [7]	0.87	120	30
LA ASM + LV ASM	0.85	15	30
FCN + LV ASM (Proposed)	**0.93**	**32**	**30**

5 Conclusion

In this paper, we propose a fully convolutional network based automated approach for left atrium segmentation, and adopt statistical shape models to make up for the lack of constraints. The experiments showed that our proposed method is comparable with state-of-the-art methods while using only a small training set. Furthermore, we preliminarily attempted to combine neural networks with statistical shape models.

The proposed method can be improved from many perspectives. For reliability and simplicity, we prefer to learn LA/LV fuzzy boundary in FCN model itself. Neural networks and shape models have respective advantages, next step we will attempt to better unify the two into the same framework. As the dataset increases, the FCN models are expected to be trained better in the future.

Acknowledgements. This work is supported by the National Natural Science Foundation of China under Grants 61622207, 61373074, 61225008, and 61572271.

References

1. Christ, P.F., Elshaer, M.E.A., et al.: Automatic liver and lesion segmentation in CT using cascaded fully convolutional neural networks and 3D conditional random fields. In: Proceedings of International Conference on Medical Image Computing and Computer-Assisted Intervention, pp. 415–423 (2016)
2. Cootes, T.F., Taylor, C.J., Cooper, D.H., Graham, J.: Active shape models-their training and application. Comput. Vis. Image Underst. **61**(1), 38–59 (1995)
3. Daoudi, A., Mahmoudi, S., Chikh, M.A.: Automatic segmentation of the left atrium on CT images. In: Proceedings of International Workshop on Statistical Atlases and Computational Models of the Heart, pp. 14–23 (2013)
4. Everingham, M., Eslami, S.M.A., Gool, L.V., Williams, C.K.I., Winn, J., Zisserman, A.: The PASCAL visual object classes challenge: a retrospective. Int. J. Comput. Vis. **111**(1), 98–136 (2015)
5. Heimann, T., Meinzer, H.: Statistical shape models for 3D medical image segmentation: a review. Med. Image Anal. **13**(4), 543–563 (2009)
6. Iglesias, J.E., Sabuncu, M.R.: Multi-atlas segmentation of biomedical images: a survey. Med. Image Anal. **24**(1), 205–219 (2015)
7. Koltun, V.: Efficient inference in fully connected CRFs with Gaussian edge potentials. Adv. Neural Inf. Proc. Syst. **2**(3), 4 (2011)
8. Margeta, J., McLeod, K., Criminisi, A., Ayache, N.: Decision forests for segmentation of the left atrium from 3D MRI. In: Proceedings of International Workshop on Statistical Atlases and Computational Models of the Heart, pp. 49–56 (2013)
9. Marrouche, N.F., Wilber, D.J., Hindricks, G., et al.: Association of atrial tissue fibrosis identified by delayed enhancement MRI and atrial fibrillation catheter ablation: The DECAAF study. J. Am. Med. Assoc. **311**(5), 498–506 (2014)
10. Noh, H., Hong, S., Han, B.: Learning deconvolution network for semantic segmentation. In: Proceedings of the IEEE International Conference on Computer Vision, pp. 1520–1528 (2015)
11. Patti, G., Pengo, V., Marcucci, R., et al.: The left atrial appendage: from embryology to prevention of thromboembolism. Eur. Heart J. (2016). doi:10.1093/eurheartj/ehw159
12. Shelhamer, E., Long, J., Darrell, T.: Fully convolutional networks for semantic segmentation. In: Proceedings of the IEEE Conference on Computer Vision and Pattern Recognition, pp. 3431–3440 (2015)
13. Tobongomez, C., Geers, A.J., Peters, J., et al.: Benchmark for algorithms segmenting the left atrium from 3D CT and MRI datasets. IEEE Trans. Med. Imaging **34**(7), 1460–1473 (2015)
14. Zheng, Y., Wang, T., John, M., Zhou, S.K., Boese, J., Comaniciu, D.: Multi-part left atrium modeling and segmentation in C-arm CT volumes for atrial fibrillation ablation. In: Proceedings of International Conference on Medical Image Computing and Computer-Assisted Intervention, pp. 487–495. Springer, New York (2011)
15. Zuluaga, M., Cardoso, M.J., Modat, M., et al.: Multi-atlas propagation whole heart segmentation from MRI and CTA using a local normalised correlation coefficient criterion. In: Proceedings of International Conference on Functional Imaging and Modeling of the Heart, pp. 174–181 (2013)

3D Randomized Connection Network
with Graph-Based Inference

Siqi Bao$^{(\boxtimes)}$, Pei Wang, and Albert C. S. Chung

Lo Kwee-Seong Medical Image Analysis Laboratory, Department of Computer
Science and Engineering, The Hong Kong University of Science and Technology,
Kowloon, Hong Kong
sbao@cse.ust.hk

Abstract. In this paper, a novel 3D deep learning network is proposed
for brain MR image segmentation with randomized connection, which
can decrease the dependency between layers and increase the network
capacity. The convolutional LSTM and 3D convolution are employed
as network units to capture the long-term and short-term 3D properties respectively. To assemble these two kinds of spatial-temporal information and refine the deep learning outcomes, we further introduce an
efficient graph-based node selection and label inference method. Experiments have been carried out on the publicly available database and
results demonstrate that the proposed method can obtain the best performance as compared with other state-of-the-art methods.

1 Introduction

The analysis of sub-cortical structures in brain Magnetic Resonance (MR)
images is crucial in clinical diagnosis, treatment plan and post-operation assessment. Recently deep learning techniques, such as Convolutional Neural Networks
(CNN), have been introduced to brain image analysis and brought significant
improvements in image labeling. To utilize CNN on 3D image analysis, the
conventional way applies the 2D CNN network on each image slice, and then
concatenates the results along third image direction. However, the temporal
information is collapsed during the 2D convolution process. To learn spatio-
temporal features, 3D convolution is recently introduced in video analysis tasks
[6,12]. Given the expensive computation cost, the size of convolution kernels is
usually set to a small number in practice, which can only capture short-term
dependencies.

For image segmentation with CNN, the classic architecture is fully convolutional network (FCN) [8]. While due to the large receptive fields and pooling
layers, FCN tends to produce segmentations that are poorly localized around
object boundaries. Therefore, the deep learning outcomes are usually combined
with probabilistic graphical models to further refine the segmentation results.
Fully connected CRF [7] is one commonly used graphic model during the FCN
post-processing [2], where each image pixel is treated as one graph node and
densely connected to the rest graph nodes. Rather than utilizing the color-based

M.J. Cardoso et al. (Eds.): DLMIA/ML-CDS 2017, LNCS 10553, pp. 47–55, 2017.
DOI: 10.1007/978-3-319-67558-9_6

affinity like fully connected CRF, boundary neural fields (BNF) [1] first predicts object boundaries with FCN feature maps and then encodes the boundary information in the pair-wise potential to enhance the semantic segmentation quality. However, given the massive pixel amount and poor contrast condition, it is different to apply these methods directly to 3D brain MR image segmentation.

To address the above challenges, in this paper, we extract long-term dependencies in spatial-temporal information with convolutional LSTM [9,14]. One novel randomized connection network is designed, which is a dynamic directed acyclic graph with symmetric architecture. Through the randomized connection, the deep network behaves like ensembles of multiple networks, which reduces the dependency between layers and increases the network capacity. To obtain the comprehensive properties for 3D brain image, both convolutional LSTM and 3D convolution are employed as the network units to capture long-term and short-term spatial-temporal information independently. Their results are assembled and refined together with the proposed graph-based node selection and label inference. Experiments have been carried out on the publicly available database and our method can obtain quality segmentation results.

2 Methodology

2.1 Convolutional LSTM and 3D Convolution

Long-Short Term Memory (LSTM) [5] is widely used for sequential modeling to capture long-term temporal information. As compared with the Recurrent Neural Network (RNN), it embraces a more complex neural network block to control information flow. The key component in LSTM is the memory cell state c_t, which carries information through the entire chain with some minor linear operations. This memory cell can be accessed and updated by three gates: forget gate f_t, input gate i_t and output gate o_t. In classic LSTM, fully-connected transformations are employed during the input-to-state and state-to-state transitions. As such, the spatial property is ignored. To gather the spatial-temporal information, convolutional LSTM (ConvLSTM) is recently proposed [9,14] to replace the fully-connected transformation with the local convolution operation.

In this paper, we utilize ConvLSTM to collect the long-term dependencies for 3D images, where the third image axis is treated as temporal dimension. The ConvLSTM for 3D image processing is illustrated in Fig. 1. To compute the pixel values in one layer, both those pixels within the corresponding local region from its last layer (at the same time stamp) and those from the current layer (at the previous time stamp) are employed as input. For example, the ConvLSTM response h_t in layer l (Purple pixel) can be estimated as follows:

$$i_t = \sigma(W_{xi} * \mathcal{X}_t + W_{hi} * \mathcal{H}_{t-1} + b_i),$$
$$f_t = \sigma(W_{xf} * \mathcal{X}_t + W_{hf} * \mathcal{H}_{t-1} + b_f),$$
$$c_t = f_t \circ c_{t-1} + i_t \circ \tanh(W_{xc} * \mathcal{X}_t + W_{hc} * \mathcal{H}_{t-1} + b_c),$$
$$o_t = \sigma(W_{xo} * \mathcal{X}_t + W_{ho} * \mathcal{H}_{t-1} + b_o),$$
$$h_t = o_t \circ \tanh(c_t), \tag{1}$$

where $*$ denotes the convolution operation, the symbol $\circ$ stands for Hadamard product, $\sigma(\cdot)$ and $\tanh(\cdot)$ refer to the sigmoid and hyperbolic tangent function. As shown in Fig. 1, $\mathcal{X}_t$ is the input from last layer at the same time stamp (Red regions) and $\mathcal{H}_{t-1}$ is the input from current layer at the previous time stamp (Green regions). $W_x.$ and $W_h.$ denote the input-to-hidden and hidden-to-hidden weight matrices, with $b.$ as the corresponding biases. Distinct with the weight matrices in classical LSTM, the input $W_x.$ and recurrent weights $W_h.$ in ConvLSTM are all 4D tensors, with a size of $n \times m \times k \times k$ and $n \times n \times k \times k$ respectively, where n is the predefined number of convolution kernels (feature maps in the current layer), m is the number of feature maps in the previous layer and k is the convolutional kernel size. ConvLSTM can be regarded as a generalized version of classic LSTM, with last two tensor dimensions equal to 1.

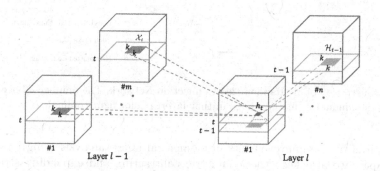

Fig. 1. Illustration of convolutional LSTM. (Color figure online)

In this paper, 3D convolution is also utilized to capture short-term dependencies for 3D images. The 3D convolutional response h_t in layer l can be estimated in the following way: $h_t = \max(W * \mathcal{X}_t + b, 0)$, where W is the convolution weight matrix, with b as the corresponding bias. $\mathcal{X}_t$ is the input from last layer, with a size of $m \times d \times k \times k$ (d is the convolutional kernel size along the third direction). In 3D convolution, the ReLU $\max(\cdot, 0)$ is employed as the non-linear activation. With convLSTM and 3D convolution settled as network unit to capture comprehensive spatial-temporal information, the next consideration is the design of the whole network architecture.

2.2   Randomized Connection Network

FCN is a classic deep learning network for image segmentation, by transforming the fully-connected layers in pre-trained classification network into convolutional layers. To extract abstract features, poolings operations are indispensable in FCN, which leads to the significant size difference between estimated probability map and the original input image. It is necessary to employ extra up-sampling or interpolation steps to make up the size difference, while the segmentation quality through one direct up-sampling can be unacceptable rough. To address this problem, the network architecture of FCN turns from a line topology into a directed acyclic graph (DAG), by adding links to append lower layers with higher resolution into the final prediction layer. U-Net [10], is another DAG with symmetric contracting and expanding architecture, which has gained great success in biomedical image segmentation. 3D U-Net [3] is recently introduced for volumetric segmentation by replacing 2D convolution with 3D convolution.

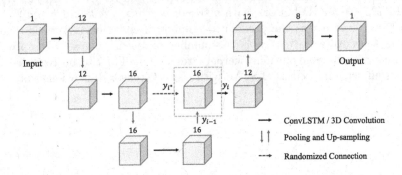

Fig. 2. Illustration of 3D Randomized Connection Network. The numbers above cubes refer to the amount of feature maps in that layer. (Color figure online)

Inspired by the improvements in biomedical image analysis using U-Net, in this paper, we also keep the symmetric contracting and expanding structure for 3D brain image segmentation, with detailed network shown in Fig. 2. The ConvLSTM/3D convolution (Black arrow) is employed to capture the long-term or short-term spatial-temporal properties. The Green arrows refer to the pooling or upsampling operations. Distinct with U-Net where all connections are fixed (static DAG), in the proposed method, the connection between contracting and expansive paths (Red arrow) is randomly established during training (dynamic DAG). To further illustrate the concept, we use one layer as an example to analyze its input and output. For the i-th layer with randomized connection (Grey dashed square) along the expansive path, its output can be estimated as:

$$y_i = \mathcal{U}(y_{i-1}) + \mathcal{R}(y_{i*}, \alpha), \tag{2}$$

where y_{i-1} the input from the previous layer along the expansive path, $\mathcal{U}(\cdot)$ is the upsampling operation, and y_{i*} the input from corresponding layer along the contracting path. $\mathcal{R}(y_{i*}, \alpha)$ is a randomized function whose result is y_{i*} with the probability α, and 0 with the probability $1 - \alpha$. During training, the input y_{i*} will be added to i-th layer with the probability α in each iteration.

Randomized connection achieves great robustness and efficiency because it reduces dependency between layers and increases the model capacity. By randomly dropping the summation connection, the layers can be fully activated and forced to learn instead of relying on the previous ones. As discussed in [13], residual network with identity skip-connections behaves like ensembles of relatively shallow networks. In the proposed method, the summation connection is randomly established in every iteration, so a number of different models are assembled implicitly during training. If there are N connections linking the two paths, then it will be 2^N model combined in the training process. In the proposed method, two randomized connection networks are trained independently, with ConvLSTM and 3D convolution to capture long-term and short-term spatial-temporal information respectively.

2.3   Graph-Based Inference

Distinct with general images, which usually are 2D images and have relatively sharp object boundaries, the size of medical image volumes is much larger and the boundary among tissues is quite blurry as a result of poor contrast condition. Although fully connected CRF and BNF can boost the segmentation performance for general images, the differences in image properties might lead to some problems if directly applying these methods on 3D medical image segmentation. It is necessary to design effective graph-based inference method for 3D brain image segmentation.

The proposed graph-based inference method involves two steps: node selection and label inference. For the sake of efficiency, it is better to prune the majority of pixels and to focus on those whose results need to be refined. The node selection and label inference are introduced based on the fundamental graph $G = (V, E)$, where node set V includes all pixels in the 3D image and edge set E corresponds to the image lattice connection. If v_i and v_j are adjacent in the 3D image, an edge e_{ij} will be set up, with w_{ij} as edge weight. As both long-term and short-term spatial-temporal information are desirable in the node selection process, the labeling results estimated by ConvLSTM and those by 3D convolution need to be employed collaboratively. Note that the examples and figures in this subsection are just for simplification to use two network results. In fact, the node selection and label inference are not limited to the number of networks.

For each node $v_i \in V$, it can be represented as $v_i = \{v_i^1, v_i^2, \cdots, v_i^K\} = \{(p_i^1, 1 - p_i^1), (p_i^2, 1 - p_i^2), \cdots, (p_i^K, 1 - p_i^K)\}$, where p_i^k refers to the probability estimated by the k-th deep learning network that v_i belongs to foreground. During node selection, two criteria are taken into consideration: the label confidence of each node and the label consistency among the neighborhood. We would like to filter out those nodes with high label confidence and consistency, so that we can focus on the rest nodes for further processing. In Fig. 3, two small image cubes are extracted from two result images for illustration. For the node v_i in the k-th result image (Yellow node), its confidence is evaluated by the contrast between foreground and background probability, with the definition as follows:

$$C_f(v_i^k) = (1 - p_i^k - p_i^k)^2 = (1 - 2p_i^k)^2. \tag{3}$$

As for the consistency, it is measured by the cosine similarity between neighboring nodes, defined as:

$$C_s(v_i^k, v_j) = \cos(v_i^k, v_j) = \frac{v_i^{k^T} v_j}{\|v_i^k\| \|v_j\|}, \tag{4}$$

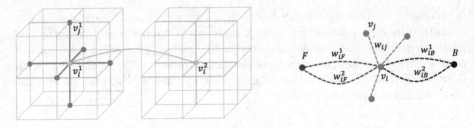

Fig. 3. Graph-based Inference Illustration. **Left**: node selection (cubes extracted from 1st and 2nd result images). **Right**: graph-based label inference for candidate nodes. (Color figure online)

where $v'_j \in \mathcal{N}(v_i^k)$, $\mathcal{N}(\cdot)$ includes the 6-nearest neighbors in k-th result image (Blue node) and the corresponding nodes v'_i in the rest images (Yellow node).

The two criteria are combined together for nodes selection and the detailed formulation is given as follows:

$$\max \sum_{v_i \in V^*} \sum_k (C_f(v_i^k) + \sum_{v'_j \in \mathcal{N}(v_i^k)} C_s(v_i^k, v'_j)), \quad s.t. \ |V^*| \leq |V| \times \theta, \quad (5)$$

where θ is the pre-defined threshold, indicating the percentage of nodes to be pruned, and $|V^*|$ is the set of confident nodes that can be pruned. The first unary term C_f measures the label confidence and the second pair-wise term C_s accesses the label consistency. Equation (5) can be solved efficiently by sorting the energy for each v_i in descending order and then set the first $|V| \times \theta$ nodes as confident nodes. The rest nodes are treated as candidate nodes and need further label inference.

The label inference is developed on a compact graph $G^C = (V^C, E^C)$, where V^C is candidate node set and E^C is the lattice edge connecting candidate nodes. The inference problem is formulated under the Random Walker framework [4], with detailed definition given as follows:

$$\min_x \sum_{v_i} \sum_k [w_{iF}^{k\ 2}(x_i - 1)^2 + w_{iB}^{k\ 2} x_i^2] + \sum_{e_{ij}} w_{ij}^2 (x_i - x_j)^2,$$

$$s.t. \quad x_F = 1, \ x_B = 0, \quad (6)$$

where x_i is the probability that node v_i belongs to the foreground, F and B refers to the foreground and background seed respectively, as shown in Fig. 3. In the first unary term, w_{iF}^k and w_{iB}^k are the priors from deep learning network, which are assigned with p_i^k and $1 - p_i^k$. In the second pairwise term, w_{ij} is the edge weight for lattice connection (Blue dashed line), which is estimated by conventional Gaussian function: $w_{ij} = \exp(-\beta(I(v_i) - I(v_j))^2)$, where $I(\cdot)$ is the intensity value and β is a tuning parameter. By minimizing Equation (6), the probability x_i for each candidate node can be obtained and the label can be then updated correspondingly: $L(v_i) = 1$ if $x_i \geq 0.5$ and $L(v_i) = 0$ otherwise.

3   Experiment

In this paper, experiments have been carried out on the publicly available brain MR image database – LPBA40 [11], which has 40 volumes with 56 structures delineated. The database was randomly divided into two equal sets for training and testing respectively. Data augmentation with elastic transformation was performed to increase the amount of training data by 20 times and the training process was set to 60 epochs, with a learning rate of 10^{-4}. The rest of the parameter settings used in the experiments are listed as follows: the probability for randomized connection $\alpha = 0.5$, the percentage to prune nodes $\theta = 0.999$ and the tuning parameter in Gaussian function $\beta = 100$.

Dice Coefficient (DC) is utilized to measure the quality of segmentation results. The quantitative results on available sub-cortical structures are given in Table 1, with the highest values shown in Red. Each sub-cortical structure has two parts (located in the left and right hemisphere), and the results are provided for the left-right part respectively. FCN was employed as the baseline during evaluation, where one patch-based classification network was first trained and then adapted to image-based segmentation network by transforming fully connected layer to convolutional layer. The intermediary results generated by randomized connection network using 3D Convolution and ConvLSTM were provided. To test the performance of randomized connection, the symmetric U-Net with fixed connection using ConvLSTM as network unit was included for comparison. As compared with conventional FCN and U-Net, randomized connection networks can obtain better results. The comparison between ConvLSTM U-Net and ConvLSTM Random Net indicates that the randomized connection can improve the labeling quality significantly by 2.15%. Through graph-based inference, the long-term and short-term information can be assembled together

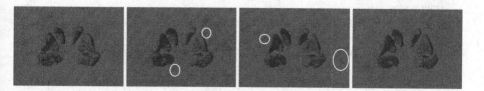

Fig. 4. Some visual results. Left to Right: Ground Truth, ConvLSTM Random Net, 3D Convolution Random Net, Graph-based Inference. (Color figure online)

Table 1. Experimental results on the LPBA40 database, measured with DC. Highest values are written in Red.

Methods	FCN	ConvLSTM U-Net	ConvLSTM Random Net	3D Convolution Random Net	Graph-based Inference
Caudate	0.8232−0.8107	0.8276−0.8204	0.8352−0.8053	0.8489−0.8455	0.8495−0.8379
Putamen	0.8358−0.8321	0.8273−0.7722	0.8362−0.8407	0.8500−0.8443	0.8533−0.8572
Hippocampus	0.7827−0.7747	0.8201−0.7826	0.8213−0.8344	0.8353−0.8165	0.8413−0.8445
Average	0.8099±0.0258	0.8084±0.0244	0.8289±0.0133	0.8401±0.0127	0.8473±0.0074

to further improve the performance. Some visual results are shown in Fig. 4, with outliers circled in Red. As displayed in the forth column, these outliers in randomized networks can be removed after graph-based inference.

4   Conclusion

In this paper, a novel deep network with randomized connection is proposed for 3D brain image segmentation, with ConvLSTM and 3D convolution network units to capture long-term and short-term spatial-temporal information respectively. To determine the label for each pixel efficiently, the graph-based node selection is introduced to prune the majority quality nodes and to focus on the nodes that really need further label inference. The long-term and short-term dependencies are encoded to the graph as priors and utilized collaboratively in the graph-based inference. Experiments carried out on the publicly available database indicate that our method can obtain the best performance as compared with other state-of-the-art methods.

References

1. Bertasius, G., Shi, J., Torresani, L.: Semantic segmentation with boundary neural fields. In: IEEE CVPR, pp. 3602–3610 (2016)
2. Chen, L.C., Papandreou, G., Kokkinos, I., et al.: Semantic image segmentation with deep convolutional nets and fully connected CRFs (2014). arxiv:1412.7062
3. Çiçek, Ö., Abdulkadir, A., Lienkamp, S.S., Brox, T., Ronneberger, O.: 3D U-net: learning dense volumetric segmentation from sparse annotation. In: Medical Image Computing and Computer-Assisted Intervention, pp. 424–432. Springer, New York (2016)
4. Grady, L.: Random walks for image segmentation. IEEE TPAMI **28**(11), 1768–1783 (2006)
5. Hochreiter, S., Schmidhuber, J.: Long short-term memory. Neural Comput. **9**(8), 1735–1780 (1997)
6. Ji, S., Xu, W., Yang, M., Yu, K.: 3D convolutional neural networks for human action recognition. IEEE TPAMI **35**(1), 221–231 (2013)
7. Koltun, V.: Efficient inference in fully connected CRFs with Gaussian edge potentials. NIPS **2**(3), 4 (2011)
8. Long, J., Shelhamer, E., Darrell, T.: Fully convolutional networks for semantic segmentation. In: IEEE CVPR, pp. 3431–3440 (2015)
9. Patraucean, V., Handa, A., Cipolla, R.: Spatio-temporal video autoencoder with differentiable memory (2015). arxiv:1511.06309
10. Ronneberger, O., Fischer, P., Brox, T.: U-net: convolutional networks for biomedical image segmentation. In: Medical Image Computing and Computer-Assisted Intervention, pp. 234–241. Springer, New York (2015)
11. Shattuck, D.W., Mirza, M., Adisetiyo, V., Hojatkashani, C., Salamon, G., Narr, K.L., Poldrack, R.A., Bilder, R.M., Toga, A.W.: Construction of a 3D probabilistic atlas of human cortical structures. NeuroImage **39**(3), 1064–1080 (2008)
12. Tran, D., Bourdev, L., Fergus, R., et al.: Learning spatiotemporal features with 3D convolutional networks. In: IEEE ICCV, pp. 4489–4497 (2015)

13. Veit, A., Wilber, M.J., Belongie, S.: Residual networks behave like ensembles of relatively shallow networks. In: NIPS, pp. 550–558 (2016)
14. Xingjian, S., Chen, Z., Wang, H., et al.: Convolutional LSTM network: a machine learning approach for precipitation nowcasting. In: NIPS, pp. 802–810 (2015)

Adversarial Training and Dilated Convolutions for Brain MRI Segmentation

Pim Moeskops[✉], Mitko Veta, Maxime W. Lafarge, Koen A.J. Eppenhof, and Josien P.W. Pluim

Medical Image Analysis Group, Department of Biomedical Engineering, Eindhoven University of Technology, Eindhoven, The Netherlands
p.moeskops@tue.nl

Abstract. Convolutional neural networks (CNNs) have been applied to various automatic image segmentation tasks in medical image analysis, including brain MRI segmentation. Generative adversarial networks have recently gained popularity because of their power in generating images that are difficult to distinguish from real images.

In this study we use an adversarial training approach to improve CNN-based brain MRI segmentation. To this end, we include an additional loss function that motivates the network to generate segmentations that are difficult to distinguish from manual segmentations. During training, this loss function is optimised together with the conventional average per-voxel cross entropy loss.

The results show improved segmentation performance using this adversarial training procedure for segmentation of two different sets of images and using two different network architectures, both visually and in terms of Dice coefficients.

Keywords: Adversarial networks · Deep learning · Convolutional neural networks · Dilated convolution · Medical image segmentation · Brain MRI

1 Introduction

Convolutional neural networks (CNNs) have become a very popular method for medical image segmentation. In the field of brain MRI segmentation, CNNs have been applied to tissue segmentation [13,14,20] and various brain abnormality segmentation tasks [3,5,8].

A relatively new approach for segmentation with CNNs is the use of dilated convolutions, where the weights of convolutional layers are sparsely distributed over a larger receptive field without losing coverage on the input image [18,19]. Dilated CNNs are therefore an effective approach to achieve a large receptive field with a limited number of trainable weights and a limited number of convolutional layers, without the use of subsampling layers.

Generative adversarial networks (GANs) provide a method to generate images that are difficult to distinguish from real images [4,15,17]. To this end,

© Springer International Publishing AG 2017
M.J. Cardoso et al. (Eds.): DLMIA/ML-CDS 2017, LNCS 10553, pp. 56–64, 2017.
DOI: 10.1007/978-3-319-67558-9_7

GANs use a discriminator network that is optimised to discriminate real from generated images, which motivates the generator network to generate images that look real. A similar adversarial training approach has been used for domain adaptation, using a discriminator network that is trained to distinguish images from different domains [2,7] and for improving image segmentations, using a discriminator network that is trained to distinguish manual from generated segmentations [11]. Recently, such a segmentation approach has also been applied in medical imaging for the segmentation of prostate cancer in MRI [9] and organs in chest X-rays [1].

In this paper we employ adversarial training to improve the performance of brain MRI segmentation in two sets of images using a fully convolutional and a dilated network architecture.

2   Materials and Methods

2.1   Data

Adult Subjects. 35 T_1-weighted MR brain images (15 training, 20 test) were acquired on a Siemens Vision 1.5T scanner at an age ($\mu \pm \sigma$) of 32.9 ± 19.2 years, as provided by the MICCAI 2012 challenge on multi-atlas labelling [10]. The images were segmented in six classes: white matter (WM), cortical grey matter (cGM), basal ganglia and thalami (BGT), cerebellum (CB), brain stem (BS), and lateral ventricular cerebrospinal fluid (lvCSF).

Elderly Subjects. 20 axial T_1-weighted MR brain images (5 training, 15 test) were acquired on a Philips Achieva 3T scanner at an age ($\mu \pm \sigma$) of 70.5 ± 4.0 years, as provided by the MRBrainS13 challenge [12]. The images were segmented in seven classes: WM, cGM, BGT, CB, BS, lvCSF, and peripheral cerebrospinal fluid (pCSF). Possible white matter lesions were included in the WM class.

2.2   Network Architecture

Two different network architectures are used to evaluate the hypothesis that adversarial training can aid in improving segmentation performance: a fully convolutional network and a network with dilated convolutions. The outputs of these networks are input for a discriminator network, which distinguishes between generated and manual segmentations. The fully convolutional nature of both networks allows arbitrarily sized inputs during testing. Details of both segmentation networks are listed in Fig. 1, left.

Fully Convolutional Network. A network with 15 convolutional layers of 32 3×3 kernels is used (Fig. 1, left), which results in a receptive field of 31×31 voxels. During training, an input of 51×51 voxels is used, corresponding to an output of 21×21 voxels. The network has 140,039 trainable parameters for $C = 7$ classes (6 plus background; adult subjects) and 140,296 trainable parameters for $C = 8$ classes (7 plus background; elderly subjects).

Fully convolutional network			
Kernel size	Dilation	Kernels	Layers
3×3	1	32	15
1×1	1	256	1
1×1	1	C	1

Dilated network			
Kernel size	Dilation	Kernels	Layers
3×3	1	32	2
3×3	2	32	1
3×3	4	32	1
3×3	8	32	1
3×3	16	32	1
3×3	1	32	1
1×1	1	C	1

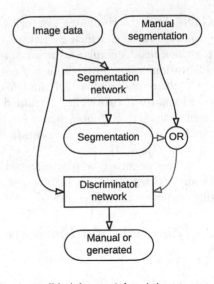

(a) Segmentation networks   (b) Adversarial training

Fig. 1. Left: Segmentation network architectures for the 17-layer fully convolutional (top) and 8-layer dilated (bottom) segmentation networks. The receptive fields are 67×67 for the dilated network and 31×31 for the fully convolutional network. No subsampling layers are used in both networks. Right: Overview of the adversarial training procedure. The red connections indicate how the discriminator loss influences the segmentation network during backpropagation. (Color figure online)

Dilated Network. The dilated network uses the same architecture as proposed by Yu et al. [19], which uses layers of 3×3 kernels with increasing dilation factors (Fig. 1, left). This results in a receptive field of 67×67 voxels using only 7 layers of 3×3 convolutions, without any subsampling layers. During training, an input of 87×87 voxels is used, which corresponds to an output of 21×21 voxels. In each layer 32 kernels are trained. The network has 56,039 trainable parameters for $C = 7$ classes (6 plus background; adult subjects) and 56,072 trainable parameters for $C = 8$ classes (7 plus background; elderly subjects).

Discriminator Network. The input to the discriminator network are the segmentation, as one-hot encoding or softmax output, and image data in the form of a 25×25 patch. In this way, the network can distinguish real from generated combinations of image and segmentation patches. The image patch and the segmentation are concatenated after two layers of 3×3 kernels on the image patch. The discriminator network further consists of three layers of 32 3×3 kernels, a 3×3 max-pooling layer, two layers of 32 3×3 kernels, and a fully connected layer of 256 nodes. The output layer with two nodes, distinguishes between manual and generated segmentations.

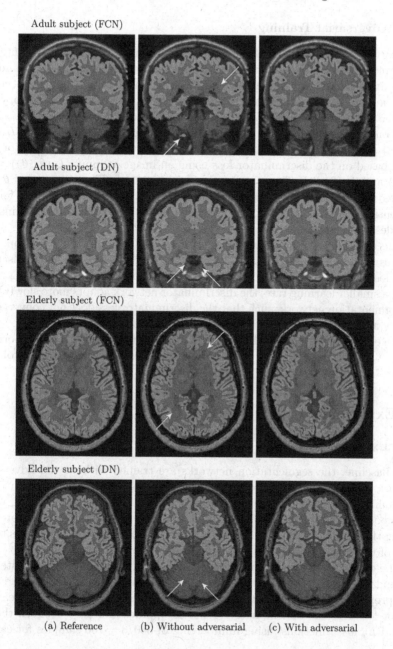

Adult subject (FCN)

Adult subject (DN)

Elderly subject (FCN)

Elderly subject (DN)

(a) Reference   (b) Without adversarial   (c) With adversarial

Fig. 2. Example segmentation results in four of the test images. From top to bottom: an adult subject using the fully convolutional network (FCN), an adult subject using the dilated network (DN), an elderly subject using the fully convolutional network (FCN), and an elderly subject using the dilated network (DN). The colours are as follows: WM in blue, cGM in yellow, BGT in green, CB in brown, BS in purple, lvCSF in orange, and pCSF in red. The arrows indicate errors that were corrected when the adversarial training procedure was used. (Color figure online)

2.3   Adversarial Training

An overview of the adversarial training procedure is shown in Fig. 1, right.

Three types of updates for the segmentation network parameters θ_s and the discriminator network parameters θ_d are possible during the training procedure: (1) an update of only the segmentation network based on the cross-entropy loss over the segmentation map, $L_s(\theta_s)$, (2) an update of the discriminator network based on the discrimination loss using a manual segmentation as input, $L_d(\theta_d)$, and (3) an update of the whole network (segmentation and discriminator network) based on the discriminator loss using an image as input, $L_a(\theta_s, \theta_d)$. Only $L_s(\theta_s)$ and $L_a(\theta_s, \theta_d)$ affect the segmentation network. The parameters θ_s are updated to maximise the discriminator loss $L_a(\theta_s, \theta_d)$, i.e. the updates for the segmentation network are performed in the direction to ascend the loss instead of to descend the loss.

The three types of updates are performed in an alternating fashion. The updates based on the segmentation loss and the updates based on the discriminator loss are performed with separate optimisers using separate learning rates. Using a smaller learning rate, the discriminator network adapts more slowly than the segmentation network, such that the discriminator loss does not converge too quickly and can have enough influence on the segmentation network.

For each network, rectified linear units are used throughout, batch normalisation [6] is used on all layers and dropout [16] is used for the 1×1 convolution layers.

3   Experiments and Results

3.1   Experiments

As a baseline, the segmentation networks are trained without the adversarial network. The updates are performed with RMSprop using a learning rate of 10^{-3} and minibatches of 300 samples. The networks are trained in 5 epochs, where each epoch corresponds to 50,000 training patches per class per image. Note that during this training sample balancing process, the class label corresponds to the label of the central voxel, even though a larger image patch is labelled.

The discriminator and segmentation network are trained using the alternating update scheme. The updates for both loss functions are performed with RMSprop using a learning rate of 10^{-3} for the segmentation loss and a learning rate of 10^{-5} for the discriminator loss. The updates alternate between the L_s, L_d and L_a loss functions, using minibatches of $300/3 = 100$ samples for each.

3.2   Evaluation

Figure 2 provides a visual comparison between the segmentations obtained with and without adversarial training, showing that the adversarial approach generally resulted in less noisy segmentations. The same can be seen from the total number of 3D components (including the background class) that compose the

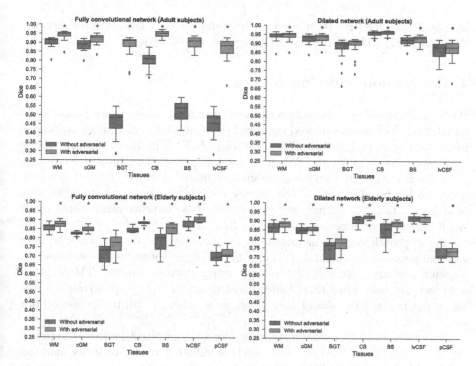

Fig. 3. Dice coefficients for the adult subjects (top row) and the elderly subjects (bottom row) for white matter (WM), cortical grey matter (cGM), basal ganglia and thalami (BGT), cerebellum (CB), brain stem (BS), lateral ventricular cerebrospinal fluid (lvCSF), and peripheral cerebrospinal fluid (pCSF), without (blue) and with (green) adversarial training. Left column: fully convolutional network. Right column: dilated network. Red stars ($p < 0.01$) and red circles ($p < 0.05$) indicate significant improvement based on paired t-tests. (Color figure online)

segmentations. For the adult subjects, the number of components per image ($\mu \pm \sigma$) decreased from 1745 ± 400 to 626 ± 247 using the fully convolutional network and from 417 ± 152 to 365 ± 122 using the dilated network. For the elderly subjects, the number of components per image ($\mu \pm \sigma$) decreased from 926 ± 134 to 692 ± 88 using the fully convolutional network and from 601 ± 104 to 481 ± 90 using the dilated network.

The evaluation results in terms of Dice coefficients (DC) between the automatic and manual segmentations are shown in Fig. 3 as boxplots. Significantly improved DC, based on paired t-tests, were obtained for each of the tissue classes, in both image sets, and for both networks. The only exception was lvCSF in the elderly subjects using the dilated network. For the adult subjects, the DC averaged over all 6 classes ($\mu \pm \sigma$) increased from 0.67 ± 0.04 to 0.91 ± 0.03 using the fully convolutional network and from 0.91 ± 0.03 to 0.92 ± 0.03 using the dilated network. For the elderly subjects, the DC averaged over all 7 classes ($\mu \pm \sigma$)

increased from 0.80 ± 0.02 to 0.83 ± 0.02 using the fully convolutional network and from 0.83 ± 0.02 to 0.85 ± 0.01 using the dilated network.

4   Discussion and Conclusions

We have presented an approach to improve brain MRI segmentation by adversarial training. The results showed improved segmentation performance both qualitatively (Fig. 2) and quantitatively in terms of DC (Fig. 3). The improvements were especially clear for the deeper, more difficult to train, fully convolutional networks as compared with the more shallow dilated networks. Furthermore, the approach improved structural consistency, e.g. visible from the reduced number of components in the segmentations. Because these improvements were usually small in size, their effect on the DC was limited.

The approach includes an additional loss function that distinguishes between real and generated segmentations and can therefore capture inconsistencies that a normal per-voxel loss averaged over the output does not capture. The proposed approach can be applied to any network architecture that, during training, uses an output in the form of an image patch, image slice, or full image instead of a single pixel/voxel.

Various changes to the segmentation network that might improve the results could be evaluated in future work, such as different receptive fields, multiple inputs, skip-connections, 3D inputs, etc. Using a larger output patch size or even the whole image as output could possibly increase the effect of the adversarial training by including more information that could help in distinguishing manual from generated segmentations. This could, however, also reduce the influence of local information, resulting in a too global decision. Further investigation is necessary to evaluate which of the choices in the network architecture and training procedure have most effect on the results.

Acknowledgements. The authors would like to thank the organisers of MRBrainS13 and the multi-atlas labelling challenge for providing the data. The authors gratefully acknowledge the support of NVIDIA Corporation with the donation of a Titan X Pascal GPU.

References

1. Dai, W., Doyle, J., Liang, X., Zhang, H., Dong, N., Li, Y., Xing, E.P.: SCAN: structure correcting adversarial network for chest X-rays organ segmentation. arXiv preprint arXiv:1703.08770 (2017)
2. Ganin, Y., Ustinova, E., Ajakan, H., Germain, P., Larochelle, H., Laviolette, F., Marchand, M., Lempitsky, V.: Domain-adversarial training of neural networks. J. Mach. Learn. Res. **17**(59), 1–35 (2016)
3. Ghafoorian, M., Karssemeijer, N., Heskes, T., Bergkamp, M., Wissink, J., Obels, J., Keizer, K., de Leeuw, F.E., van Ginneken, B., Marchiori, E., Platel, B.: Deep multi-scale location-aware 3D convolutional neural networks for automated detection of lacunes of presumed vascular origin. NeuroImage. Clin. **14**, 391–399 (2017)

4. Goodfellow, I., Pouget-Abadie, J., Mirza, M., Xu, B., Warde-Farley, D., Ozair, S., Courville, A., Bengio, Y.: Generative adversarial nets. In: NIPS, pp. 2672–2680 (2014)
5. Havaei, M., Davy, A., Warde-Farley, D., Biard, A., Courville, A., Bengio, Y., Pal, C., Jodoin, P.M., Larochelle, H.: Brain tumor segmentation with deep neural networks. Med. Image Anal. **35**, 18–31 (2017)
6. Ioffe, S., Szegedy, C.: Batch normalization: accelerating deep network training by reducing internal covariate shift. In: ICML (2015)
7. Kamnitsas, K., et al.: Unsupervised domain adaptation in brain lesion segmentation with adversarial networks. In: Niethammer, M., Styner, M., Aylward, S., Zhu, H., Oguz, I., Yap, P.-T., Shen, D. (eds.) IPMI 2017. LNCS, vol. 10265, pp. 597–609. Springer, Cham (2017). doi:10.1007/978-3-319-59050-9_47
8. Kamnitsas, K., Ledig, C., Newcombe, V.F., Simpson, J.P., Kane, A.D., Menon, D.K., Rueckert, D., Glocker, B.: Efficient multi-scale 3D CNN with fully connected CRF for accurate brain lesion segmentation. Med. Image Anal. **36**, 61–78 (2017)
9. Kohl, S., Bonekamp, D., Schlemmer, H.P., Yaqubi, K., Hohenfellner, M., Hadaschik, B., Radtke, J.P., Maier-Hein, K.: Adversarial networks for the detection of aggressive prostate cancer. arXiv preprint arXiv:1702.08014 (2017)
10. Landman, B.A., Ribbens, A., Lucas, B., Davatzikos, C., Avants, B., Ledig, C., Ma, D., Rueckert, D., Vandermeulen, D., Maes, F., Erus, G., Wang, J., Holmes, H., Wang, H., Doshi, J., Kornegay, J., Manjon, J., Hammers, A., Akhondi-Asl, A., Asman, A.J., Warfield, S.K.: MICCAI 2012 Workshop on Multi-Atlas Labeling. CreateSpace Independent Publishing Platform, Nice (2012)
11. Luc, P., Couprie, C., Chintala, S., Verbeek, J.: Semantic segmentation using adversarial networks. In: NIPS Workshop on Adversarial Training (2016)
12. Mendrik, A.M., Vincken, K.L., Kuijf, H.J., Breeuwer, M., Bouvy, W.H., de Bresser, J., Alansary, A., de Bruijne, M., Carass, A., El-Baz, A., Jog, A., Katyal, R., Khan, A.R., van der Lijn, F., Mahmood, Q., Mukherjee, R., van Opbroek, A., Paneri, S., Pereira, S., et al.: MRBrainS challenge: online evaluation framework for brain image segmentation in 3T MRI scans. Comput. Intel. Neurosci. **2015** (2015). Article No. 813696
13. Moeskops, P., Viergever, M.A., Mendrik, A.M., de Vries, L.S., Benders, M.J., Išgum, I.: Automatic segmentation of MR brain images with a convolutional neural network. IEEE Trans. Med. Imag. **35**(5), 1252–1261 (2016)
14. Moeskops, P., Wolterink, J.M., Velden, B.H.M., Gilhuijs, K.G.A., Leiner, T., Viergever, M.A., Išgum, I.: Deep learning for multi-task medical image segmentation in multiple modalities. In: Ourselin, S., Joskowicz, L., Sabuncu, M.R., Unal, G., Wells, W. (eds.) MICCAI 2016. LNCS, vol. 9901, pp. 478–486. Springer, Cham (2016). doi:10.1007/978-3-319-46723-8_55
15. Radford, A., Metz, L., Chintala, S.: Unsupervised representation learning with deep convolutional generative adversarial networks. In: ICLR (2016)
16. Srivastava, N., Hinton, G., Krizhevsky, A., Sutskever, I., Salakhutdinov, R.: Dropout: a simple way to prevent neural networks from overfitting. J. Mach. Learn. Res. **15**(1), 1929–1958 (2014)
17. Wolterink, J.M., Leiner, T., Viergever, M.A., Išgum, I.: Generative adversarial networks for noise reduction in low-dose CT. IEEE Trans. Med. Imag. (2017). https://doi.org/10.1109/TMI.2017.2708987

18. Wolterink, J.M., Leiner, T., Viergever, M.A., Išgum, I.: Dilated convolutional neural networks for cardiovascular MR segmentation in congenital heart disease. In: Zuluaga, M.A., Bhatia, K., Kainz, B., Moghari, M.H., Pace, D.F. (eds.) RAMBO/HVSMR -2016. LNCS, vol. 10129, pp. 95–102. Springer, Cham (2017). doi:10.1007/978-3-319-52280-7_9
19. Yu, F., Koltun, V.: Multi-scale context aggregation by dilated convolutions. In: ICLR (2016)
20. Zhang, W., Li, R., Deng, H., Wang, L., Lin, W., Ji, S., Shen, D.: Deep convolutional neural networks for multi-modality isointense infant brain image segmentation. NeuroImage **108**, 214–224 (2015)

CNNs Enable Accurate and Fast Segmentation of Drusen in Optical Coherence Tomography

Shekoufeh Gorgi Zadeh[1]([✉]), Maximilian W.M. Wintergerst[2], Vitalis Wiens[1],
Sarah Thiele[2], Frank G. Holz[2], Robert P. Finger[2], and Thomas Schultz[1]

[1] Department of Computer Science, University of Bonn, Bonn, Germany
gorgi@cs.uni-bonn.de
[2] Department of Ophthalmology, University of Bonn, Bonn, Germany

Abstract. Optical coherence tomography (OCT) is used to diagnose
and track progression of age-related macular degeneration (AMD).
Drusen, which appear as bumps between Bruch's membrane (BM) and
the retinal pigment epithelium (RPE) layer, are among the most important biomarkers for staging AMD. In this work, we develop and compare
three automated methods for Drusen segmentation based on the U-Net
convolutional neural network architecture. By cross-validating on more
than 50,000 annotated images, we demonstrate that all three approaches
achieve much better accuracy than a current state-of-the-art method.
Highest accuracy is achieved when the CNN is trained to segment the
BM and RPE, and the drusen are detected by combining shortest path
finding with polynomial fitting in a post-process.

1 Introduction

Age-related macular degeneration (AMD) is the most common cause of irre-
versible vision loss for people over the age of 50 in the developed countries [7].
Drusen, i.e., focal deposits of acellular debris between the retinal pigment epithelium layer and Bruch's membrane, are usually the first clinical sign of AMD.
Their size, number, and location can serve as biomarkers for disease progression.

Optical Coherence Tomography (OCT) is a fast and non-invasive way of
obtaining three-dimensional images of the retina, and is increasingly used to
monitor the onset and progression of AMD [1]. We would like to use OCT in
large epidemiological studies, which requires detecting and quantifying drusen
in tens of thousands of eyes, and is infeasible by manual analysis.

Even though considerable progress has been made on (semi-)automated
quantitative OCT image analysis, in recent years [14], current approaches still do
not achieve sufficient accuracy, or require too much interaction, to be practical
for use in large-scale studies. The fact that convolutional neural networks (CNNs)
have recently achieved excellent results in related biomedical image analysis tasks
[11] motivates us to develop and compare three CNN-based approaches to drusen
segmentation. In an evaluation on more than 50,000 semiautomatically annotated B-scans (i.e., two-dimensional cross-sectional images acquired by OCT),
we demonstrate that all three approaches outperform a current state-of-the-art

M.J. Cardoso et al. (Eds.): DLMIA/ML-CDS 2017, LNCS 10553, pp. 65–73, 2017.
DOI: 10.1007/978-3-319-67558-9_8

technique for drusen segmentation. Highest accuracy is achieved by combining CNNs with application-specific post-processing.

2 Related Work

Among the many approaches for drusen segmentation [2,4,6,8], we identified the method by Chen et al. [2] as a state-of-the-art reference, based on the reported accuracy and its successful use in a longitudinal study [13]. In Sect. 4.2, we will show that our CNN-based results compare very favorably to this method.

Even though we are not aware of any prior work that would have applied deep learning to the segmentation of drusen or retinal layers, two very recent works by Lee et al. [10] and Zheng et al. [15] have used convolutional neural networks to classify OCT images as either healthy or having AMD. The advantage of a segmentation, as it is achieved in our work, is that it can be used to derive intuitive measures such as size and number of drusen, which can be entered into progression models [12,13], or correlated with genetic or lifestyle variables.

3 Method

3.1 Data Preparation

In each OCT scan, a three-dimensional volume is covered by a varying number of noncontiguous two-dimensional slice images, so-called B-scans. Our data set is generated as part of the MODIAMD (Molecular Diagnostics of Age-related Macular Degeneration) study, an observational cohort study on intermediate AMD, and it consists of 52,377 such B-scans (512 × 496 pixels) that belong to 682 OCT scans from 98 different subjects, each having a different number of followup scans taken. Each eye is scanned with a density of either 19 or 145 B-scans that cover approximately 5–6 mm. Due to this varying density, we segment B-scans independently; accounting for 3D context is left for future work.

Layer segmentation map Complex segmentation map Drusen segmentation map

Fig. 1. Different label images overlaid on the corresponding B-scan. The layer segmentation map has different classes for RPE (red) layer and BM (yellow) layer. (Color figure online)

The annotations provided with the data are segmentation maps of the RPE and BM, as shown in Fig. 1 (left). For each B-scan, a medical expert performed a careful manual correction of an initial segmentation that was created with a proprietary algorithm, spending about two minutes per B-scan on average.

Since a gold standard segmentation of the drusen themselves was not available, we generated reference drusen maps from the RPE and BM curves. To this end, we rectified the images by vertically shifting each column so that the BM forms a straight horizontal line. We then estimated the normal RPE layer by fitting a third degree polynomial to the segmented RPE in the rectified image, which was then warped back into the original image coordinates, where drusen were detected as areas in which the segmented RPE is elevated by more than two pixels above the normal RPE, cf. Fig. 1 (right). An experienced rater confirmed that this procedure led to plausible drusen segmentation masks. Figure 2 illustrates how rectification helps to avoids false positive detections.

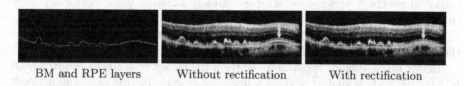

BM and RPE layers   Without rectification   With rectification

Fig. 2. Rectification helps to avoid false drusen detections. The blue curve is the RPE layer, the green curve is the estimated normal RPE and the red areas show the drusen. (Color figure online)

3.2 Network Architecture and Training

Because of its good performance in related segmentation tasks, we chose the U-Net architecture [11] as the basis of our methods. It consists of two symmetric paths, a contracting one that is used to capture image context, and an expanding one to recover the original resolution.

Since background is a dominant class in all our maps, it is crucial to assign spatially varying pixel weights. Similar to [11], we define weights according to

$$w(x) = w_c(x) + w_0 G_\sigma(x), \tag{1}$$

where w_c is defined as the overall number of pixels divided by those from class c, in order to account for class frequencies. In addition, training is focused on the exact boundaries in the segmentation masks by centering two-dimensional Gaussian weights $G_\sigma(x)$ with standard deviation σ on them. We keep the maximum weight in case weights from neighboring Gaussians overlap.

Since the number of foreground and boundary pixels differs greatly between our three approaches, we had to adapt the values of w_0 and σ accordingly. We will mention the exact values that were used in each experiment. They were found empirically in a small pilot experiment that involved only a small fraction of all data (3 out of 682 scans for training, one for testing).

Stochastic gradient descent from the caffe framework [9] is used for training. The momentum is set to 0.99. The initial learning rate is 10^{-3}, and is reduced at every 200,000 steps by a factor of 0.1. We stopped the training when the validation accuracy plateaued, i.e., at 603,000 iteration. We used the U-Net implementation provided by the authors, including their augmentation layer, and used the same overlap-tile strategy with 2 tiles per image. We added batch normalization layers after ReLu layers to avoid internal covariate shift [5].

3.3   Three Approaches to Drusen Segmentation

The most straightforward way of applying the U-Net to our problem is **direct segmentation,** i.e., training on the drusen maps derived in Sect. 3.1. We used weights according to Eq. (1) with parameters $w_0 = 150$ and $\sigma = 30$. As will be shown in Sect. 4.2, this simple strategy already achieves much better accuracy than a state-of-the-art method. Our main technical contribution is to explore two more complex approaches, which improve accuracy further by combining the CNN with application-specific post-processing.

In our **layer based approach,** we train the U-Net to segment the RPE and BM layers, and estimate the drusen from the results. In this case, the network is set up to output four class probabilities for each pixel, i.e., background, RPE layer, BM layer, and a class that captures cases in which the two layers overlap. We used weights according to Eq. (1) with parameters $w_0 = 100$ and $\sigma = 15$.

We recover continuous curves that represent the RPE layer and the BM via a shortest path algorithm. To this end, we convert the combined probability p^{c+o} of class c (i.e., RPE or BM) and the overlap class o, to a cost χ^{c+o} so that, in each image column, the pixel with highest probability can be traversed at zero cost. This is achieved by defining

$$\chi^{c+o} = -\log\left(\frac{p_{ij}^{c+o}}{\max(p_{:j}^{c+o})}\right) \tag{2}$$

where $\max(p_{:j}^{c+o})$ is the maximum combined probability in column j. Dijkstra's algorithm is used to find continuous curves that connect the left and right image boundaries with the lowest accumulated cost. To obtain smoother curves, we favor non-diagonal moves over diagonal ones by multiplying the step size to the local path cost.

Based on these layer estimates, we proceed to detect drusen via rectification, polynomial fitting and thresholding, in the same way as described in Sect. 3.1.

In our **RPE+drusen complex based approach,** the U-Net segments the area between RPE layer and BM, which has been termed RPE+drusen complex in [3], and which is shown in the center image of Fig. 1. We simply perform a hard segmentation into RPE+drusen complex or background. We used weights according to Eq. (1) with parameters $w_0 = 30$ and $\sigma = 15$.

In almost all cases, the largest component of the segmentation connects the left and right image boundaries, and its upper and lower boundaries can be used as RPE layer and BM, respectively. In the case of a discontinuity, which is usually due to a layer atrophy, the respective boundaries of all components are kept as candidates, and a shortest path is found that connects the left and right image boundaries by adding the lowest possible number of additional pixels. From the resulting curves, drusen are detected in the same way as in the layer based approach.

4   Experiments and Results

4.1   Cross-Validation Setup

We used five-fold cross-validation to evaluate the methods on all available data. To ensure unbiased results, all scans from the same subject were placed in the same fold. For each of the three approaches described in Sect. 3.3, this resulted in five runs with $\sim$70 subjects ($\sim$37, 500 images) for training, 8 subjects ($\sim$4, 000 images) for validation, and $\sim$20 subjects ($\sim$10, 500 images) for testing. The U-Net was trained end-to-end with random initial weights that were sampled from a Gaussian distribution with a standard deviation of size $\sqrt{2/N}$, with N being the number of input nodes of each neuron. On a Titan X GPU, each of these 15 experiments took about three days.

4.2   Quantitative Evaluation

We evaluated the drusen segmentation with two quality measures that were previously established for this application [2]. The overall area covered by drusen is an important summary of drusen load, and the absolute drusen area difference (ADAD) measures the error in estimating it. Formally, $\text{ADAD}(Y^k, X^k) = |\text{Area}(Y^k) - \text{Area}(X^k)|$, where Y and X are the ground truth and the predicted segmentation, and k is the B-scan index. Following [2], the mean and standard deviation of ADAD are computed only over the columns that contain a druse.

Since the ADAD does not capture the accuracy of spatially localizing the drusen, it is complemented by the overlapping ratio (OR), which is defined as $OR(Y^k, X^k) = \frac{X^k \cap Y^k}{X^k \cup Y^k}$, i.e., the ratio of pixels with true positive drusen detections over the sum of true positives, false positives, and false negatives. Again, mean and standard deviation are computed.

In addition to the final drusen segmentation, our layer based and RPE+drusen complex based approaches produce a segmentation of the RPE layer and BM. We compare these two approaches with respect to the accuracy of this intermediate result by also computing the OR for the area between these two layers.

Table 1 compares the errors (ADAD) and accuracies (OR) of our three approaches over all 52, 377 images. As a baseline, it also includes results from our reimplementation of the state-of-the-art method by Chen et al. [2], which performs bilateral filtering, detection and removal of the retinal nerve fiber layer (i.e., the bright layer that can be seen at the top in Fig. 2), thresholding based segmentation of the RPE, detection of drusen by comparing the RPE segmentation to the result of a polynomial fit, and finally false positive elimination using

Table 1. All CNN-based methods achieve much better results than Chen et al. [2], the previous state of the art. Lowest absolute drusen area difference (ADAD), and highest overlap ratio (OR), have been achieved by our layer based approach. The bottom two rows show the results of after an additional false positive elimination (FPE) step.

Measure	Chen et al	Direct	Layer based	Complex based
ADAD	92.33 ± 460.95	28.45 ± 74.08	**7.19 ± 36.04**	11.14 ± 52.92
OR (drusen)	$20.48\% \pm 26.87$	$41.35\% \pm 31.3$	**$55.88\% \pm 33.85$**	$47.42\% \pm 33.45$
OR (complex)	-	-	**$82.6\% \pm 7.26$**	$76.74\% \pm 6.9$
ADAD (FPE)	55.67 ± 277.97	15.08 ± 44.76	**4.92 ± 29.69**	9.58 ± 50.87
OR (FPE,drusen)	$24.20\% \pm 31.52$	$47.90\% \pm 33.23$	**$64.24\% \pm 36.09$**	$58.66\% \pm 35.63$

an *en face* projection image. The top three rows of the table are the evaluation results without the false positive elimination (FPE) step, and the two bottom rows are the results after using the FPE step with similar parameters as in [2].[1] Clearly, the FPE step has a positive overall effect on all methods.

We note that our results from the Chen et al. method are significantly worse than those reported in [2]. We believe that the main reason for this is the much lower axial resolution of our B-scans (496 pixels) compared to theirs (1024 pixels). As a result, layers that they could easily distinguish via thresholding often got merged in our data. Table 1 clearly shows that, at the image resolution available to us, all three CNN-based approaches were able to segment drusen much more accurately than the previous state of the art, both with or without the FPE step. Further improvements are obtained by our customized layer and complex based approaches, which use higher-level knowledge, e.g., that drusen can be recognized as deviations from a smooth normal RPE layer, which runs in parallel to the BM. The fact that the layer based approach achieved highest accuracy might, in part, be due to the fact that it makes use of continuous CNN-derived class probabilities.

Without the FPE step, all methods take less than a second per B-scan: 0.31 s for Chen's method; 0.34 s for the direct drusen segmentation; 0.57 s for the complex based and 0.81 s for the layer based approach.

4.3   Robustness to Additional Pathology

In more advanced stages of AMD, it is important that methods for automated drusen detection are robust to the presence of additional types of pathology, such as geographic atrophy (GA), which can be seen from the relative hyperreflectance in the choroidal layer, at the center of all images in Fig. 3. Our training data included correctly annotated images with such pathology, so the CNN should have learned how to deal with it.

---

[1] The selective *en face* projection relies on an estimate of the RPE layer, which the direct drusen segmentation does not provide. Thus, only those steps of the FPE that do not rely on the *en face* could be applied in case of the direct segmentation.

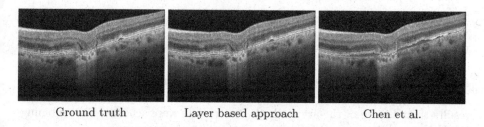

| Ground truth | Layer based approach | Chen et al. |

Fig. 3. This visual comparison indicates that drusen segmentation using our approach is more robust to additional pathology such as GA than the method by Chen et al. [2].

We computed the same measures as in Table 1 for a subset of 49 OCT scans that contained GA. The results were similar, with or without the FPE step. For the layer based approach, results were ADAD = 2.84 ± 8.28, OR = $49.87\% \pm 28.08$, and for the complex based approach, ADAD = 5.66 ± 19.74 and $40.94\% \pm 24.76$. The fact that these numbers are similar to those in the overall dataset illustrates the robustness of these approaches in the presence of GA.

4.4   3D Visualization of Results

As an additional visual check of the segmentation quality, Fig. 4 shows summaries of the full 3D OCT data in an eye for which a high-resolution scan was available. Chen et al. [2] propose a selective volume projection that reduces such data to a 2D *en face* image in which the drusen stand out as bright spots. Figure 4 (left) shows such an image with the ground truth and CNN-derived drusen segmentations overlaid as colored curves. It can be seen that they strongly overlap.

The center and right subfigures visualize the ground truth and segmented RPE layers in the same eye as 3D surfaces, which have been rectified based on the BM for better visualization. A high degree of continuity between adjacent B-scans is observed despite the fact that we process each B-scan independently.

| *En face* image | Ground truth | Layer based approach |

Fig. 4. On the *en face* image, red curves delineate the ground truth drusen segmentations, blue curves our segmentation, purple curves regions where the two overlap. Surface renderings illustrate that the RPE layer has been reliably detected. (Color figure online)

5   Conclusion

We present the first three CNN-based approaches for a fully automated segmentation of drusen in OCT images, which is an important task for diagnosing age-related macular degeneration and modeling disease progression. On the $52,377$ annotated images available to us, all three approaches produced much better results than a state-of-the-art method. Best results were achieved by combining the U-Net architecture with application-specific post-processing.

After training, segmentations are obtained fast enough for use in epidemiological studies, and they have been shown to be robust to the additional presence of geographic atrophy. In future work, we would like to investigate how to best account for three-dimensional context that has a highly variable resolution.

References

1. Abràmoff, M.D., Garvin, M.K., Sonka, M.: Retinal imaging and image analysis. IEEE Rev. Biomed. Eng. **3**, 169–208 (2010)
2. Chen, Q., Leng, T., Zheng, L., Kutzscher, L., Ma, J., de Sisternes, L., Rubin, D.L.: Automated drusen segmentation and quantification in SD-OCT images. Med. Image Anal. **17**(8), 1058–1072 (2013)
3. Chiu, S.J., Izatt, J.A., O'Connell, R.V., Winter, K.P., Toth, C.A., Farsiu, S.: Validated automatic segmentation of AMD pathology including drusen and geographic atrophy in SD-OCT images. Invest. Opthalmol. Vis. Sci. **53**(1), 53 (2012)
4. Farsiu, S., Chiu, S.J., Izatt, J.A., Toth, C.A.: Fast detection and segmentation of drusen in retinal optical coherence tomography images. In: Proceedings of SPIE, Ophthalmic Technologies XVIII, vol. 6844, p. 68440D (2008)
5. Ioffe, S., Szegedy, C.: Batch normalization: accelerating deep network training by reducing internal covariate shift. arXiv preprint arXiv:1502.03167 (2015)
6. Iwama, D., Hangai, M., Ooto, S., Sakamoto, A., Nakanishi, H., Fujimura, T., Domalpally, A., Danis, R.P., Yoshimura, N.: Automated assessment of drusen using three-dimensional spectral-domain optical coherence tomography. Invest. Ophthalmol. Vis. Sci. **53**(3), 1576–1583 (2012)
7. Jager, R.D., Mieler, W.F., Miller, J.W.: Age-related macular degeneration. New Engl. J. Med. **358**(24), 2606–2617 (2008)
8. Jain, N., Farsiu, S., Khanifar, A.A., Bearelly, S., Smith, R.T., Izatt, J.A., Toth, C.A.: Quantitative comparison of drusen segmented on SD-OCT versus drusen delineated on color fundus photographs. Invest. Ophthalmol. Vis. Sci. **51**(10), 4875–4883 (2010)
9. Jia, Y., Shelhamer, E., Donahue, J., Karayev, S., Long, J., Girshick, R., Guadarrama, S., Darrell, T.: Caffe: convolutional architecture for fast feature embedding. arXiv preprint arXiv:1408.5093 (2014)
10. Lee, C.S., Baughman, D.M., Lee, A.Y.: Deep learning is effective for classifying normal versus age-related macular degeneration optical coherence tomography images. Ophthalmol. Retina **1**(4), 322–327 (2017)
11. Ronneberger, O., Fischer, P., Brox, T.: U-Net: convolutional networks for biomedical image segmentation. In: Navab, N., Hornegger, J., Wells, W.M., Frangi, A.F. (eds.) MICCAI 2015. LNCS, vol. 9351, pp. 234–241. Springer, Cham (2015). doi:10.1007/978-3-319-24574-4_28

12. Schmidt-Erfurth, U., Bogunovic, H., Klimscha, S., Hu, X., Schlegl, T., Sadeghipour, A., Gerendas, B.S., Osborne, A., Waldstein, S.M.: Machine learning to predict the individual progression of AMD from imaging biomarkers. In: Proceedings of Association for Research in Vision and Ophthalmology, p. 3398 (2017)
13. de Sisternes, L., Simon, N., Tibshirani, R., Leng, T., Rubin, D.L.: Quantitative SD-OCT imaging biomarkers as indicators of age-related macular degeneration progressionpredicting AMD progression using SD-OCT features. Invest. Ophthalmol. Vis. Sci. **55**(11), 7093–7103 (2014)
14. Sonka, M., Abràmoff, M.D.: Quantitative analysis of retinal OCT. Med. Image Anal. **33**, 165–169 (2016)
15. Zheng, Y., Williams, B.M., Pratt, H., Al-Bander, B., Wu, X., Zhao, Y.: Computer aided diagnosis of age-related macular degeneration in 3D OCT images by deep learning. In: Proceedings of Association for Research in Vision and Ophthalmology, p. 824 (2017)

Region-Aware Deep Localization Framework for Cervical Vertebrae in X-Ray Images

S.M. Masudur Rahman Al Arif[1]([✉]), Karen Knapp[2], and Greg Slabaugh[1]

[1] City, University of London, London, England
S.Al-Arif@city.ac.uk
[2] University of Exeter, Exeter, England

Abstract. The cervical spine is a flexible anatomy and vulnerable to injury, which may go unnoticed during a radiological exam. Towards building an automatic injury detection system, we propose a localization framework for the cervical spine in X-ray images. The proposed framework employs a segmentation approach to solve the localization problem. As the cervical spine is a single connected component, we introduce a novel region-aware loss function for training a deep segmentation network that penalises disjoint predictions. Using data augmentation, the framework has been trained on a dataset of 124 images and tested on another 124 images, all collected from real life medical emergency rooms. The results show a significant improvement in performance over the previous state-of-the-art cervical vertebrae localization framework.

Keywords: Cervical spine · X-ray · Localization · FCN · Region-aware

1 Introduction

The cervical spine is a critical part of human body and is vulnerable to high-impact collisions, sports injuries and falls. Roughly 20% of the injuries remain unnoticed in X-ray images and 67% of these missed injuries end in tragic consequences [1,2]. Computer-aided-detection has the potential to reduce the number of undetected injuries on radiological images. Towards this goal, we propose a robust spine localization framework for lateral cervical X-ray radiographs. We reformulated the localization problem as a segmentation problem at a lower resolution. Given a set of high-resolution images and manually segmented vertebrae ground truth, at a lower resolution, the ground truth becomes a single connected region. We train a deep segmentation network to predict this region. To force the network to predict a single connected region, we introduce a novel term in the loss function which penalizes small disjoint areas and encourages single region prediction. This novel loss has produced significant improvement in localization performance. Previous work in vertebrae localization includes generalized Hough transform based approaches [3,4] and more recent random forest based approaches [5–7]. The state-of-the-art (SOTA) work on cervical vertebrae localization [7], uses a sliding window technique to extract patches from the images.

© Springer International Publishing AG 2017
M.J. Cardoso et al. (Eds.): DLMIA/ML-CDS 2017, LNCS 10553, pp. 74–82, 2017.
DOI: 10.1007/978-3-319-67558-9_9

A random forest classifier decides which patches belong to the spinal area. Then, a rectangular bounding box is generated to localize the spinal region. In contrast, the proposed framework can produce localization map of arbitrary shape in a one-shot process and provides a localisation result that models the cervical spine better than a rectangular box. We have trained our framework on a dataset of 124 images using data augmentation and tested on a separate 124 images having different shapes, sizes, ages and medical conditions. An average pixel level accuracy of 99.1% and sensitivity of 93.6% was achieved. There are two key contributions of this paper. First, a novel loss function which constrains the segmentation to form a single connected region and second, the adaptation and application of deep segmentation networks to cervical spine localization in real-life emergency room X-ray images. The networks learn from a small dataset and robustly outperform the SOTA both quantitatively and qualitatively.

2   Data

Our dataset contains 248 lateral view emergency room X-ray images collected from Royal Devon and Exeter Hospital. Image size, orientation, resolution, patient position, age, medical conditions and scanning systems all vary greatly in the dataset. Some images can be seen in Fig. 1. Along with the images, our medical partners have provided us with the manual segmentation of the cervical vertebrae, C3–C7. The top two vertebrae, C1–C2, were excluded from the study as ground truth was only available for C3 to C7. The segmentation (green) and localization (blue) ground truth (GT) for the images are highlighted in Fig. 1.

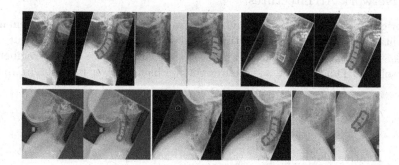

Fig. 1. Examples of X-ray images and corresponding ground truth. (Color figure online)

3   Methodology

We have approached the localization problem as a segmentation problem at a lower resolution. The X-ray images are converted into square images by padding an appropriate number zeros in the smaller dimension and the square images are resized to a lower resolution using bicubic interpolation. This resolution can vary based on the available memory and size of the training networks. For our case, we chose this resolution to be 100 × 100 pixel. The corresponding binary segmentations of the vertebrae are also resized to the same resolution. At this resolution, the provided vertebrae segmentation becomes a single localized area encompassing the spine (blue region in Fig. 1). For this work, we have experimented with three different deep segmentation architectures: fully convolutional network (FCN) [8], deconvolutional network (DeConvNet) [9] and UNet [10]. In this work, we train the networks from scratch. The networks take an input X-ray image of 100 × 100 pixels and produce a probabilistic binary segmentation map of the same resolution.

3.1   Localization Ground Truth

As stated earlier, our target is to localize the spinal area in a cervical X-ray image. For this purpose, we convert our manual vertebra segmentations to a localization ground truth. As our networks are designed to produce an output localization map of 100 × 100 pixels, we create our localization ground truth in these dimensions. Since our original image sizes are approximately in the range of 1000 to 5000 pixels, a simple bicubic interpolation based resize of the vertebra segmentation produces a connected localization ground truth in the smaller dimension. To visualize the ground truth, it can be transformed back to the original dimensions. The blue overlay in Fig. 1 shows how much area the localization ground truth covers apart from the actual vertebrae (green).

3.2   Network Architectures

The original FCN [8] and DeConvNet [9] were designed to tackle a semantic segmentation problem having multiple classes on natural images. Since our task here is to localize the spinal region, we essentially have a binary segmentation problem. Thus, we use a shallower version having fewer parameters. In our

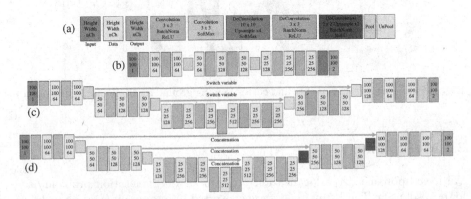

Fig. 2. (a) Legends (b) FCN (c) DeConvNet (d) UNet.

implementation, the FCN network has six convolutional layers and two pooling layers (size 2×2, stride 2). The two stages of pooling reduce the dimension from 100×100 to 25×25 thus creating an activation map of smaller size. The final deconvolutional layer upsamples the 25×25 activations to 100×100 pixels, producing an output map of the input size. Instead of upsampling from the lower resolution to the input resolution in a single step, DeConvNet uses a deconvolutional network which expands the activations step by step using a series of deconvolutional and unpooling layers. The expanding path forms a mirrored version of the contracting convolutional path. The UNet follows a similar structure but instead of an unpooling layer, it uses deconvolution to upsample the input. Both UNet and DeConvNet use information from the contracting path in the expanding path. DeConvNet does this through switch variables from the pooling layers and UNet uses concatenation of data. Figure 2 shows the network diagrams that include data sizes after each layer for a single input image. The number of filters in each layer can be tracked from the number of channels in the data blocks. In total, our FCN has 1,199,042 parameters whereas DeConvNet and UNet have 4,104,194 and 6,003,842 parameters, respectively.

3.3 Training

We have a small dataset of only 248 manually segmented images. We divide our data randomly into 124 training and 124 test images. In order to train any network with a large number of parameters, 124 images are not enough. In order to increase the number of training data, we have augmented the images by rotating each image from $5°$ to $355°$ with a step of $5°$. This results in a training set of 8928 images. It also made the framework rotation invariant. Our choice for data augmentation was only limited to rigid transformations since non-rigid transformation will affect the natural appearance of the spine in the image. All the networks were trained from randomly initialized parameters using a mini-batch gradient descent optimization algorithm from this augmented training dataset.

Given a dataset of training image (x)-segmentation label (y) pairs, training a deep segmentation network means finding a set of parameters $\boldsymbol{W}$ that minimizes a loss function, L_t. The simplest form of the loss function for segmentation problem is the pixel-wise log loss.

$$\hat{\boldsymbol{W}} = \arg\min_{\boldsymbol{W}} \sum_{n=1}^{N} L_t(\{x^{(n)}, y^{(n)}\}; \boldsymbol{W}) \tag{1}$$

where N is the number of training examples and $\{x^{(n)}, y^{(n)}\}$ represents n-th example in the training set with corresponding manual segmentation. The pixel-wise segmentation loss per image can be defined as:

$$L_t(\{x, y\}; \boldsymbol{W}) = -\sum_{i \epsilon \Omega_p} \sum_{j=1}^{M} y_i^j \log P(y_i^j = 1 | x_i; \boldsymbol{W}) \tag{2}$$

$$P(y_i^j = 1 | x_i; \boldsymbol{W}) = \frac{\exp(a_j(x_i))}{\sum_{k=1}^{M} \exp(a_k(x_i))} \qquad (3)$$

where $a_j(x_i)$ is the output of the penultimate activation layer of the network for the pixel x_i, Ω_p represents the pixel space, M is the number of class labels and P are the corresponding class probabilities. However, this term doesn't constrain the predicted maps to be connected. Since the objective of the localization problem is to find a single connected region encompassing the spine area, we add a novel region-aware term in the loss function to force the network to learn to penalize small and disconnected regions.

3.4 Region-Aware Term

We translate our domain knowledge into the training by adding a region based term, L_r. This term forces the network to produce a single region by penalizing small disjoint regions. This term can be defined as:

$$L_r(\{x, y\}; \boldsymbol{W}) = \frac{1}{2} \sum_{i \in \Omega_p} \sum_{j=1}^{M} y_i^j E_i P^2(y_i^j = 1 | x_i; \boldsymbol{W}) \qquad (4)$$

$$E_i = \begin{cases} \max(N_r - N_t, 0) \frac{A_{max_t} - A_q}{A_{max_t}} & \text{if } i \epsilon R_q \\ 0 & \text{otherwise} \end{cases} \qquad (5)$$

where N_r is the number of regions predicted as spine regions, N_t is the number of target regions we are looking for, A_q is the area of the q-th region, A_{max_t} is area of the t-th largest region, R_q is the set of pixels in the region q, and q represents the regions having area less than A_{max_t}. In our case, $N_t = 1$. Notice that, if N_r is equal to or less than N_t and/or $A_{max_t} = A_q$, no region based error will be added with the loss function of Eq. 1.

3.5 Updated Loss Function

Finally, the loss function of Eq. 1 can be extended as:

$$\hat{\boldsymbol{W}} = \arg \min_{\boldsymbol{W}} \sum_{n=1}^{N} L_t(\{x^{(n)}, y^{(n)}\}; \boldsymbol{W}) + L_r(\{x^{(n)}, y^{(n)}\}; \boldsymbol{W}) \qquad (6)$$

The contribution of each term in the total loss can be controlled by introducing a weight parameter in Eq. 6. However, in our case, the best performance was achieved when both terms contributed equally.

3.6 Experiments and Inference

The networks are trained on a server with two NVidia Quadro M4000 GPUs. Training each network for 30 epochs took 22 to 30 h. Batch size during training was selected as 10 images and RMSprop [11] version of mini-batch gradient descent algorithm was used to update the parameters in every epoch. We have three different networks and two versions of the loss function, with and without the region-aware term. In total six networks have been trained: FCN, DeConvNet, UNet and FCN-R, DeConvNet-R, UNet-R, '-R' signifying if the region-aware term of Eq. 6 has been used.

When testing, a test image is padded with zeros to form a square, resized to 100×100 pixels and fed forward through the network to produce probabilistic localization map. This map is converted into a single binary map and compared with the corresponding localization ground truth. Pixel level accuracy, object level Dice, sensitivity and specificity are computed. These metrics demonstrate the performance of the trained networks at the lower resolution at which the network generates the prediction. From a practical point of view, the performance of the localization should also be computed at the original resolution with the manually segmented vertebrae ground truth. In order to achieve this, the predicted localization map is transformed (resized and unpadded) back to the original image dimension and sensitivity and specificity are computed by comparing them with vertebrae segmentation.

4   Results and Discussions

The mean and standard deviation of the metrics over 124 test images at lower and original resolutions are reported in Table 1. In all cases and all metrics (other than specificity), inclusion of the region-aware term in the loss function improves the performance. The improvements are statistically significant for most of the metrics according to a paired t-test at a 5% significance level (bold numbers signify statistical significance for that metric over the other version of the same network architecture in the table). It can be noted that as the sensitivity increases, the specificity may decrease. This is because when the predicted region increases in size to cover more spinal regions, it may also start to encompass some other regions. This effect can also be seen in the qualitative results in Fig. 4. However, the specificity is always in the high range of 97.2% to 97.7%. Quantitatively, FCN performs better than UNet and DeConvNet. But qualitatively UNet and DeConvNet produce finer localization maps (Fig. 4b, d). The coarser map for FCN can be attributed to the single stage upsampling strategy of the network. Figure 4 shows some of the difficult images in the test dataset: osteoporosis (a), image artefacts (b) and severe degenerative change (c, e). In most of these images, our region-aware term has been able to produce better results. It also decreases the standard deviation of the metrics (Table 1 and Fig. 3) proving its usefulness in regularizing the localized maps. However, outliers in the box plot of Fig. 3 show that there are images where all methods fail. Most of these images have severe clinical issues. One example of a complete failure of our algorithm for an image with bone implants is shown in Fig. 4f.

To compare with the previous state-of-the-art (SOTA) in cervical vertebra localization [7], we have implemented and trained the random forest based framework on our dataset. Our algorithm produces a 17.1% relative improvement in average sensitivity with a drop of only 0.9% in specificity. In terms of time

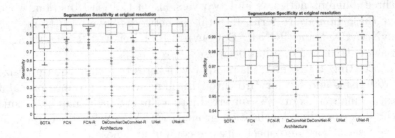

Fig. 3. Box plot of quantitative metrics.

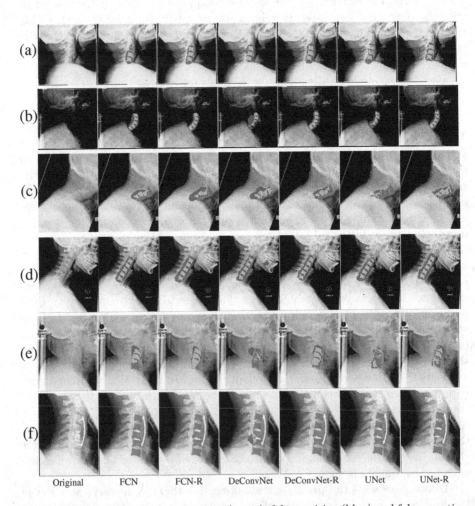

Fig. 4. Qualitative results: true positive (green), false positive (blue) and false negative (red) based on manual vertebrae segmentation. (Color figure online)

Table 1. Quantitative results (%).

Networks	Lower resolution				Original resolution	
	Pixel accuracy	Dice	Sensitivity	Specificity	Sensitivity	Specificity
SOTA [7]	Not available				79.9 ± 16.7	98.1 ± 1.1
FCN	98.9 ± 1.0	83.8 ± 15.6	80.2 ± 18.1	**99.7 ± 0.3**	91.5 ± 18.4	**97.5 ± 0.7**
FCN-R	**99.1 ± 1.0**	**85.8 ± 14.7**	**85.0 ± 17.6**	99.6 ± 0.3	**93.6 ± 17.6**	97.2 ± 0.8
DeConvNet	98.6 ± 1.1	79.7 ± 16.4	77.2 ± 19.1	99.5 ± 0.5	88.2 ± 20.4	97.3 ± 0.8
DeConvNet-R	**99.1 ± 1.0**	**85.7 ± 15.6**	81.0 ± 18.1	**99.8 ± 0.2**	91.5 ± 18.3	**97.7 ± 0.7**
UNet	98.9 ± 1.1	84.1 ± 18.5	79.9 ± 21.9	99.8 ± 0.3	87.5 ± 23.3	**97.6 ± 0.8**
UNet-R	99.0 ± 1.0	85.1 ± 17.0	**82.5 ± 20.8**	99.7 ± 0.2	**89.6 ± 21.1**	97.4 ± 0.8

required for the algorithm to produce a result, our slowest framework (UNet) is approximately 60 times faster than [7]. To prove the robustness, we have tested the proposed framework on 275 cervical X-ray images of NHANES-II dataset [12] and even without any adaptation or transfer learning on the networks, it showed promising capability of generalization in localizing the cervical spine. However, due to insufficient ground truth information, quantitative results are not available. A few qualitative localization results on this dataset are shown in Fig. 5.

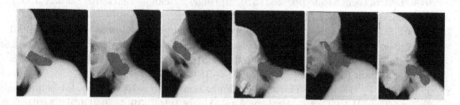

Fig. 5. Localization results (blue overlay) on NHANES-II dataset. (Color figure online)

5 Conclusion

In this paper, we have proposed a framework for spine localization in cervical X-ray images. The localization problem has been reformulated as a binary segmentation problem in a lower resolution. Based on the domain knowledge, a novel region-aware term was added to the loss function to produce a single region as localized output. Three segmentation networks were investigated and the novel loss function improved the performance of all these networks significantly. A maximum average sensitivity of 93.6% and specificity of 97.7% was achieved. Currently, we are adapting the proposed method for precise vertebrae segmentation. In future work, we plan to build a fully automatic computer-aided detection system for cervical spine injuries.

References

1. Platzer, P., Hauswirth, N., Jaindl, M., Chatwani, S., Vecsei, V., Gaebler, C.: Delayed or missed diagnosis of cervical spine injuries. J. Trauma Acute Care Surg. **61**(1), 150–155 (2006)
2. Morris, C., McCoy, E.: Clearing the cervical spine in unconscious polytrauma victims, balancing risks and effective screening. Anaesthesia **59**(5), 464–482 (2004)
3. Tezmol, A., Sari-Sarraf, H., Mitra, S., Long, R., Gururajan, A.: Customized Hough transform for Robust segmentation of cervical vertebrae from X-ray images. In: Proceedings of Fifth IEEE Southwest Symposium on Image Analysis and Interpretation, pp. 224–228. IEEE (2002)
4. Larhmam, M.A., Mahmoudi, S., Benjelloun, M.: Semi-automatic detection of cervical vertebrae in X-ray images using generalized Hough transform. In: 2012 3rd International Conference on Image Processing Theory, Tools and Applications (IPTA), pp. 396–401. IEEE (2012)
5. Glocker, B., Feulner, J., Criminisi, A., Haynor, D.R., Konukoglu, E.: Automatic localization and identification of vertebrae in arbitrary field-of-view CT scans. In: Ayache, N., Delingette, H., Golland, P., Mori, K. (eds.) MICCAI 2012. LNCS, vol. 7512, pp. 590–598. Springer, Heidelberg (2012). doi:10.1007/978-3-642-33454-2_73
6. Glocker, B., Zikic, D., Konukoglu, E., Haynor, D.R., Criminisi, A.: Vertebrae localization in pathological spine CT via dense classification from sparse annotations. In: Mori, K., Sakuma, I., Sato, Y., Barillot, C., Navab, N. (eds.) MICCAI 2013. LNCS, vol. 8150, pp. 262–270. Springer, Heidelberg (2013). doi:10.1007/978-3-642-40763-5_33
7. Al Arif, S.M.M.R., Gundry, M., Knapp, K., Slabaugh, G.: Global localization and orientation of the cervical spine in X-ray images. In: Yao, J., Vrtovec, T., Zheng, G., Frangi, A., Glocker, B., Li, S. (eds.) CSI 2016. LNCS, vol. 10182, pp. 3–15. Springer, Cham (2016). doi:10.1007/978-3-319-55050-3_1
8. Shelhamer, E., Long, J., Darrell, T.: Fully convolutional networks for semantic segmentation. IEEE Trans. Pattern Anal. Mach. Intell. **39**(4), 640–651 (2016)
9. Noh, H., Hong, S., Han, B.: Learning deconvolution network for semantic segmentation. In: Proceedings of the IEEE International Conference on Computer Vision, pp. 1520–1528 (2015)
10. Ronneberger, O., Fischer, P., Brox, T.: U-Net: convolutional networks for biomedical image segmentation. In: Navab, N., Hornegger, J., Wells, W.M., Frangi, A.F. (eds.) MICCAI 2015. LNCS, vol. 9351, pp. 234–241. Springer, Cham (2015). doi:10.1007/978-3-319-24574-4_28
11. Ruder, S.: An overview of gradient descent optimization algorithms, arXiv preprint arXiv:1609.04747 (2016)
12. NHANES-II Dataset. https://ceb.nlm.nih.gov/proj/ftp/ftp.php. Accessed 19 Feb 2017

Domain-Adversarial Neural Networks to Address the Appearance Variability of Histopathology Images

Maxime W. Lafarge[✉], Josien P.W. Pluim, Koen A.J. Eppenhof,
Pim Moeskops, and Mitko Veta

Medical Image Analysis Group, Department of Biomedical Engineering,
Eindhoven University of Technology, Eindhoven, The Netherlands
m.w.lafarge@tue.nl

Abstract. Preparing and scanning histopathology slides consists of several steps, each with a multitude of parameters. The parameters can vary between pathology labs and within the same lab over time, resulting in significant variability of the tissue appearance that hampers the generalization of automatic image analysis methods. Typically, this is addressed with ad-hoc approaches such as staining normalization that aim to reduce the appearance variability. In this paper, we propose a systematic solution based on domain-adversarial neural networks. We hypothesize that removing the domain information from the model representation leads to better generalization. We tested our hypothesis for the problem of mitosis detection in breast cancer histopathology images and made a comparative analysis with two other approaches. We show that combining color augmentation with domain-adversarial training is a better alternative than standard approaches to improve the generalization of deep learning methods.

Keywords: Domain-adversarial training · Histopathology image analysis

1 Introduction

Histopathology image analysis aims at automating tasks that are difficult, expensive and time-consuming for pathologists to perform. The high variability of the appearance of histopathological images, which is the result of the inconsistency of the tissue preparation process, is a well-known observation. This hampers the generalization of image analysis methods, particularly to datasets from external pathology labs.

The appearance variability of histopathology images is commonly addressed by standardizing the images before analysis, for example by performing staining normalization [6,7]. These methods are efficient at standardizing colors while

© Springer International Publishing AG 2017
M.J. Cardoso et al. (Eds.): DLMIA/ML-CDS 2017, LNCS 10553, pp. 83–91, 2017.
DOI: 10.1007/978-3-319-67558-9_10

keeping structures intact, but are not equipped to handle other sources of variability, for instance due to differences in tissue fixation.

We hypothesize that a more general and efficient approach in the context of deep convolutional neural networks (CNNs) is to impose constraints that disregard non-relevant appearance variability with domain-adversarial training [3]. We trained CNN models for mitosis detection in breast cancer histopathology images on a limited amount of data from one pathology lab and evaluated them on a test dataset from different, external pathology labs. In addition to domain-adversarial training, we investigated two additional approaches (color augmentation and staining normalization) and made a comparative analysis. As a main contribution, we show that domain-adversarial neural networks are a new alternative for improving the generalization of deep learning methods for histopathology image analysis.

2   Materials and Methods

2.1   Datasets

This study was performed with the TUPAC16 dataset [1] that includes 73 breast cancer cases annotated for mitotic figures. The density of mitotic figures can be directly related to the tumor proliferation activity, and is an important biomarker for breast cancer prognostication.

The cases come from three different pathology labs (23, 25 and 25 cases per lab) and were scanned with two different whole-slide image scanners (the images from the second two pathology labs were scanned with the same scanner). All CNN models were trained with eight cases (458 mitoses) from the first pathology lab. Four cases were used as a validation set (92 mitoses). The remaining 12 cases (533 mitoses) from the first pathology lab were used as an internal test set[1] and the 50 cases from the two other pathology labs (469 mitoses) were used to evaluate inter-lab generalization performance.

2.2   The Underlying CNN Architecture

The most successful methods for mitosis detection in breast cancer histopathology images are based on convolutional neural networks (CNN). These methods train models to classify image patches based on mitosis annotations resulting from the agreement of several expert pathologists [1,2,9].

The baseline architecture that is used in all experiments of this study is a 6-layer neural network with four convolutional and two fully connected layers that takes a 63×63 image patch as an input and produces a probability that there is a mitotic figure in the center of the patch as an output. The first convolutional layer has 4×4 kernels and the remaining three convolutional layers have 3×3 kernels. All convolutional layers have 16 feature maps. The first fully connected layer has 64 neurons and the second layer serves as the output layer with softmax activation. Batch normalization, max-pooling and ReLU nonlinearities

---

[1] This test set is identical to the one used in the AMIDA13 challenge [9].

are used throughout. This architecture is similar to the one proposed in [2]. The neural network can be densely applied to images in order to produce a mitosis probability map for detection.

2.3   Three Approaches to Handling Appearance Variability

Poor generalization occurs when there is a discrepancy between the distribution of the training and testing data. Increasing the amount of training data can be of help, however, annotation of histology images is a time-consuming process that requires scarce expertise. More feasible solutions are needed, therefore we chose to investigate three approaches.

One straightforward alternative is to artificially produce new training samples. Standard data augmentation methods include random spatial and intensity/color transformation (e.g. rotation, mirroring and scaling, and color shifts). In this study, we use spatial data augmentation (arbitrary rotation, mirroring and ±20% scaling) during the training of all models. Since the most prominent source of variability in histopathology images is the staining color appearance, the contribution of color augmentation (CA) during training is evaluated separately.

The opposite strategy is to reduce the appearance variability of all the images as a pre-processing step before training and evaluating a CNN model. For hematoxylin and eosin (H&E) stained slides, staining normalization (SN) methods can be used [6,7].

A more direct strategy is to constrain the weights of the model to encourage learning of mitosis-related features that are consistent for any input image appearance. We observed that the features extracted by a baseline CNN mitosis classifier carry information about the origin of the input patch (see Sect. 3). We expect that better generalization can be achieved by eliminating this information from the learned representation with domain-adversarial training [3].

Finally, in addition to the three individual approaches, we also investigate all possible combinations.

Color Augmentation. Color variability can be increased by applying random color transformations to original training samples. We perform color augmentation by transforming every color channels $I_c \leftarrow a_c \cdot I_c + b_c$, where a_c and b_c are drawn from uniform distributions $a_c \sim U\,[0.9, 1.1]$ and $b_c \sim U\,[-10, +10]$.

Staining Normalization. The RBG pixel intensities of H&E-stained histopathology images can be modeled with the Beer-Lambert law of light absorption: $I_c = I_0 \exp\left(-\mathbf{A}_{c,*} \cdot \mathbf{C}\right)$. In this expression $c = 1, 2, 3$ is the color-channel index, $\mathbf{A} \in [0, +\infty]^{3 \times 2}$ is the matrix of absorbance coefficients and $\mathbf{C} \in [0, +\infty]^2$ are the stain concentrations [7]. We perform staining normalization with the method described in [6]. This is an unsupervised method that decomposes any image with estimates of its underlying $\mathbf{A}$ and $\mathbf{C}$. The appearance variability over the dataset can then be reduced by recomposing all the images using some fixed absorbance coefficients (Fig. 1).

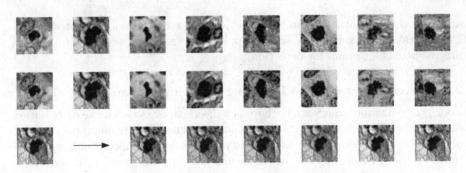

Fig. 1. Illustration of the variability of histological images with 8 patches from different slides (first row), and their transformed version after staining normalization (second row). The third row illustrates the range of color variation induced by color augmentation. (Color figure online)

Domain Adversarial Neural-Network. Since every digital slide results from a unique combination of preparation parameters, we assume that all the image patches extracted from the same slide come from the same unique data distribution and thus constitute a domain. Domain-adversarial neural networks (DANN) allow to learn a classification task, while ensuring that the domain of origin of any sample of the training data cannot be recovered from the learned feature representation [3]. Such a domain-agnostic representation improves the cross-domain generalization of the trained models.

Any image patch **x** extracted from the training data can be given two labels: its class label y (assigned to "1" if the patch is centered at a mitotic figure, "0" otherwise) and its domain label d (a unique identifier of the slide that is the origin of the patch).

The training of the mitosis classifier introduced in Sect. 2.2 is performed by minimizing the cross-entropy loss $\mathcal{L}_M(\mathbf{x}, y; \boldsymbol{\theta}_M)$, where $\boldsymbol{\theta}_M$ are the parameters of the network.

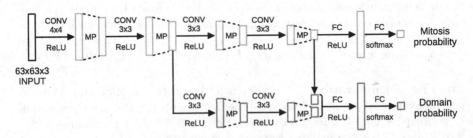

Fig. 2. Architecture of the domain-adversarial neural network. The domain classification (red) bifurcates from the baseline network (blue) at the second and forth layers. (Color figure online)

The DANN is made of a second CNN that takes as input the activations of the second and fourth layers of the mitosis classifier and predicts the domain identifier d. This network is constructed in parallel to the mitosis classifier, with the same corresponding architecture (Fig. 2). Multiple bifurcations are used to make domain classification possible from different levels of abstraction and to improve training stability as in [4]. The cross-entropy loss of the domain classifier is $\mathcal{L}_D(\mathbf{x}, d; \boldsymbol{\theta}_M, \boldsymbol{\theta}_D)$, where $\boldsymbol{\theta}_D$ are the parameters of the bifurcated network (note however that the loss is also a function of $\boldsymbol{\theta}_M$).

The weights of the whole network are optimized via gradient back-propagation during an iterative training process that consists of three successive update rules:

Optimization of the mitosis classifier
$$\boldsymbol{\theta}_M \leftarrow \boldsymbol{\theta}_M - \lambda_M \frac{\partial \mathcal{L}_M}{\partial \boldsymbol{\theta}_M} \quad (1)$$
with learning rate λ_M:

Optimization of the domain classifier
$$\boldsymbol{\theta}_D \leftarrow \boldsymbol{\theta}_D - \lambda_D \frac{\partial \mathcal{L}_D}{\partial \boldsymbol{\theta}_D} \quad (2)$$
with learning rate λ_D:

Adversarial update of the mitosis classifier:
$$\boldsymbol{\theta}_M \leftarrow \boldsymbol{\theta}_M + \alpha \lambda_D \frac{\partial \mathcal{L}_D}{\partial \boldsymbol{\theta}_M} \quad (3)$$

The update rules (1) and (2) work in an adversarial way: with (1), the parameters $\boldsymbol{\theta}_M$ are updated for the mitosis detection task (by minimizing $\mathcal{L}_M$), and with (3), the same parameters are updated to prevent the domain of origin to be recovered from the learned representation (by maximizing $\mathcal{L}_D$). The parameter $\alpha \in [0, 1]$ controls the strength of the adversarial component.

2.4 Evaluation

The performances of the mitosis detection models were evaluated with the F1-score as described in [1,2,9]. We used the trained classifiers to produce dense mitosis probability maps for all test images. All local maxima above an operating point were considered detected mitotic figures. The operating point was determined as the threshold that maximizes the F1-score over the validation set.

We used the t-distributed stochastic neighbor embedding (t-SNE) [5] method for low-dimensional feature embedding, to qualitatively compare the domain overlap of the learned feature representation for the different methods.

3 Experiments and Results

For every possible combination of the three approaches developed in Sect. 2.3, we trained three convolutional neural networks with the same baseline architecture, under the same training procedure, but with random initialization seeds to assess the consistency of the approaches.

Baseline Training. Training was performed with stochastic gradient descent with momentum and with the following parameters: batch size of 64 (with balanced class distribution), learning rate λ_M of 0.01 with a decay factor of 0.9 every 5000 iterations, weight decay of 0.0005 and momentum of 0.9. The training was stopped after 40000 iterations.

Because the training set has a high class imbalance, hard negative mining was performed as previously described [2,8]. To this purpose, an initial classifier was trained with the baseline CNN model. A set of hard negative patches was then obtained by probabilistically sampling the probability maps produced by this first classifier (excluding ground truth locations). We use the same set of hard-negative samples for all experiments.

Domain-Adversarial Training. Every training iteration of the DANN models involves two passes. The first pass is performed in the same manner as the baseline training procedure and it involves the update (1). The second pass uses batches balanced over the domains of the training set, and is used for updates (2) and (3). Given that the training set includes eight domains, the batches for the second pass are therefore made of 8 random patches from each training case. The learning rate λ_D for these updates was fixed at 0.0025.

As remarked in [3,4], domain-adversarial training is an unstable process. Therefore we use a cyclic scheduling of the parameter α involved in the adversarial update (3). This allows alternating between phases in which both branches learn their respective tasks without interfering, and phases in which domain-adversarial training occurs. In order to avoid getting stuck in local maxima and to ensure that domain information is not recovered over iterations in the main branch, the weights of the domain classifier θ_D are reinitialized at the beginning of every cycle.

Performance. The F1-scores for all three approaches and their combinations are given in Table 1. t-SNE embeddings of the feature representations learned by the baseline model and the three investigated approaches are given in Fig. 3.

Table 1. Mean and standard deviation of the F1-score over the three repeated experiments. Every column of the table represents the performance of one method on the internal test set (ITS; from the same pathology lab) and the external test sets (ETS; from different pathology labs). The squares indicate the different investigated methods. Multiple squares indicate a combination of methods.

CA		■			■	■		■
SN			■		■		■	■
DANN				■		■	■	■
ITS	.61 ± .02	.61 ± .01	.57 ± .06	.61 ± .02	.55 ± .01	.62 ± .02	.61 ± .01	.57 ± .01
ETS	.33 ± .08	.58 ± .03	.46 ± .02	.55 ± .05	.48 ± .08	.62 ± .00	.51 ± .02	.53 ± .03

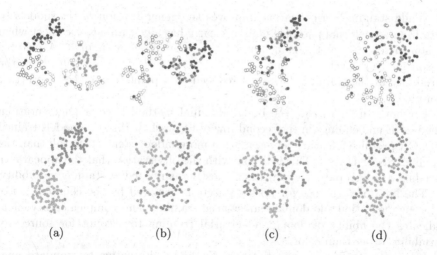

Fig. 3. t-SNE embeddings of 80 patches represented by the learned features at the fourth layer of the mitosis classifier. First row: patches are balanced across classes (mitosis: disk, non-mitosis: circle) and are equally sampled from two different slides of the training set (red/blue). Second row: patches of mitotic figures sampled from slides of the internal (orange) and external test set (green). Each column corresponds to one approach: (a) baseline, (b) SN, (c) CA, (d) DANN. (Color figure online)

Although the t-SNE embeddings of the first row only show two domains for clarity, the same observations can be made for almost all pairs of domains.

4 Discussion and Conclusions

On the internal test set, all methods and combinations have good performance in line with previously reported results [1,2,8,9]. The combination of color augmentation and domain-adversarial training has the best performance (F1-score of 0.62 ± 0.02). The staining normalization method and combinations of staining normalization with other methods have the worst performance (F1-scores lower than the baseline method).

As with the internal test set, the best performance on the external test set is achieved by the combination of color augmentation and domain-adversarial training (F1-score of 0.62 ± 0.00). On the external test set, all three investigated methods show improvement since the baseline method has the worst performance (F1-score of 0.33 ± 0.08).

The intra-lab t-SNE embeddings presented in the first row of Fig. 3 show that the baseline model learns a feature representation informative of the domain, as shown by the presence of well-defined clusters corresponding to the domains of the embedded image patches. In contrast, each of the three approaches produces some domain confusion in the model representation, since such domain clusters are not produced by t-SNE under the same conditions.

While staining normalization improves the generalization of the models to data from an external pathology lab, it clearly has a general adverse effect when combined to other methods, compared to combinations without it. A possible reason for this effect could be that by performing staining normalization, the variability of the training dataset is reduced to a point that makes overfitting more likely.

For both test datasets, the best individual method is color augmentation. The t-SNE embeddings in the second row of Fig. 3 show that the models trained with CA produce a feature representation more independent of the lab than the baseline, SN or DANN. This is in line with the observation that the appearance variability in histopathology images is mostly manifested as staining variability.

The best performance for both datasets is achieved by the combination of color augmentation and domain-adversarial training. This complementary effect indicates the ability of domain-adversarial training to account for sources of variability other than color.

In conclusion, we investigated DANNs as an alternative to standard augmentation and normalization approaches, and made a comparative analysis. The combination of color augmentation and DANNs had the best performance, confirming the relevance of domain-adversarial approaches in histopathology image analysis. This study is based on the performances for a single histopathology image analysis problem and only one staining normalization method was investigated. These are limiting factors, and further confirmation of the conclusions we make is warranted.

References

1. Tumor proliferation assessment challenge 2016. http://tupac.tue-image.nl
2. Cireşan, D.C., Giusti, A., Gambardella, L.M., Schmidhuber, J.: Mitosis detection in breast cancer histology images with deep neural networks. In: Mori, K., Sakuma, I., Sato, Y., Barillot, C., Navab, N. (eds.) MICCAI 2013. LNCS, vol. 8150, pp. 411–418. Springer, Heidelberg (2013). doi:10.1007/978-3-642-40763-5_51
3. Ganin, Y., Ustinova, E., Ajakan, H., Germain, P., Larochelle, H., Laviolette, F., Marchand, M., Lempitsky, V.: Domain-adversarial training of neural networks. J. Mach. Learn. Res. **17**(59), 1–35 (2016)
4. Kamnitsas, K., Baumgartner, C., Ledig, C., Newcombe, V., Simpson, J., Kane, A., Menon, D., Nori, A., Criminisi, A., Rueckert, D., Glocker, B.: Unsupervised domain adaptation in brain lesion segmentation with adversarial networks. In: Niethammer, M., Styner, M., Aylward, S., Zhu, H., Oguz, I., Yap, P.-T., Shen, D. (eds.) IPMI 2017. LNCS, vol. 10265, pp. 597–609. Springer, Cham (2017). doi:10.1007/978-3-319-59050-9_47
5. van der Maaten, L., Hinton, G.: Visualizing data using t-SNE. J. Mach. Learn. Res. **9**, 2579–2605 (2008)
6. Macenko, M., Niethammer, M., Marron, J., Borland, D., Woosley, J.T., Guan, X., Schmitt, C., Thomas, N.E.: A method for normalizing histology slides for quantitative analysis. In: IEEE ISBI 2009, pp. 1107–1110 (2009)
7. Ruifrok, A.C., Johnston, D.A., et al.: Quantification of histochemical staining by color deconvolution. Anal. Quant. Cytol. **23**(4), 291–299 (2001)

8. Veta, M., van Diest, P.J., Jiwa, M., Al-Janabi, S., Pluim, J.P.: Mitosis counting in breast cancer: object-level interobserver agreement and comparison to an automatic method. PloS one **11**(8), e0161286 (2016)
9. Veta, M., Van Diest, P.J., Willems, S.M., Wang, H., Madabhushi, A., Cruz-Roa, A., Gonzalez, F., Larsen, A.B., Vestergaard, J.S., Dahl, A.B., et al.: Assessment of algorithms for mitosis detection in breast cancer histopathology images. Med. Image Anal. **20**(1), 237–248 (2015)

Accurate Lung Segmentation via Network-Wise Training of Convolutional Networks

Sangheum Hwang$^{(\boxtimes)}$ and Sunggyun Park

Lunit Inc., Seoul, Korea
{shwang,sgpark}@lunit.io

Abstract. We introduce an accurate lung segmentation model for chest radiographs based on deep convolutional neural networks. Our model is based on atrous convolutional layers to increase the field-of-view of filters efficiently. To improve segmentation performances further, we also propose a multi-stage training strategy, *network-wise training*, which the current stage network is fed with both input images and the outputs from pre-stage network. It is shown that this strategy has an ability to reduce falsely predicted labels and produce smooth boundaries of lung fields. We evaluate the proposed model on a common benchmark dataset, JSRT, and achieve the state-of-the-art segmentation performances with much fewer model parameters.

Keywords: Lung segmentation · Network-wise trainnig · Atrous convolution

1 Introduction

Accurate lung boundaries provide valuable image-based information such as total lung volume or shape irregularities, but it also has an important role as a prerequisite step for developing computer-aided diagnosis (CAD) system. However, an automated segmentation of lung fields is a challenging problem due to high variations in shape and size among different chest radiographs.

For automatic detection of lung fields, a lot of methods have been proposed over the past decade [1,3,10,12]. The early segmentation methods can be partitioned into *rule-based, pixel classification-based, deformable model-based,* and *hybrid* methods [3]. Recently, deep neural network-based approaches [10,12][1] have been proposed due to the success of deep learning in various computer vision tasks including object classification [8], localization [13], and segmentation [2,11].

For semantic segmentation, the encoder-decoder architecture is commonly used [11]. In this architecture, encoder is a typical convolutional neural network

---

[1] In [10], the authors propose a hybrid model combined distance regularized level sets with a deep learning model for lung segmentation. This model shows high overlap scores but it requires good initial guesses. Therefore, we exclude this model from our comparison.

© Springer International Publishing AG 2017
M.J. Cardoso et al. (Eds.): DLMIA/ML-CDS 2017, LNCS 10553, pp. 92–99, 2017.
DOI: 10.1007/978-3-319-67558-9_11

(CNN), while decoder consists of transposed convolutions and upsampling operations. The role of decoder is to restore the abstracted feature map by learning how to densify the sparse activations. The final output of decoder is a probability map with the same size as that of the ground-truth masks, and pixel-wise cross entropy loss is employed for training. Such encoder-decoder architecture has also been shown its promising performances in various medical imaging problems [12,14]. For example, U-Net [14], a variant of the encoder-decoder architecture, shows the impressive results on segmentation of neuronal structures in electron microscopic stacks. For the task of lung segmentation, the authors of [12] present U-Net-based CNN architecture for automated segmentation of anatomical organs (e.g., lung, cavicles and heart) in chest radiographs. They also propose a modified loss function to deal with the multi-class segmentation problem.

Another succesful approach for semantic segmentation is to employ atrous convolutional layers by replacing some convolutional layers [2]. It is known that atrous convolution effectively enlarges the global receptive field of CNN [9], and therefore larger context information can be efficiently utilized for prediction of pixel-wise labels.

In this paper, we introduce an accurate lung segmentation model for chest radiographs based on deep CNN with atrous convolutions. The proposed model is designed to have a deep-and-thin architecture, which has much fewer parameters compared to other CNN-based lung segmentation models. To improve further, we propose a multi-stage training strategy, *network-wise training*, which the current stage network is fed with both input images and the outputs of pre-stage network. It is shown that this strategy has an ability to reduce falsely predicted labels (i.e. false positives and false negatives) and produce smooth boundaries of segmented lung fields.

We evaluate the proposed method on a common benchmark dataset, the Japanese Society of Radiological Technology (JSRT) [15], and achieve the state-of-the-art results under four popular segmentation metrics: the Jaccard similarity coefficient, Dice's coefficient, average contour distance, and average surface distance. To investigate generalization capability of our method, we test on another dataset, the Montgomery County (MC) [6]. It is observed that performances on this dataset are comparable in terms of mean values, but have high variances since there is some degree of a shift between training (JSRT) and test (MC) distributions.

2 Methods

2.1 Lung Segmentation with Atrous Convolutions

We present a deep-and-thin CNN architecture based on residual learning [5] which has skip connections to prevent the gradient vanishing problem. Dense prediction problems should consider large context to predict class labels of pixels. Simple way for larger context is increasing the global receptive fields of network

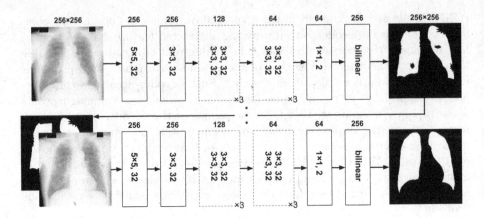

Fig. 1. Simplified framework of the proposed method. In network, the boxes in black represent convolutions and bilinear interpolation, and the dotted boxes in red denote residual blocks consisting of consecutive two convolutional layers. The number in each box means the size and the number of filters, respectively, and that above each box represents the resolution of feature maps. Our model is trained in a network-wise manner by feeding the outputs from pre-stage network. (Color figure online)

by stacking more convolution layers or using downsampling operations (e.g., pooling or strided convolution) [9].

Recently, it is known that atrous convolution is useful to enlarge the field-of-view (i.e. receptive fields) of filters. This enlargement is particularly effective for segmentation task since it should consider the context around the location where we want to predict class labels [2]. Atrous convolution contains 'holes' between weights of filters so that it involves larger fields to compute activations. Given a filter $\mathbf{k} = [k(m, n)]$ for $m, n = 1, ..., 2k + 1$ and the input $\mathbf{x} = [x(i, j)]$ at location (i, j), atrous convolution with rate r computes the output $\mathbf{y} = [y(i, j)]$ as follows:

$$y(i, j) = \sum_{m=-k}^{k} \sum_{n=-k}^{k} x(i + rm, j + rn)k(m, n) \tag{1}$$

Note that if $r = 1$, Eq. 1 stands for standard convolution operation. Therefore, the global receptive field of network can be controlled via rate r while maintaining the number of weights.

Figure 1 shows an architecture of our network for lung segmentation task. It consists of 3 convolutional layers and 6 residual blocks, i.e. 15 convolutional layers. We employ atrous convolutions with $r = 3$ for the end of two residual blocks. Batch normalization layer is followed by every convolutional layer. The global stride of our network is 4, i.e. 2 convolutional layers at the beginning of particular residual blocks (the first layer in each red block in Fig. 1) operate convolutions with stride 2 (i.e. 2-strided convolution). To calculate pixel-wise cross entropy loss with the groud-truth mask, we upsample network outputs by bilinear interpolation.

The advantage of the proposed deep-and-thin architecture is that it has much fewer model parameters compared to other CNN-based lung segmentation models. For examples, our model has 120,672 weights (26 times fewer parameters) while the encoder-decoder network like U-Net has 3,140,771 weights [12].

2.2 Network-Wise Training of CNN

Generally, CNN with atrous convolutions and bilinear interpolation has some limitations. First, it may produce small false positive or false negative areas. This is mainly caused due to pixel-wise cross entropy loss dealing with every pixel independently. Second, it outputs blurry object boundaries, which is inevitable if we use a bilinear interpolation to upsample the downsampled feature maps. To overcome these issues, postprocessing via conditional random fields [7] is widely used to smooth such noisy segmentation maps [2].

We propose another strategy, network-wise training, to refine segmentation results. It is designed as a repeated training pipeline which has an output of pre-stage model as an input (see Fig. 1). At the first stage (namely stage 1), a network is trained using only input chest radiographs. After training it, both input chest radiographs and network outputs from trained model at stage 1 are fed into the second network. Specifically, input chest radiographs and the corresponding output from pre-stage network are concatenated across the channel dimension. From relatively coarse segmentation outputs, a network can more focus on the details to learn accurate boundaries of lung fields. This procedure is iterated until validation performance is saturated. Note that this strategy can be considered as iterative cascading, an extended version of the cascaded network [4].

3 Computational Experiments

We use a common benchmark dataset, the Japanese Society of Radiological Technology (JSRT) dataset [15], to evaluate lung segmentation performance of our model. JSRT database contains 247 the posterior-anterior (PA) chest radiographs, 154 have lung nodules and 93 have no nodules. The ground-truth lung masks can be obtained in the Segmentation in Chest Radiographs (SCR) database [3].

Following previous practices in literatures, JSRT dataset is split in two folds: one contains 124 odd numbered and the other contain 123 even numbered chest radiographs. Then, one fold is used for training[2] and the other fold used for testing, and vice versa. Final performances are computed by averaging results from both cases. Also, all training images are resized to 256×256 as in the literatures. The network is trained via stochastic gradient descent with momentum 0.9. For learning rate scheduling, we set initial learning rate to 0.1 and it is decreased to 0.01 after training 70 epochs.

---

[2] After the search of hyperparameters with randomly selected 30% training data, the network is re-trained with the entire training data.

We use Montgomery County (MC) dataset [6] as another testset to investigate generalization capability of our model. MC dataset contains PA chest radiographs collected from National Library of Medicine, National Institutes of Health, Bethesda, MD, USA. It consists of 80 normal and 58 abnormal cases with manifestations of tuberculosis. It is interesting to see segmentation performances on this dataset since it has different characteristics compared to training set (JSRT): image acquisition equipment, abnormal diseases, nationality of patients, etc.

3.1   Performance Metrics

We use four commonly used metrics in the literatures: the Jaccard similarity coefficient(JSC), Dice's coefficient (DC), average contour distance (ACD), average surface distance (ASD)[3]. JSC and DC are similar in that they only consider the number of true positives, false positives and false negatives. Therefore, they are metrics ignoring predicted locations. On the other hand, ACD and ASD are distance-based metrics. They penalize if the minimum distance of a particular pixel predicted as lung boundaries to the ground-truth boundaries is large. Therefore, performance from these metrics may vary even if JSC and DC are almost the same.

Let s_i, $i = 1, ..., n_S$, and g_j, $i = 1, ..., n_G$, be the pixels on the segmented boundary S and the ground-truth boundary G. The minimum distance of s_i on S to G is defined as $d(s_i, G) = \min_j \|g_j - s_i\|$. Then, ACD and ASD are computed as follows:

$$\mathrm{ACD(S,G)} = \frac{1}{2} \left(\frac{\sum_i d(s_i, G)}{n_S} + \frac{\sum_j d(g_i, S)}{n_G} \right)$$

$$\mathrm{ASD(S,G)} = \frac{1}{n_S + n_G} \left(\sum_i d(s_i, G) + \sum_j d(g_i, S) \right).$$

(2)

3.2   Quantatitive and Qualititive Results

Table 1 summarizes segmentation performances of our model compared to previous methods[4]. First, we evaluate the models at stage 1 and 3, which are trained without any preprocessing method such as histogram equalization and data augmentation techniques to exclude other potential factors that may affect performances. These results show that segmentation performances are continously improved through a network-wise training, and those from stage 3 model outperforms other methods.

The left side in Fig. 2 shows the effect of the proposed network-wise training, false positive and negative reduction and boundary smoothing. The top row

---

[3] Average surface distance is also known as symmetric mean absolute surface distance [12].

[4] Note that JSC and DC numbers in Candemir [1] are incorrect since DC should be 2JSC/(1+JSC).

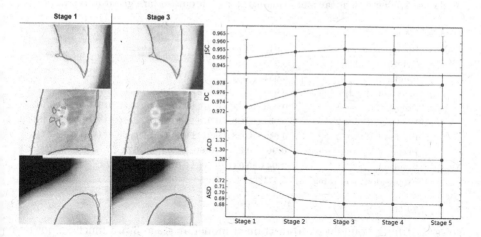

Fig. 2. The effect of network-wise training at each stage. (1) Left figure shows some regions improved through network-wise training. The ground-truth and prediction is depicted in blue and red line, respectively. (2) Right plot shows performances at each stage. It is shown that they are saturated at stage 3. (Color figure online)

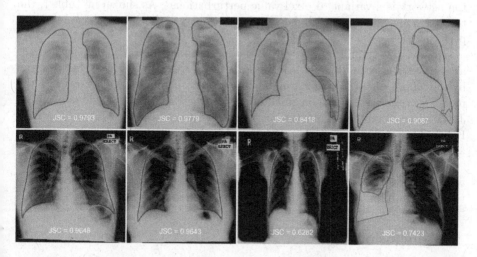

Fig. 3. Segmentation examples for JSRT and MC datasets. The ground-truth masks are depicted in blue and segmentation boundaries from our model are in red. The first and second row show the results from JSRT and MC, respectively. Left two columns and right two columns present the best and worst two examples in terms of JSC. (Color figure online)

Table 1. Mean and standard deviations of segmentation performances for JSRT and MC datasets. The best mean performances for each dataset are given in bold.

Dataset	Methods	JSC	DC	ACD (mm)	ASD (mm)
JSRT	Human observer [3]	0.946±0.018	–	1.64±0.69	–
	Hybrid voting [3]	0.949±0.020	–	1.62±0.66	–
	Candemir [1]	0.954±0.015	0.967±0.008	1.321±0.316	–
	InvertedNet [12]	0.950	0.973	–	0.69
	Proposed (Stage 1)	0.950±0.023	0.974±0.012	1.347±0.919	0.724±0.163
	Proposed (Stage 2)	0.954±0.020	0.976±0.011	1.295±0.846	0.690±0.151
	Proposed (Stage 3)	0.956±0.018	0.977±0.010	1.283±0.814	0.683±0.145
	Proposed w/ aug (Stage 3)	**0.961±0.015**	**0.980±0.008**	**1.237±0.702**	**0.675±0.122**
MC	Candemir [1]	**0.941±0.034**	0.960±0.018	**1.599±0.742**	–
	Proposed w/ aug (Stage 3)	0.931±0.049	**0.964±0.028**	2.186±1.795	0.915±0.258

shows that lung boundaries from trained model at stage 3 are much smoother than those from the model at stage 1. Also, the second and third rows support that false positives and false negatives can be supressed as stage goes. The performance plot in the right side in Fig. 2 shows the change of performances according to stages. It is observed that they are saturated at stage 3, so we report the performances from the model at stage 3.

In addition, we investigate the effect of data augmentation. For this, we adjust pixel values randomly through adjusting brightness and contrast so that the network is invariant to pixel value perturbations[5]. As shown in Table 1, the trained model at stage 3 with data augmentation gives much better segmentation performances.

However, it should be noted that the performances on MC dataset are not as good as those on JSRT. Mean performances are slightly lower than the the hybrid approach in [1], but standard deviations are much higher even if the model is trained with data augmentation. It means that our model traind using JSRT gives unstable segmentation results on some cases in MC as shown in Fig. 3. This is due to the presence of a shift between distributions of training and test datasets, which needs to solve *Domain Adaptation* problem.

The samples of segmented lung boundaries are visualized in Fig. 3. Left two columns show the best two results in terms of JSC, and right two columns show the worst two for JSRT and MC datasets.

4   Conclusion

In this paper, we present an accurate lung segmentation model based on CNN with atrous convolutions. Furthermore, a novel multi-stage training strategy, network-wise training, to refine the segmentation results is also proposed. Computational experiments on benchmark dataset, JSRT, show that the proposed

---

[5] Cropping, horizontal flipping and rotation were not effective. This is because lung segmentation network does not need to be invariant to such transformations.

architecture and the network-wise training are very effective to obtain the accurate segmentation model for lung fields. We also evaluate the trained model on MC dataset, which raises the task for us to develop the model insensitive to domain shift.

References

1. Candemir, S., Jaeger, S., Palaniappan, K., Musco, J.P., Singh, R.K., Xue, Z., Karargyris, A., Antani, S., Thoma, G., McDonald, C.J.: Lung segmentation in chest radiographs using anatomical atlases with nonrigid registration. IEEE Trans. Med. Imaging **33**(2), 577–590 (2014)
2. Chen, L.C., Papandreou, G., Kokkinos, I., Murphy, K., Yuille, A.L.: DeepLab: semantic image segmentation with deep convolutional nets, atrous convolution, and fully connected CRFs (2016). arXiv:1606.00915
3. van Ginneken, B., Stegmann, M.B., Loog, M.: Segmentation of anatomical structures in chest radiographs using supervised methods: a comparative study on a public database. Med. Image Anal. **10**(1), 19–40 (2006)
4. Havaei, M., Davy, A., Warde-Farley, D., Biard, A., Courville, A., Bengio, Y., Pal, C., Jodoin, P.M., Larochelle, H.: Brain tumor segmentation with deep neural networks. Med. Image Anal. **35**, 18–31 (2017)
5. He, K., Zhang, X., Ren, S., Sun, J.: Deep residual learning for image recognition. In: CVPR, pp. 770–778 (2016)
6. Jaeger, S., et al.: Automatic tuberculosis screening using chest radiographs. IEEE Trans. Med. Imaging **33**(2), 233–245 (2014)
7. Krähenbühl, P., Koltun, V.: Efficient inference in fully connected CRFs with Gaussian edge potentials. In: NIPS, pp. 109–117 (2011)
8. Krizhevsky, A., Sutskever, I., Hinton, G.E.: Imagenet classification with deep convolutional neural networks. In: NIPS, pp. 1097–1105 (2012)
9. Luo, W., Li, Y., Urtasun, R., Zemel, R.: Understanding the effective receptive field in deep convolutional neural networks. In: NIPS, pp. 4898–4906 (2016)
10. Ngo, T.A., Carneiro, G.: Lung segmentation in chest radiographs using distance regularized level set and deep-structured learning and inference. In: 2015 IEEE International Conference on Image Processing, pp. 2140–2143 (2015)
11. Noh, H., Hong, S., Han, B.: Learning deconvolution network for semantic segmentation. In: ICCV, pp. 1520–1528 (2015)
12. Novikov, A.A., Major, D., Lenis, D., Hladůvka, J., Wimmer, M., Bühler, K.: Fully convolutional architectures for multi-class segmentation in chest radiographs (2017). arXiv:1701.08816
13. Ren, S., He, K., Girshick, R., Sun, J.: Faster R-CNN: towards real-time object detection with region proposal networks. In: NIPS, pp. 91–99 (2012)
14. Ronneberger, O., Fischer, P., Brox, T.: U-net: convolutional networks for biomedical image segmentation. In: Navab, N., Hornegger, J., Wells, W.M., Frangi, A.F. (eds.) MICCAI 2015. LNCS, vol. 9351, pp. 234–241. Springer, Cham (2015). doi:10.1007/978-3-319-24574-4_28
15. Shiraishi, J., Katsuragawa, S., Ikezoe, J., Matsumoto, T.: Kobayashi, T., ichi Komatsu, K., Matsui, M., Fujita, H., Kodera, Y., Doi, K.: Development of a digital image database for chest radiographs with and without a lung nodule: receiver operating characteristic analysis of radiologists detection of pulmonary nodules. Am. J. Roentgenol. **174**(1), 71–74 (2000)

Deep Residual Recurrent Neural Networks for Characterisation of Cardiac Cycle Phase from Echocardiograms

Fatemeh Taheri Dezaki[1(✉)], Neeraj Dhungel[1], Amir H. Abdi[1],
Christina Luong[2], Teresa Tsang[2], John Jue[2], Ken Gin[2], Dale Hawley[2],
Robert Rohling[1], and Purang Abolmaesumi[1(✉)]

[1] The University of British Columbia, Vancouver, BC, Canada
{fatemeht,purang}@ece.ubc.ca
[2] Vancouver General Hospital's Cardiology Laboratory,
Vancouver, BC, Canada

Abstract. Characterisation of cardiac cycle phase in echocardiography data is a necessary preprocessing step for developing automated systems that measure various cardiac parameters. Accurate characterisation is challenging, due to differences in appearance of the cardiac anatomy and the variability of heart rate in individuals. Here, we present a method for automatic recognition of cardiac cycle phase from echocardiograms by using a new deep neural networks architecture. Specifically, we propose to combine deep residual neural networks (ResNets), which extract the hierarchical features from the individual echocardiogram frames, with recurrent neural networks (RNNs), which model the temporal dependencies between sequential frames. We demonstrate that such new architecture produces results that outperform baseline architecture for the automatic characterisation of cardiac cycle phase in large datasets of echocardiograms containing different levels of pathological conditions.

Keywords: Deep residual neural networks · Recurrent neural networks · Long short term memory · Echocardiograms · Frame identification

1 Introduction

According to the World Health Organization[1] millions of people worldwide suffer from the heart-related disease, a major cause of mortality. The 2-D echocardiography (echo) examination is a widely used imaging modality for early diagnosis. Echo

---

F.T. Dezaki and N. Dhungel—Contributed equally.

T. Tsang is the Director of the Vancouver General Hospital and University of British Columbia Echocardiography Laboratories, and Principal Investigator of the CIHR-NSERC grant supporting this work.

[1] Global status report on noncommunicable diseases, 2014.

© Springer International Publishing AG 2017
M.J. Cardoso et al. (Eds.): DLMIA/ML-CDS 2017, LNCS 10553, pp. 100–108, 2017.
DOI: 10.1007/978-3-319-67558-9_12

can be used to estimate several cardiac parameters such as stroke volume, end-diastolic volume, and ejection fraction [1]. These parameters are generally measured by identifying end-systolic and end-diastolic frames from cine echos [1,2]. Current detection is either manual or semi-automatic [3,4], relying primarily on the availability of electrocardiogram (ECG) data, which can be challenging in point-of-care and non-specialized clinics. Moreover, such detection techniques add to the workload of medical personnel and are subject to inter-user variability. Automatic detection of cardiac cycle phase in echo can potentially alleviate these issues. However, such detection is challenging due to low signal-to-noise ratio in echo data, subtle differences between the consecutive frames, the variability of cardiac phases, and temporal dependencies between these frames.

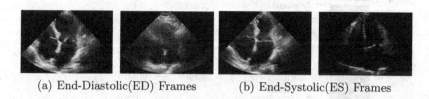

(a) End-Diastolic(ED) Frames   (b) End-Systolic(ES) Frames

Fig. 1. Examples of end-diastolic and end-systolic frames in 2-D echocardiograms.

There have been a number of attempts for automatic detection and labeling of frames in echo [2,4,5]. The most common approach is to use a segmentation strategy, such as level-sets or graph-cuts, for identifying the boundary of the left ventricle in an echo sequence [2,5]. Boundaries that correspond to largest and smallest ventricular areas are regarded as end-diastolic (ED) and end-systolic (ES) frames [5] (Fig. 1). A major drawback of those segmentation-based approaches is the requirement for a good initialization and localization of the left ventricle. In another approach [4], information from cine frames was projected onto a low-dimensional 2-D manifold, after which the difference of distance between points on the manifold was used to determine ED and ES frames. However, this approach does not consider the complex temporal dependencies between the frames, which may subsequently lead to sub-optimal results. Recently, methods based on convolutional neural networks (CNNs) combined with recurrent neural networks (RNNs) have produced state-of-the-art results in many computer vision problems [6–9]. A combination of CNNs and RNNs has been successfully applied in various problem domains such as detecting frames from videos, natural language processing, object detection and tracking [3,8,10,11]. However, our experience is that such network structure cannot be easily extended to the analysis of echo data, which suffer from low signal-to-noise ratio, where anatomical boundaries may not be as clearly visible compared to other imaging modalities.

Here, we formulate a very deep CNN architecture using residual neural networks (ResNets) [7] for extracting deep hierarchical features, which are then passed to RNNs containing two blocks of long short term memory (LSTM) [12] to decode the temporal dependencies between these features. The primary motivation of using a ResNets with LSTMs is that layers in ResNets are reformulated for learning the residual function with respect to input layers for countering the vanishing or exploding gradient problem in CNN while going deeper [7]. In addition to this, we minimize the structured loss function that mimics the temporal changes in left ventricular volume during the systolic and diastolic phase [3]. We demonstrate that the proposed method using ResNets and LSTMs with structured loss function produces state-of-the-art accuracy for detecting cardiac phase cycle in echo data without the need of segmentation.

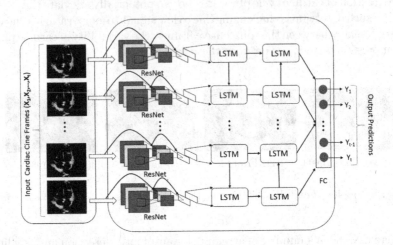

Fig. 2. The proposed method for characterisation of cardiac cycle phase from echocardiograms using deep residual recurrent neural networks (RRNs), which contains residual nets (ResNets), followed by two blocks of long term short term memory (LSTM) and a fully connected layer (FC).

2   Methods

2.1   Dataset

Let $\mathcal{D} = \{(\mathbf{c}^{(i)}, \mathbf{y}^{(i)})_j\}_{j=1}^{|\mathcal{D}|}$ represent the dataset, where $i \in \{1, 2, .., N\}$ indexes the number of cardiac cycles for an individual study j, $\mathbf{c} = \{\mathbf{x}_1, \mathbf{x}_2, .., \mathbf{x}_t\}$ represents the collection of frames $\mathbf{x}$ for each patient such that $\mathbf{x}_t : \Omega \to \mathbb{R}$ with $\Omega \subset \mathbb{R}^2$, and $\mathbf{y} = \{y_1, y_2, .., y_t\}$ denotes the class label for individual frames computed using the following function [3]:

$$y_t = \begin{cases} \left(\frac{|t-T_{ES}^i|}{|T_{ES}^i-T_{ED}^i|}\right)^{\tau}, & \text{if } T_{ED}^i < t < T_{ES}^i \\ \\ \left(\frac{|t-T_{ES}^i|}{|T_{ES}^i-T_{ED}^{i+1}|}\right)^{\frac{1}{\tau}}, & \text{if } T_{ES}^i < t < T_{ED}^{i+1} \end{cases} \tag{1}$$

where T_{ES}^i and T_{ED}^i are the locations of ES and ED frames in the ith cardiac cycle and τ is an integer constant. The expression in Eq. 1 mimics the ventricular volume changes during the systole and diastole phases of a cardiac cycle [3].

2.2 Deep Residual Recurrent Neural Networks (RRNs)

We propose deep residual recurrent neural networks (RRNs) to decode the phase information in echo data. RRNs constitute of a stack of residual units for extracting the hierarchical spatial features from echo data, RNNs that use LSTMs for decoding the temporal information, and a fully connected layer at the end. As shown in Fig. 3(a), each residual unit is made by the addition of a residual function and an identity mapping, and can be expressed in a general form by the following expression [7]:

$$\mathbf{x}_L^{(t)} = \mathbf{x}_l^{(t)} + \sum_{l=1}^{L-1} f_{\text{RES}}(\mathbf{x}_l^{(t)}; \mathcal{W}_l), \tag{2}$$

where $\mathbf{x}_l^{(t)}$ is the input feature to the $l \in \{1, ..., L\}^{th}$ residual unit (for $l = 1$, $\mathbf{x}_l = \mathbf{x}_t$), $\mathcal{W}_l = \mathbf{w}_{l,k}$ is the set of weights for the lth residual unit, $k \in \{1, ..., K\}$ represents the numbers of layers in the residual unit, $f_{\text{RES}}(.)$ is called the residual function represented by a convolutional layer (weight) [6,13], batch normalization (BN) [14] and rectilinear unit (ReLU) [7,15] (Fig. 3(a)).

We use RNNs for decoding the temporal dependencies within the echo sequence. LSTM network as shown in Fig. 3(b) is a type of RNNs with the addition of memory cell that allows the network to learn to remember or forget the hidden states. In particular, LSTM updates the hidden states $\mathbf{h}_t$ and its memory units $\mathbf{c}_t$ with given sequential input $\mathbf{x}_t$ using the following expressions [8]:

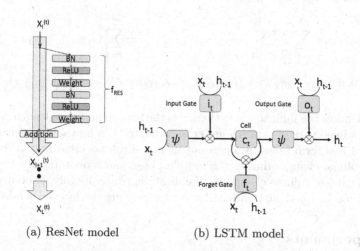

(a) ResNet model   (b) LSTM model

Fig. 3. Two basic building blocks of our proposed method: (a) ResNet, and (b) LSTM. A typical ResNet model contains residual units with BN, ReLU, and convolutional (weight) layers, stacked together and identity mapping (addition). An LSTM model is made of memory units (cells) which maintain a cell state using input, output, and forget gates.

$$\mathbf{i}_t = \phi\left(\mathbf{w}_{xi}\mathbf{x}_t + \mathbf{w}_{hi}\mathbf{h}_{t-1} + \mathbf{b}_i\right);$$
$$\mathbf{f}_t = \phi\left(\mathbf{w}_{xf}\mathbf{x}_t + \mathbf{w}_{hf}\mathbf{h}_{t-1} + \mathbf{b}_f\right); \tag{3}$$
$$\mathbf{o}_t = \phi\left(\mathbf{w}_{xo}\mathbf{x}_t + \mathbf{w}_{ho}\mathbf{h}_{t-1} + \mathbf{b}_o\right);$$

$$\mathbf{c}_t = \mathbf{f}_t \odot \mathbf{c}_{t-1} + \mathbf{i}_t \odot \psi\left(\mathbf{w}_{xi}\mathbf{x}_t + \mathbf{w}_{hi}\mathbf{h}_{t-1} + \mathbf{b}_i\right),$$
$$\mathbf{h}_t = \mathbf{o}_t \odot \psi(\mathbf{c}_t), \tag{4}$$

where $\odot$ is element-wise product, ψ is the hyperbolic tangent function, ϕ is the sigmoid function, and $\mathbf{w}_{xi}$, $\mathbf{w}_{hi}$, and $\mathbf{b}_i$ are the weights and biases between input $\mathbf{i}_t$ and hidden state $\mathbf{h}_{t-1}$; $\mathbf{w}_{xf}$, $\mathbf{w}_{hf}$, and $\mathbf{b}_f$ are the weights and biases between forget state $\mathbf{f}_t$ and hidden state $\mathbf{h}_{t-1}$; and $\mathbf{w}_{xo}$, $\mathbf{w}_{ho}$, and $\mathbf{b}_o$ are the weights and biases between output $\mathbf{o}_t$ and hidden state $\mathbf{h}_{t-1}$. Our proposed RRNs (Fig. 2) can be expressed as the following function:

$$\{\tilde{y}_1, \tilde{y}_2, .., \tilde{y}_t\} = f_{\text{RRNs}}\left(\mathbf{x}_1, \mathbf{x}_2, .., \mathbf{x}_t; \mathcal{W}_{\text{RRNs}}\right), \tag{5}$$

where f_{RRNs} takes the sequential echo images as input to RRNs and outputs the predicted values $\tilde{y}_t$ with parameters of RRNs as $\mathcal{W}_{\text{RRNs}} = [\mathcal{W}_{\text{ResNets}}, \mathcal{W}_{\text{LSTM}}, \mathbf{w}_{\text{fc}}]$. $\mathcal{W}_{\text{ResNets}}$ represents the parameters of ResNets, $\mathcal{W}_{\text{LSTM}}$ represents the parameters of LSTM, and $\mathbf{w}_{\text{fc}}$ is associated with weights of the final fully connected layer. We train the proposed RRNs in an end-to-end fashion with stochastic gradient decent to minimise the following loss function [3] containing a first term as L^2 norm and a second term as structured loss (for notation simplicity, we dropped index i denoting the echo sequences):

$$\ell(\mathcal{W}_{\text{RRNs}}) = \sum_{j=1}^{|\mathcal{D}|}\left[\frac{\alpha}{T}\sum_{t=1}^{T}||y_{(j,t)} - \tilde{y}_{(j,t)}||^2 + \frac{2\beta}{T}\sum_{t=2}^{T}\left(\mathbb{I}(y_{(j,t)} > y_{(j,t-1)})\right.\right.$$
$$\left.\left. \max(0, \tilde{y}_{(j,t-1)} - \tilde{y}_{(j,t)}) + \mathbb{I}(y_{(j,t)} < y_{(j,t-1)}) \max(0, \tilde{y}_{(j,t)} - \tilde{y}_{(j,t-1)}))\right], \right. \tag{6}$$

where $\mathbb{I}$ denotes the indicator function, T is the maximum frame length, and α, β are user defined variables which are cross validated during the experiment. The structured loss term takes into account monotonically decreasing nature of the cardiac volume changes during the systole phase and monotonically increasing nature of cardiac volume changes in the diastolic phase. Finally, inference in the purposed method is done in a feed-forward way using the learned model.

3   Experiments

We carried out the experiments on a set of 2-D apical 4 chambers (AP4) cine echoes obtained at Vancouver General Hospital (VGH) with ethics approval from the Clinical Medical Research Ethics Board of the Vancouver Coastal Health (VCH) (H13-02370). The dataset used in this project consists of 1,868 individual patient studies containing a range of pathological conditions. The data were obtained using different types of ultrasound probes from various manufacturers, which consists of both normal and abnormal cases. Experiments were run by randomly dividing these cases into mutually exclusive subsets, such that 60% of the cases were available for training, 20% for validation, and 20% for the test. The average length of echo sequence in each study was about 30 frames. The electrocardiogram (ECG) signals, synchronized with cine clips, were also available. We considered the ECG signal to generate the ground-truth labels for ED and ES frames by identifying R and end of T points in that signal. Subsequently, the intermediate labels for all frames in each echo sequence were derived using Eq. (1), where the value of τ is selected to be 3 and labels are normalized in the range of [0–1] (Note that 0 represents ES and 1 represents ED). The experiment was conducted on the image resolution of 120×120, where we sub-sample the original image resolution using bi-cubic interpolation.

Our proposed RRNs, shown in Fig. 2, takes as input a cardiac sequence containing 30 frames irrespective of its location in the cardiac phase. Each frame is passed through the convolutional layer (weights) plus a ReLU, where the convolutional layer contains eight filters of size 3×3 followed by nine subsequent residual units. Each residual unit is made up of batch normalization (BN) plus ReLU plus weights. Each convolutional layer in the first three residual units contained the same eight filters (size 3×3), the fourth, fifth and sixth residual units contained 16 filters of size 3×3, and the seventh, eighth and ninth units had 32 filters of size 3×3. In the second to the last layer, we concatenated the 32 output features from each ResNet to form 192 features (32×6), followed by two blocks of LSTM layer and a final fully connected layer containing 30 neurons, to generate a label for each frame of the cardiac cycle. For comparison, we also used the shallow CNN model, Zeiler-Fergus (ZF), [3] in combination with different loss functions. The performance of our approach is calculated based on three metrics, the coefficient of determination (R^2 score), average absolute frame detection error, and computation time of each sample. R^2 score is defined by $R^2 = 1 - \frac{\sum(y_t - \tilde{y}_t)^2}{\sum(y_t - \bar{y})^2}$, where $\bar{y}_t$ is the mean of true labels. Average absolute frame detection error is calculated as $Err_S = \frac{1}{|\mathcal{D}|} \sum |T_S - \tilde{T}_S|$, where S is either the ED or the ES frame. All experiments were performed on a system with an Intel(R) Core(TM) i7-2600k 3.40 GHz×8 CPU with 8 GB RAM equipped with the NVIDIA GeForce GTX 980Ti graphics card.

4 Results and Discussion

Table 1 shows the result of the proposed approach using two different loss functions: a) L_2 loss, and b) L_2 loss + structured loss [3]. The R^2 values for these two settings are 0.36 and 0.66, respectively. Similarly, (Err_{ED}, Err_{ES}) for these two settings are (4.4, 4.7) and (3.7, 4.1), respectively. The baseline method [3] using CNN with the same setting produces the R^2 score of 0.13, whereas the average absolute frame detection error for both ED and ES error is 6.3 and 7.3 in setting (a), and 6.4 and 7.3 in setting (b).

Results in Table 1 show that the proposed RRN method outperforms the baseline method in both settings with a large margin. The advantage of the proposed method lies in its ability to go deeper with 126 layers in ResNets compared to the shallow architecture in the baseline model (CNN). Furthermore, we note that performance of the proposed RRNs model improves with the use of structured loss with higher R^2 score and lower frame detection error compared to the use of L_2 loss only. This is due to fact that structured loss introduces a structured constraint in the loss function which helps with smoothing the predicted labels, thus increasing the overall accuracy. It is also important to note that R^2 score of our proposed model is far better than that of baseline model (0.66 vs. 0.13), indicating that our approach is a better function approximation in the global context (with respect to assigning the correct label to individual frames). The method takes only 80 ms to characterise each cardiac cycle showing its efficiency in terms of computation cost. Figure. 4 shows some visual results of frame characterisation in a sequence of 30 frames as a function of labels represented in Eq. (1). In particular, Fig. 4(a and b) shows cases where the first frame in the sequence starts as an ED frame, and Fig. 4(c and d) shows cases where there is a shift in the location of the ED frame. Similarly, in Fig. 5, we also show one fairly accurate visual example of detection of ED and ES frames using the proposed method in the test set.

Table 1. Performance of the proposed and state-of-the-art methods on the test set.

Method	R^2 Score	Err_{ED}	Err_{ES}
ResNet+LSTM+L_2 loss (Proposed)	0.36	4.4	4.7
ResNet+LSTM+L_2 loss+struct loss (Proposed)	**0.66**	**3.7**	**4.1**
CNN+LSTM+L_2 loss [3]	0.13	6.3	7.3
CNN+LSTM+L_2 loss+struct loss [3]	0.13	6.4	7.3

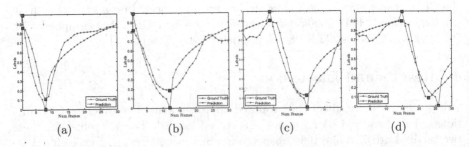

(a)   (b)   (c)   (d)

Fig. 4. Label approximation of the proposed method (blue) plotted against the ground-truth labels (red) on some sample cases, where the ES and ED frames (denoted by rectangular boxes) are correctly identified within an error margin of 3 to 4 frames. (Color figure online)

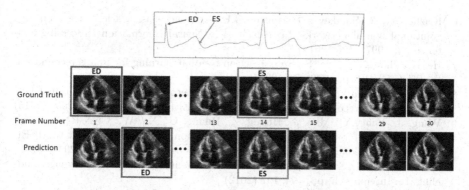

Fig. 5. A sample echo cine, consisting of 30 frames, alongside its ECG signal. The results of the proposed method in terms of ED and ES frame detection are demonstrated and compared with their ground-truth labels.

5 Conclusion and Future Works

In this paper, we proposed a deep residual recurrent neural net (RRN) for automated identification of cardiac cycle phases (ED and ES) from echocardiograms. We also showed that the proposed method produces results that outperform a baseline method using CNN with a large margin in detecting ED and ES frames. We achieved the R^2 score of 0.66 and the average absolute frame detection error of 3.7 and 4.1 for ED and ES, respectively. Our results suggest that the method has the potential to be used in clinical setting and is robust to all sorts of the pathological condition of the patient. In future, we plan to use the method as a pre-processing step for assessing several cardiac function parameters, including the ejection fraction.

References

1. Barcaro, U., Moroni, D., Salvetti, O.: Automatic computation of left ventricle ejection fraction from dynamic ultrasound images. Pattern Recogn. Image Anal. **18**(2), 351 (2008)
2. Abboud, A.A., Rahmat, R.W., et al.: Automatic detection of the end-diastolic and end-systolic from 4D echocardiographic images. JCS **11**(1), 230–240 (2015)
3. Kong, B., Zhan, Y., Shin, M., Denny, T., Zhang, S.: Recognizing end-diastole and end-systole frames via deep temporal regression network. In: Ourselin, S., Joskowicz, L., Sabuncu, M.R., Unal, G., Wells, W. (eds.) MICCAI 2016. LNCS, vol. 9902, pp. 264–272. Springer, Cham (2016). doi:10.1007/978-3-319-46726-9_31
4. Gifani, P., Behnam, H., Shalbaf, A., Sani, Z.A.: Automatic detection of end-diastole and end-systole from echocardiography images using manifold learning. Physiol. Meas. **31**(9), 1091 (2010)
5. Darvishi, S., Behnam, H., Pouladian, M., Samiei, N.: Measuring left ventricular volumes in two-dimensional echocardiography image sequence using level-set method for automatic detection of end-diastole and end-systole frames. Res. Cardiovasc. Med. **2**(1), 39 (2013)

6. Krizhevsky, A., Sutskever, I., Hinton, G.E.: Imagenet classification with deep convolutional neural networks. In: Advances in Neural Information Processing Systems, pp. 1097–1105 (2012)
7. He, K., Zhang, X., Ren, S., Sun, J.: Deep residual learning for image recognition. In: IEEE CVPR, pp. 770–778 (2016)
8. Donahue, J., Anne Hendricks, L., et al.: Long-term recurrent convolutional networks for visual recognition and description. In: IEEE CVPR, pp. 2625–2634 (2015)
9. Wang, J., Yang, Y., Mao, J., Huang, Z., Huang, C., Xu, W.: CNN-RNN: a unified framework for multi-label image classification. In: Proceedings of the IEEE Conference on Computer Vision and Pattern Recognition, pp. 2285–2294 (2016)
10. Sundermeyer, M., Schlüter, R., Ney, H.: LSTM neural networks for language modeling. In: Interspeech, pp. 194–197 (2012)
11. Milan, A., Rezatofighi, S.H., Dick, A., Reid, I., Schindler, K.: Online multi-target tracking using recurrent neural networks, arXiv preprint arXiv:1604.03635 (2016)
12. Hochreiter, S., Schmidhuber, J.: Long short-term memory. Neural Comput. $9(8)$, 1735–1780 (1997)
13. LeCun, Y., Bengio, Y.: Convolutional networks for images, speech, and time series. In: Arbib, M.A. (ed.) Handbook of Brain Theory and Neural Networks, vol. 3361. MIT Press (1995)
14. Ioffe, S., Szegedy, C.: Batch normalization: Accelerating deep network training by reducing internal covariate shift, arXiv preprint arXiv:1502.03167 (2015)
15. Nair, V., Hinton, G.E.: Rectified linear units improve restricted boltzmann machines. In: ICML, pp. 807–814 (2010)

Computationally Efficient Cardiac Views Projection Using 3D Convolutional Neural Networks

Matthieu Le[✉], Jesse Lieman-Sifry, Felix Lau, Sean Sall, Albert Hsiao, and Daniel Golden

Arterys Inc., San Francisco, USA
matthieu@arterys.com

Abstract. 4D Flow is an MRI sequence which allows acquisition of 3D images of the heart. The data is typically acquired volumetrically, so it must be reformatted to generate cardiac long axis and short axis views for diagnostic interpretation. These views may be generated by placing 6 landmarks: the left and right ventricle apex, and the aortic, mitral, pulmonary, and tricuspid valves. In this paper, we propose an automatic method to localize landmarks in order to compute the cardiac views. Our approach consists of first calculating a bounding box that tightly crops the heart, followed by a landmark localization step within this bounded region. Both steps are based on a 3D extension of the recently introduced ENet. We demonstrate that the long and short axis projections computed with our automated method are of equivalent quality to projections created with landmarks placed by an experienced cardiac radiologist, based on a blinded test administered to a different cardiac radiologist.

1 Introduction

Cardiac pathologies are often best evaluated along the principle axes of the heart, on long and short axis views. Because 4D Flow scans are typically acquired volumetrically, these planar views must be reconstructed for interpretation by a cardiac radiologist. In order to generate the standard cardiac views, the user may manually place 6 cardiac landmarks: the left and right ventricle apex (LVA, and RVA), and the aortic, mitral, pulmonary, and tricuspid valves (AV, MV, PV, and TV). With training, locating the landmarks using an interactive clinical tool takes a trained radiologist roughly 3 min. The landmark positions are then used to compute the 2, 3, and 4 chamber views, as well as the short axis stack (Fig. 1). Although these views can be computed for the left and right ventricles, for the sake of conciseness, we focus on the left ventricle cardiac views. The 4 chamber view is defined as the plane going through the TV, MV, and LVA, with the LVA oriented on the top of the image. The 3 chamber view is defined as the plane going through the AV, MV, and LVA, with the LVA placed on the left of the image. The 2 chamber view is defined as the plane bisecting the large angle of the 3 and 4 chamber planes, with the LVA placed on the left of the image. The SAX

© Springer International Publishing AG 2017
M.J. Cardoso et al. (Eds.): DLMIA/ML-CDS 2017, LNCS 10553, pp. 109–116, 2017.
DOI: 10.1007/978-3-319-67558-9_13

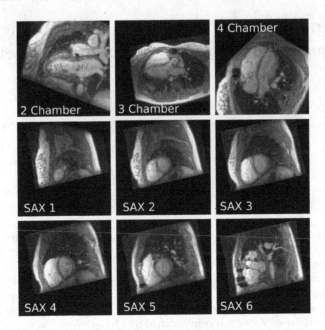

Fig. 1. First row: 2 chamber, 3 chamber, and 4 chamber views; second and third rows, left to right, then top to bottom: 6 slices of the SAX views from the apex to the base for the left ventricle.

(short axis stack views) are generated by sampling planes orthogonal to the axes defined by the LVA and MV, with the left ventricle in the center of the image, and the right ventricle on the left of the image.

Methods that utilize deep learning techniques to determine keypoints in 2D or 3D images have recently shown state of the art results. Payer et al. [1] proposed a method to detect landmarks in 2D and 3D medical images by regressing heat maps of the landmark location. Other approaches have been suggested for landmark detection such as using reinforcement learning [2], where an agent travels through the image and stops when a landmark is found. Some approaches proposed to detect cardiac landmarks starting with left ventricle segmentations [3].

In this paper, we build on [1,4] and introduce a method to automatically compute cardiac views based on a landmark detection network. Our approach may obviate manual placement of landmarks, to fully automate computation of cardiac views. We demonstrate that the long and short axis projections computed with our automated method are of equivalent quality to those created by expert annotators, based on a blinded test administered to a board-certified cardiac radiologist. We finally show that our model can be applied to other cardiac MRI sequences, such as volumetrically-acquired 3D-cine single breath hold MRI.

2   Methods

Pipeline. We utilize a two-step approach to locate the six landmarks that define the cardiac views. First, we predict a tight bounding box around the heart.

Original Image   Preprocessed   Cropped   Landmark
and Bounding Box   Preprocessed Image   Prediction

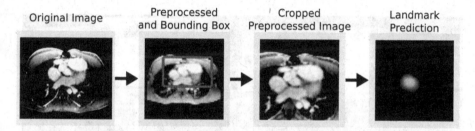

Fig. 2. From left to right: original input, preprocessed image for bounding box prediction, cropped preprocessed input, landmark detection (here the aortic valve).

Second, we use the cropped image to predict the landmark location (Fig. 2). For both steps, we use a 3D ENet architecture based on the original 2D implementation from [5]. For bounding box prediction, the input is the 3D image, and the output is a semantic segmentation map of the volume within the bounding box (Fig. 3). We convert the bounding box segmentation map to bounding box coordinates by taking, per axis, the location of the 5th and 95th percentiles of the cumulative sum over the non-considered axis. For landmark prediction, the input of the network is a crop around the heart of the original 3D image and the output is a 6 channel heat map (one channel per landmark), where the location of the peak corresponds to the predicted landmark position (Fig. 4).

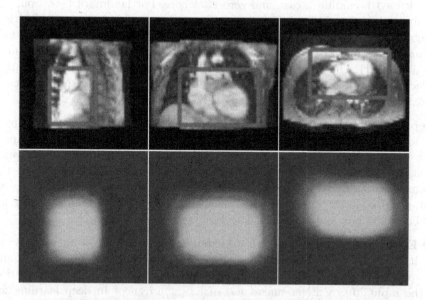

Fig. 3. Bounding box prediction examples for one 3D image of the test set. The anatomy image is on top, the output of the network on the bottom. The red lines outline the inferred bounding box from the output map. From left to right, sagittal, coronal, and axial views. (Color figure online)

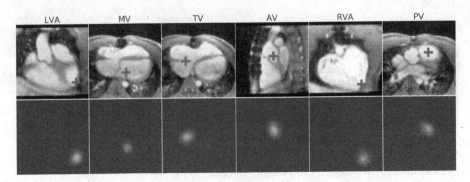

Fig. 4. Example of prediction of the 6 landmarks for one 3D image of the test set. The anatomy image is on top, the output of the network on the bottom. For each landmark, we show one 2D slice of the 3D input and output. The red cross on the anatomy image localizes the peak of the output map of the network for the different landmarks. From left to right, the views are coronal, axial, axial, sagittal, coronal, and axial. (Color figure online)

Data. For bounding box detection, the database comprises 81 unique patients. For each patient, an average of 2.9 time points were manually annotated with a bounding box enclosing the heart, resulting in 234 distinct 3D images. The ground truth for the network is a binary mask equal to one within the manually defined bounding boxes, and zero elsewhere. For landmark detection, the database comprises 310 unique patients. For each patient, an average of 2.1 time points were annotated by expert cardiac radiologists, resulting in 664 distinct 3D images. Note that for each image, not all six landmarks are necessarily annotated (Table 1). The ground truth is a 6-channel 3D heat map where the location of each landmark is encoded by a 3D isotropic Gaussian with a fixed variance of 4 voxels.

Preprocessing. Before both bounding box and landmark detection, we apply the following preprocessing: (i) resize the image to an isotropically re-sampled $64 \times 64 \times 64$ cube, padding with zero if necessary, (ii) clip the image pixel intensities between the 1st and 99th percentile intensities, (iii) normalize the image intensity to be between 0 and 1, (iv) apply adaptive histogram [6], (v) center and scale the image. Figure 2 shows the result of the pre-processing for the bounding box and landmark detection steps.

3D ENet. In both steps we use a 3D extension of the 2D ENet neural network [5]. It is a fully convolutional network composed of a downsampling path and an upsampling path that produces segmentation maps at the same resolution as the input image. ENet makes use of recent advances in deep learning for improved computational efficiency [5]. The network architecture utilizes early downsampling and a smaller expanding path than contracting path. It also makes use of bottleneck modules, which are convolutions with a small receptive field that are applied in order to project the feature maps into a lower dimensional space in which larger kernels can be applied [7]. Finally, throughout the network, ENet leverages a diversity of low cost convolution operations such as asymmetric $(1 \times 1 \times n, 1 \times n \times 1,$ and $n \times 1 \times 1)$ convolutions and dilated convolutions [8].

Training. We split the data such that 80% of patients are in the training set, 10% in the validation set, and 10% in the test set, such that a patient images are only present in one set. For the bounding box ENet, we use the parameters described in [5]. We train it using the Adam optimizer [9] with pixelwise cross-entropy loss. We train the landmark detection models with the Adam optimizer and an l2 loss. We only compute the loss corresponding to the landmarks present in the considered annotations. This allows to train for all 6 landmarks using every 3D image even though not all images have ground truth for all 6 landmarks. We run a hyperparameter search in which we train 50 models for 40 epochs and pick the 3 models with the lowest l2 loss. We then re-train the three best models for 100 epochs. We then designate the best model as the one with the lowest median error over all landmarks. We search over: the level of the applied distortions (whether or not to flip the images, the intensity of the Gaussian noise, the elastic distortions, affine transformations, brightness, contrast, and blur parameters), the size of the asymmetric convolutions, the number of 1.x bottlenecks, the number of repeated Sect. 2, the number of initial filters, the type of pooling (max or average), the projection scale at which the dimensionality is reduced in a bottleneck, the learning rate, and the amount of dropout (see [5] for details on the naming conventions). We additionally tested the use of skip connections and pyramidal scene parsing module [10], but did not find it improved the results.

3 Results

Bounding Box. The goal of the bounding box is to be able to crop a region around the heart containing the landmarks. This allows to feed higher resolution images to the subsequent landmark detection network. For this reason, and because the exact extent of a "good" bounding box is subject to interpretation, the classical measures of accuracy for semantic segmentation problems are not well suited. Rather we are interested in making sure that the landmarks are within the bounding boxes, and that they are cropping the original image by a sufficient amount. On the landmark database, the bounding boxes are on average 18% the volume of the original image, and 98% of all ground truth landmarks are enclosed within the bounding boxes. The bounding box validation per pixel accuracy is 95%, and the validation Dice coefficient is 0.83.

Figure 3 shows the results of the bounding box prediction for one patient, and how we compute bounding box coordinates (outlined in red on the top row) from the output segmentation map (bottom row). Note that because the image is then resized to a cube with isotropic spacing for landmark prediction, the final cropped image is often larger than the predicted bounding box.

Landmarks. The results of the designated best model on the test set are presented in Table 1. The per landmark median error (distance between the ground truth and the prediction), averaged over all 6 landmarks, is 8.9 mm, and the median error over all prediction is 8.8 mm for the test set. Cadaveric studies by [11] find that the average radius of the smallest valve (the PV) is 10.8 mm. Our average localization error is therefore less than the average radius of the smallest cardiac valve. Figure 4 presents the landmark detection results for one patient. The inference time for one study is on average a little less than a second on an NVIDIA Maxwell Titan X, depending on the original size of the input image.

Table 1. Result of landmarks detection error, and number of annotations in the database. The error is defined as the distance between the ground truth and the prediction in mm.

	Train (nbr of annot.)	Validation (nbr of annot.)	Test (nbr of annot.)
LVA	6.6 (403)	5.5 (50)	7.1 (62)
MV	8.4 (514)	8.8 (59)	9.5 (68)
AV	6.3 (284)	6.2 (35)	6.5 (34)
RVA	7.5 (453)	6.0 (48)	9.3 (45)
TV	8.9 (513)	10.5 (64)	10.6 (67)
PV	8.8 (326)	13.6 (34)	10.2 (34)
Average median error	7.7	8.4	8.9
Median error	7.7	7.9	8.8

Cardiac Views. Fig. 1 shows the predicted cardiac views for a single time point for the left ventricle. Although the model is trained at sparsely annotated time points of the cardiac cycle (mainly the time points corresponding to the end diastole and the end systole), we can predict the landmarks location for each time point.

In order to compute the cardiac views, we predict each landmark position at each time point, and then take the median position over time in order to compute the projection with fixed landmark (otherwise, the projected cardiac views would move during the cycle, which can be confusing for the clinician). To evaluate the quality of the cardiac views, we blindly presented them to a board-certified cardiac radiologist who was not involved in the ground truth collection or any other aspect of this work. We presented the radiologist with 2D cine movies over the cardiac cycle of projections computed from both the ground truth annotations and the predicted landmarks for the full test set. The order of the movies were randomized so the clinician did not know whether any given movie came from ground truth or predicted landmarks. We asked the clinician to grade each projections (2/3/4 chamber and SAX views) from 1 (bad projection unsuitable for diagnosis) to 5 (clinically relevant cardiac views). Table 2 shows the results. The grade shows that the quality of the predicted cardiac views is on par with the ground truth views with an average grade difference of 0.5 between the two, in favor of the predicted cardiac views. The failure mode of the cardiac view computations mainly include the presence of part of the right ventricle in the left ventricle 2 chamber views.

Other MR Sequences. We show on Fig. 5 the results of the cardiac views computation using a 3D-cine single breath hold MRI, which is a different sequence than 4D flow. We can see that the learned model generalizes well to similar MRI modalities.

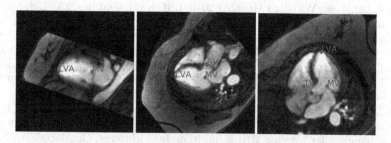

Fig. 5. Cardiac views results for a 3D-cine single breath hold MRI. Left: 4 chamber view where the prediction is successful. Middle: 3 chamber view at a time point where the prediction is successful. Right: 3 chamber view at a time point where the prediction is failing.

Table 2. Grade of the cardiac view evaluation by a board-certified cardiac radiologist.

	1 - Automated views	2 - Radiologist views	Difference 1 - 2
2 Chamber	4.0	3.6	0.4
3 Chamber	4.7	4.6	0.1
4 Chamber	4.8	4.1	0.7
SAX	4.1	4.1	0.0
Average	4.4	4.1	0.5

4   Conclusion

We presented a method to automatically compute long and short axis views for volumetrically acquired cardiac MRI. We showed that our error approaches anatomical variability, and we confirmed the efficacy of our approach with quality ratings in a blinded comparison test from a board-certified cardiac radiologist. The presented approach may obviate manual placement of landmarks, to fully automate computation of cardiac views. Future work could explore the use of non-cubic cropped images fed to the landmark detection network since it is fully convolutional. We will also investigate the incorporation of flow information to help localize the landmarks. This additional data would likely be most helpful to locate the valves. However, flow data requires careful handling because it can be particularly noisy.

References

1. Payer, C., Štern, D., Bischof, H., Urschler, M.: Regressing heatmaps for multiple landmark localization using CNNs. In: Ourselin, S., Joskowicz, L., Sabuncu, M.R., Unal, G., Wells, W. (eds.) MICCAI 2016. LNCS, vol. 9901, pp. 230–238. Springer, Cham (2016). doi:10.1007/978-3-319-46723-8_27
2. Ghesu, F.C., Georgescu, B., Mansi, T., Neumann, D., Hornegger, J., Comaniciu, D.: An artificial agent for anatomical landmark detection in medical images. In: Ourselin, S., Joskowicz, L., Sabuncu, M.R., Unal, G., Wells, W. (eds.) MICCAI 2016. LNCS, vol. 9902, pp. 229–237. Springer, Cham (2016). doi:10.1007/978-3-319-46726-9_27
3. Lu, X., Jolly, M.-P., Georgescu, B., Hayes, C., Speier, P., Schmidt, M., Bi, X., Kroeker, R., Comaniciu, D., Kellman, P., Mueller, E., Guehring, J.: Automatic view planning for Cardiac MRI acquisition. In: Fichtinger, G., Martel, A., Peters, T. (eds.) MICCAI 2011. LNCS, vol. 6893, pp. 479–486. Springer, Heidelberg (2011). doi:10.1007/978-3-642-23626-6_59
4. Pfister, T., Charles, J., Zisserman, A.: Flowing convnets for human pose estimation in videos. In: ICCV, pp. 1913–1921 (2015)
5. Paszke, A., et al.: Enet: a deep neural network architecture for real-time semantic segmentation arXiv:1606.02147 (2016)
6. Zuiderveld, K.: Contrast limited adaptive histogram equalization. In: Graphics gems IV. Academic Press Professional Inc., pp. 474–485 (1994)
7. He, K., Zhang, X., Ren, S., Sun, J.: Deep residual learning for image recognition. In: Proceedings of the IEEE CVPR, pp. 770–778 (2016)
8. Yu, F., Koltun, V.: Multi-scale context aggregation by dilated convolutions arXiv:1511.07122 (2015)
9. Kingma, D., Ba, J.: Adam: a method for stochastic optimization arXiv:1412.6980 (2014)
10. Zhao, H., Shi, J., Qi, X., Wang, X., Jia, J.: Pyramid scene parsing network arXiv:1612.01105 (2016)
11. Ilankathir, S.: A cadaveric study on adult human heart valve annular circumference and its clinical significance. IOSR-JDMS 1(14), 60–64 (2015)

Non-rigid Craniofacial 2D-3D Registration Using CNN-Based Regression

Yuru Pei[1(✉)], Yungeng Zhang[1], Haifang Qin[1], Gengyu Ma[2], Yuke Guo[3],
Tianmin Xu[4], and Hongbin Zha[1]

[1] Key Laboratory of Machine Perception (MOE), Department of Machine
Intelligence, Peking University, Beijing, China
Peiyuru@cis.pku.edu.cn
[2] USens Inc., San Jose, USA
[3] Luoyang Institute of Science and Technology, Luoyang, China
[4] School of Stomatology, Peking University, Beijing, China

Abstract. The 2D-3D registration is a cornerstone to align the inter-treatment X-ray images with the available volumetric images. In this paper, we propose a CNN regression based non-rigid 2D-3D registration method. An iterative refinement scheme is introduced to update the reference volumetric image and the digitally-reconstructed-radiograph (DRR) for convergence to the target X-ray image. The CNN-based regressor represents the mapping between an image pair and the in-between deformation parameters. In particular, the short residual connections in the convolution blocks and long jump connections for the multi-scale feature map fusion facilitate the information propagation in training the regressor. The proposed method has been applied to 2D-3D registration of synthetic X-ray and clinically-captured CBCT images. Experimental results demonstrate the proposed method realizes an accurate and efficient 2D-3D registration of craniofacial images.

1 Introduction

The long-term orthodontic treatments often take a few years, during which several images are captured for the purpose of structural morphology assessments due to treatments and growth, especially for adolescent patients. Before the wide use of volumetric images, such as CBCT images, in clinical orthodontics, 2D X-ray cephalograms are the only medium for recording the craniofacial morphology. The non-rigid 2D-3D registration is a crucial step to obtain the 3D volumetric images from inter-treatment X-ray images.

Traditional 2D-3D registration techniques rely on iterative optimizations to minimize the difference between the digitally-reconstructed-radiographs (DRR) and the target X-ray image to solve the transformations, which is known to be time-consuming due to the enormous online DRR evaluations [7,12,15]. The time complexity of the non-rigid 2D-3D registration is even higher with more parameters to be solved. Several 3D image registration metrics have been adapted for the intensity-based 2D-3D registration, including the stochastic rank correlation-based metric [9], the sparse histogramming based metric [15], and the variational mutual information metric [12]. In order to relieve the time complexity in metric

© Springer International Publishing AG 2017
M.J. Cardoso et al. (Eds.): DLMIA/ML-CDS 2017, LNCS 10553, pp. 117–125, 2017.
DOI: 10.1007/978-3-319-67558-9_14

evaluation, the feature-based methods have been used for 2D-3D registration by using the corners, lines and segmentations-based metrics [1]. However, the accurate detection of geometric features in its own is challenging considering the blurry images caused by structure overlapping [4]. Moreover, the optimization-based registration is prone to propagate feature detection errors.

The statistical surface and intensity models can reduce the parameter space in the 2D-3D registration [10,11,13]. However, it is not a trivial task to fine tune surface models by limited parameters for a close fitting of complex structures, such as the skull. Moreover, the statistical-model-based registration still relies on a large amount of online DRRs. The regression models bridge the gap between 2D X-ray images and 3D transformation parameters without online DRRs [5,6,14]. Zheng et al. [14] proposed a 3D volumetric image reconstruction technique based on the partial least squares regression (PLSR). However, the system required additional 3D surrogate meshes for model learning. The regression forest acted as a mapping function between the feature space of X-ray images and that of volumetric images, though the model relied on feature extraction [6]. Miao et al. [5] proposed a CNN regression-based system for a rigid 2D-3D registration with limited transformation parameters, where the one-shot transformation estimation is limited to handle fine-grained structures in craniofacial registration.

In this paper, we propose a CNN-regression-based method for the non-rigid 2D-3D registration between 2D cephalograms and 3D volumetric CBCT images (Fig. 1). The CNN has shown amazing performance in a variety of computer vision and medical image processing tasks [2,8]. Here we utilize the CNN-based regressor to bridge an image pair to the corresponding in-between non-rigid deformation parameters. In particular, the short and long residual connections are introduced to the neural network to reinforce the image information propagation. The randomly sampled deformation pairs from a subspace spanned by the training volumetric images, as well as the corresponding synthetic DRRs, are used to train the CNN-based regressor. Instead of the one-shot registration estimation, a feedback scheme is introduced to the regression-based 2D-3D registration, where the deformation parameters are fine-tuned iteratively. The reference image of the input image pair is iteratively updated by the CNN-based regressor, until the newly-generated DRR is consistent with the target X-ray image. The proposed method enables a reliable and efficient online nonrigid 2D-3D registration by an integration of a mixed residual CNN and an iterative refinement scheme. We show that 2D-3D registration results of cephalograms and CBCT images can be improved by the proposed multi-scale feature fusion of the mixed residual CNN and the iterative refinement compared with conventional approaches.

2   Methods

2.1   Regression-Based 2D-3D Registration

We denote the target X-ray image as I_{tar}, and reference volumetric image as V_{ref}. The goal of the 2D-3D registration is to derive the transformation of V_{ref} to make the deformed volumetric image consistent to I_{tar} in the sense of DRR projection. The input of the proposed system is an image pair (I_{tar}, I_{ref}), where I_{ref} denotes the DRR projection of V_{ref}. The output is transformation parameters defined as the 3D displacements vector at the control grid of a B-spline based deformation model. The 3D tensor product of cubic B-splines is used to define the non-rigid deformations, where the control grid G is defined on the V_{ref}. The objective function is defined as $g = d(I_{tar}, P(M(V_{ref}, G \circ \delta T)))$. The transformation $\delta T = f(I_{tar}, I_{ref})$, where f denotes the CNN-based regression function. M denotes the B-spline based deformation, and $M = \sum_{i=0}^{3} \sum_{j=0}^{3} \sum_{k=0}^{3} B_i(u)B_j(v)B_k(w)(G \circ \delta T)$. B denotes the base function of cubic B-splines. P refers to the DRR projection function, which is used to generate the 2D X-ray image from the deformed volumetric image. d denotes the mutual information based 2D image metric.

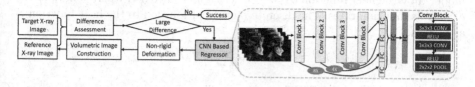

Fig. 1. The flowchart of the proposed method.

Iterative Refinement. As shown in Fig. 1, an iterative feedback scheme is introduced to handle the fine-grained shape variations of the craniofacial structures, instead of one-shot regression based registration. To begin with, the target X-ray image I_{tar} combined with I_{ref} of the reference CBCT is fed to the regressor. The output deformation parameters are imposed to the reference CBCT for a volumetric image consistent with I_{tar}. In case the difference between the target X-ray image I_{tar} and newly-generated DRR I'_{ref} is large enough, the input image pair is replaced by (I_{tar}, I'_{ref}), and fed to the regressor once again. The process terminates when the iteratively updated DRR is consistent with I_{tar}. The objective function is reformulated as

$$g^{(t)} = d\left(I_{tar}, P\left(M\left(V_{ref}^{(t-1)}, G^{(t)} \circ \delta T^{(t)}\right)\right)\right), \tag{1}$$

where $\delta T^{(t)} = f(I_{tar}, I_{ref}^{(t-1)})$. At the t-th iteration, the control grid G is determined by the deformation parameters obtained in previous iterations, and $G^{(t)} = G^{(0)} \circ \sum_{i=0}^{t-1} \delta T^{(i)}$. The final volumetric image $V^{(t)}$ after t iterations is determined by deformation parameters obtained in previous iterations.

$$V^{(t)} = M\left(V_{ref}^{(t-1)}, G^{(t)} \circ \delta T^{(t)}\right). \tag{2}$$

Due to the iterative refinement, the system also consists of online DRRs and 2D metric evaluations, though limited iterations are enough for a convergence as stated in experiments.

Fig. 2. Five levels of anisotropic diffusion and related gradient magnitude channels.

Input Channels. Different from the traditional feature-based 2D-3D registration methods, the entire X-ray images are used as the input of the CNN-based regressor without feature extraction, such as corners and line segments. Considering that the X-ray image is a mono-channel gray image, we enforce the input X-ray image by a set of anisotropic diffusion filtered channels [7], accompanied by related gradient magnitudes (see Fig. 2). The anisotropic diffusion process is modeled by the heat function $\partial I/\partial t = c(x, y, t)\triangle I + \triangledown c \triangledown I$, where flux function c controls diffusion rates. The discrete numerical solution to the heat function is as follows:

$$I^{(t)} = I^{(t-1)} + \alpha \sum_{r=1}^{4} \beta_r \nabla_r I^{(t-1)}, \tag{3}$$

where the constant α is related to the stability of the numerical solution, and set at 0.1. The coefficient β is set according to the gradient magnitude, and $\beta = \exp(-\| \nabla I \|^2/\kappa)$. The constant κ is set at 0.3. The value of r varies from 1 to 4 corresponding to four directions, including up, down, left, and right, in gradient estimation. As shown in Fig. 2, the anisotropic diffusion filtered images $I^{(t)}$, together with the gradient magnitude $I_G^{(t)}$ are used as the input of the regressor, where $t \in \{0, \dots, 8\}$. Thus, the input image pairs $\mathcal{I}$ of the regressor have 18 channels, and $\mathcal{I} = [I^{(t)}, I_G^{(t)}|t = 0, \dots, 8]$.

2.2 CNN Based Regressor

As illustrated in Fig. 1, the regression network consists of a convolutional part and a fully connected (FC) part. The first part serves as encoders for the automatic feature map extraction. The basic unit of the convolutional part, i.e. the Conv-block, is set following the common pattern of the residual convolutional neural networks [2]. The Conv-block is composed of two convolutional layers followed by the rectified linear unit (RELU). The size of the reception fields of all convolutional filters is set at 3 with a stride of one. One $2 \times 2 \times 2$ pooling layer follows the convolution layer. The short residual connection is added between the input and the output of the duplicated convolutional layers to reinforce the information propagation.

The FC part consists of three fully connected layers for the transformation parameter prediction. The long short-cut connections have been applied in recent works, and was demonstrated to facilitate the network training [11]. In this work, the long-jump residual connections are appended between the multi-scale Conv-blocks and the first FC layer, for the purpose of accurate transformation prediction by virtue of the multi-scale feature map fusion. The feature maps from Conv-block 1, 2, and 3 are down-sampled by the rates of $1/8$, $1/4$, $1/2$ respectively, and combined with the output of Conv-block 4, serving as the input of the FC part. The long short-cut connections are illustrated as side arrows in Fig. 1, which are implemented by a convolution with a stride of 8, 4, and 2 respectively. The fusion of four FC layers corresponding to four Conv-blocks is followed by two other FC layers. The loss function $\mathcal{L}$ of the CNN regressor is simply defined as the Euclidean distance between the predicted and the ground truth transformation parameters, denoted as δT^p and δT^g respectively, and $\mathcal{L} = \|\delta T^p - \delta T^g\|^2$.

3   Experiments

Data Set. The data set consists of 120 CBCT images captured by NewTom with a 12-inch field of view. The CBCT images are resampled to a resolution of $200 \times 200 \times 190$ with a voxel size is of $1 \times 1 \times 1\,mm^3$. The 2D X-ray cephalograms are generated by the DRR technique with a resolution of 200×200. It is not a trivial task to get the ground truth data for the craniofacial 2D-3D image registration. In experiments, we use the synthetic data for both training and testing of the proposed model. We employ the non-rigid B-spline-based registration of Plastimatch toolbox to obtain the deformation parameters compared with a randomly-selected reference image. The control grid of the B-spline deformation is set at $20 \times 20 \times 20$ with an 8000×3-dimensional deformation vector associated with each volumetric image. The principal component analysis is used for a low dimensional subspace representation of the non-rigid deformations parameters. In our experiments, 80 principal components are retained covering more than 99.5% variances. We randomly sample 30k point pairs $(T_s, T_{s'})$ in the subspace, with corresponding synthesized X-ray image pair $[P(M(V_{ref}, T_s)), P(M(V_{ref}, T_{s'}))]$. The anisotropic diffusion and gradient magnitude channels (Sect. 2.1) of the synthesized image pair, together with the in-between transformation $\delta T = T_{s'} - T_s$, are used to train the CNN regressor.

Training Details. The training of the CNN regressor is performed using Caffe [3] on a machine with one Titan X GPU. The batch size is set at 1000. The momentum is set at 0.9. The learning rate is set at 1e-4, and divided by 10 after every 10000 epochs. The weight decay is set at 5e-4. The batch normalization is performed to avoid the internal covariance shifts by making the input data of each layer to be of zero mean and unit variance.

3.1   Qualitative Assessment

The proposed 2D-3D registration method is evaluated by three metrics on the synthetic testing data set with the ground truth available. The first metric is the contour distance e_C between the 2D target X-ray image and the DRR of the output deformed volumetric image, and $e_C = 1/n_c \sum_{i=1}^{n_c} \|p_i - p_i^g\|$. p_i and p_i^g denote n_c evenly-sampled points on the contours of the DRR and the target X-ray image respectively. Our experiments report e_C of three kinds of structures, i.e. the anterior cranial base (ACB), the maxilla (MAX), and the mandible (MAD). The second metric is the mean intensity difference e_I between the output volumetric image and the ground truth, and $e_I = 1/n_v \sum_{i=1}^{n_v} |v_i - v_i^g|$, where n_v is the voxel number. v_i and v_i^g denote the intensity values of the reconstructed and ground truth volumetric images. We also measure the distance e_T between the estimated deformation parameters and the ground truth, and $e_T = 1/n_g \sum_{i=1}^{n_g} \|T_i - T_i^g\|$, where T_i and T_i^g are the estimated and the ground truth displacement vectors of the i-th control point of the B-spline-based deformation model. We compare with the traditional intensity-based 2D-3D registration methods using the mean squared metric (Inten-MS) [9] and the mutual information metric (Inten-MI) [12]. We also perform the ablation experiments to assess the short and long residual connections. We compare with one-shot regression based on the CNN network without residual connections (CNN)[5], the one with short residual connection in the Conv-blocks (Short) [2], the one with the long residual connection (Long) [11], and the one with both short and long residual connections (Both). Moreover, we compare with the partial least squares regression (PLSR) [14].

Table 1. Comparisons of the 2D-3D registration by the proposed method with iterative refinements (Refine), the Inten-MS, the Inten-MI, and the regression models based on the one-shot PLSR and variants of the CNN with and without long and short residual connections, in terms of e_C (mm), e_I (HU), and e_T (mm).

		Inten-MI	Inten-MSD	PLSR	CNN	Short	Long	Both	Refine
e_C	CB	0.57 ±0.23	0.55±0.21	9.7±1.0	0.95±0.35	0.59±0.18	0.54±0.15	0.42±0.08	**0.41±0.12**
	MAD	0.64±0.23	0.55±0.19	5.8±1.1	0.79±0.20	0.54±0.12	0.54±0.13	0.50±0.14	**0.48±0.17**
	MAX	0.99 ±0.42	0.74±0.34	7.7±0.44	0.63±0.43	0.56±0.14	0.56±0.12	0.42±0.12	**0.35±0.08**
$e_I (\times 10^2)$		2.3 ±0.32	2.2±0.26	6.2 ±0.37	1.1±0.44	1.0±0.38	1.05±0.65	0.94±0.31	**0.92±0.30**
e_T		3.5±0.43	3.4±0.51	16±1.0	0.83±0.36	0.77±0.32	0.76±0.35	0.73±0.33	**0.60±0.33**

As shown in Table 1, the proposed method (Refine) outperforms the compared methods in both the contour distance on 2D X-ray images and the intensity difference of the volume images. In the ablation experiments, the proposed CNN regression model with both long and short residual connections produces improvements to other CNN variants. We think the reason is that the residual connections reinforce the information propagation, especially the fine-grained structures accommodated in multi-level feature maps. The proposed method even outperforms the traditional intensity based 2D-3D registration methods with thousands of DRR evaluations in each iteration. We think the reason is that the learning-based regression model accounts for the craniofacial shape distribution in both the 2D X-ray and 3D volumetric images. Moreover, the iterative refinement scheme tends to acquire the accurate deformation parameters. The CNN regression model addresses the nonlinearity in the mapping between the X-ray image pair and transformation parameters, and shows better performance

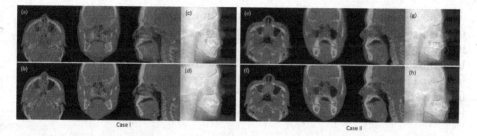

Case I   Case II

Fig. 3. The semi-transparent axial, sagittal, and coronal cross-sectional slices overlappings before (a, e) and after (b, f) non-rigid 2D-3D registration (gray-ground truth, red-reference before and after registration). The contour overlapping of the anterior cranial base, the maxilla, and the mandible on 2D X-ray images before (c, g) and after (d, h) registration (yellow-reference, red-target, blue-deformed reference). (Color figure online)

than the PLSR. The overlapping of the ground truth and the reference volumetric images before and after the 2D-3D registration of two cases are illustrated in Fig. 3. As we can see, the output deformed volumetric image is consistent with the ground truth. Moreover, the contours of the ACB, the MAX, and the MAD on the DRR of the output volumetric image coincide with those on the target X-ray images (see Fig. 3(d, h)). We also illustrate boxplots of e_I and e_T on 30 testing cases by the proposed method and CNN variants in Fig. 4(a, b).

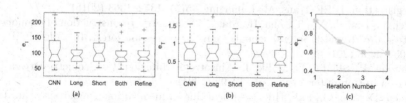

Fig. 4. Boxplots of (a) e_I and (b) e_T on 30 testing cases. (c) e_T after different numbers of iterations by the proposed method.

The proposed method utilizes the iterative refinement scheme. In each iteration, one time-consuming 3D non-rigid deformation and one DRR projection are needed to update the reference X-ray image. Fortunately, only a few iterations are enough for a successful registration as shown in Fig. 4(c). In our experiments, approximately three iterations are used to get desired deformation parameters.

4   Conclusion

We propose a CNN regression based non-rigid 2D-3D registration method. The CNN regressor accounts for the nonlinear mapping between the image pairs and the in-between transformations defined on the control grids of the B-spline-based deformation model. The refinement scheme updates the reference DRR iteratively for convergence to the target X-ray image. The proposed method has been applied to the 2D-3D registration of the synthetic X-ray and clinical CBCT images. The proposed method outperforms the compared conventional methods in terms of the contour consistency on the X-ray images and the intensity difference of volumetric images.

Acknowledgments. This work was supported by National Natural Science Foundation of China under Grant 61272342.

References

1. Groher, M., Bender, F., Hoffmann, R.T., Navab, N.: Segmentation-driven 2D–3D registration for abdominal catheter interventions. In: Medical image computing and computer-assisted intervention-MICCAI, pp. 527–535 (2007)
2. He, K., Zhang, X., Ren, S., Sun, J.: Deep residual learning for image recognition. In: CVPR, pp. 770–778 (2016)
3. Jia, Y., Shelhamer, E., Donahue, J., Karayev, S., Long, J., Girshick, R., Guadarrama, S., Darrell, T.: Caffe: convolutional architecture for fast feature embedding. In: ACM Multimedia, pp. 675–678 (2014)
4. Mahfouz, M.R., Hoff, W.A., Komistek, R.D., Dennis, D.A.: Effect of segmentation errors on 3D-to-2D registration of implant models in X-ray images. J. Biomech. **38**(2), 229–239 (2005)
5. Miao, S., Wang, Z.J., Liao, R.: A CNN regression approach for real-time 2D/3D registration. IEEE Trans. Med. imaging **35**(5), 1352–1363 (2016)
6. Pei, Y., Dai, F., Xu, T., Zha, H., Ma, G.: Volumetric reconstruction of craniofacial structures from 2D lateral cephalograms by regression forest. In: IEEE ICIP, pp. 4052–4056 (2016)
7. Perona, P., Shiota, T., Malik, J.: Anisotropic diffusion. In: Geometry-Driven Diffusion in Computer Vision, pp. 73–92 (1994)
8. Shen, D., Wu, G., Suk, H.I.: Deep learning in medical image analysis. Ann. Rev. Biomed. Eng. **19**, 221–248 (2017)
9. Steininger, P., Neuner, M., Birkfellner, W., Gendrin, C., Mooslechner, M., Bloch, C., Pawiro, S., Sedlmayer, F., Deutschmann, H.: An ITK-based implementation of the stochastic rank correlation (SRC) metric. Insight J. **11**, 5 (2010)
10. Whitmarsh, T., Humbert, L., De Craene, M., Luis, M., Fritscher, K., Schubert, R., Eckstein, F., Link, T., Frangi, A.F.: 3D bone mineral density distribution and shape reconstruction of the proximal femur from a single simulated DXA image: an in vitro study. In: SPIE Medical Imaging, pp. 76234U–76234U (2010)
11. Yu, W., Tannast, M., Zheng, G.: Non-rigid free-form 2D–3D registration using a B-spline-based statistical deformation model. Pattern Recogn. **63**, 689–699 (2017)
12. Zheng, G.: Effective incorporating spatial information in a mutual information based 3D–2D registration of a CT volume to X-ray images. Comput. Med. Imaging Graph. **34**(7), 553–562 (2010)
13. Zheng, G.: Personalized X-ray reconstruction of the proximal femur via intensity-based non-rigid 2D–3D registration. In: Medical Image Computing and Computer-Assisted Intervention-MICCAI, pp. 598–606 (2011)

14. Zheng, G.: 3D volumetric intensity reconsturction from 2D X-ray images using partial least squares regression. In: IEEE ISBI, pp. 1268–1271 (2013)
15. Zöllei, L.: 2D–3D rigid-body registration of X-ray Fluoroscopy and CT images. Ph.D. thesis, Harvard Medical School (2001)

A Deep Level Set Method for Image Segmentation

Min Tang[✉], Sepehr Valipour, Zichen Zhang, Dana Cobzas,
and Martin Jagersand

Department of Computing Science, University of Alberta, Edmonton, Canada
min.tang@ualberta.ca

Abstract. This paper proposes a novel image segmentation approach that integrates fully convolutional networks (FCNs) with a level set model. Compared with a FCN, the integrated method can incorporate smoothing and prior information to achieve an accurate segmentation. Furthermore, different than using the level set model as a post-processing tool, we integrate it into the training phase to fine-tune the FCN. This allows the use of unlabeled data during training in a semi-supervised setting. Using two types of medical imaging data (liver CT and left ventricle MRI data), we show that the integrated method achieves good performance even when little training data is available, outperforming the FCN or the level set model alone.

Keywords: Image segmentation · Level set · Deep learning · FCN · Semi-supervised learning · Shape prior

1 Introduction

Image segmentation is one of the central problems in medical imaging. It is often more challenging than natural image segmentation as the results are expected to be highly accurate, but at the same time, little training data is provided.

To address these issues, often strong assumptions and anatomical priors are imposed on the expected segmentation results. For quite a few years, the field was dominated by energy-based approaches, where the segmentation task is formulated as an energy minimization problem. Different types of regularizers and priors [1] can be easily incorporated into such formulations. Since the seminal work of Chan and Vese [2], the level set has been one of the preferred models due to its ability to handle topological changes of the segmentation function. Nevertheless, there are some limitations of traditional level set approaches: they rely on a good contour initialization and a good guess of parameters involved in the model. Additionally, they often have a relatively simple appearance model despite some progress [1,3].

The recently introduced deep neural net architectures address some of these issues by automatically learning appearance models from a large annotated

© Springer International Publishing AG 2017
M.J. Cardoso et al. (Eds.): DLMIA/ML-CDS 2017, LNCS 10553, pp. 126–134, 2017.
DOI: 10.1007/978-3-319-67558-9_15

dataset. Moreover, FCNs [4] have been proved successful in many segmentation tasks including medical imaging [5,6]. Despite their success, FCNs have a few limitations compared to traditional energy-based approaches: they have no explicit way of incorporating regularization and prior information. The network often requires a lot of training data and tends to produce low-resolution results due to subsampling in the strided convolutional and pooling layers.

We address these limitations by proposing an integrated FCN-levelset model that iteratively refines the FCN using a level set module. We show that (1) the integrated model achieves good performance even when little training data is available, outperforming the FCN or level set alone and (2) the unified iterative model trains the FCN in a semi-supervised way, which allows an efficient use of the unlabeled data. In particular, we show that using only a subset of the training data with labels, the jointly-trained FCN achieves comparable performance with the FCN trained with the whole training set.

Few other works address the problem of introducing smoothness into convolutional nets by using explicit regularization terms in the cost function [7] or by using a conditional random field (CRF) either as a post-processing step [8,9] or jointly trained [10] with the FCN or CNN. However, only specific graphical models can be trained with the FCN pipeline, and they cannot easily integrate shape priors. A joint deep learning and level set approach has also been recently proposed [11,12], but their work considers a generative model (deep belief network DBM) that is not trained by the joint model.

2 Methods

An overview of the proposed FCN-levelset framework is shown in Fig. 1. The FCN is pre-trained with a small dataset with labels. Next, in the semi-supervised training stage, the integrated FCN and level set model is trained with both labeled (top half of Fig. 1) and unlabeled (bottom half of Fig. 1) data. The segmentation is gradually refined by the level set at each iteration. Based on the refined segmentation the loss is computed and back-propagated through the network to improve the FCN. With the FCN trained in this manner, inference is done in the traditional way. A new image is fed to the network and a probability map of the segmentation is obtained. The output can be further refined by the level set if desired. The following subsections show details on the level set model as well as the integration with FCN.

2.1 The Level Set Method

Following traditional level set formulations [2,13], an optimal segmentation is found by minimizing a functional of the following form:

$$E(\Phi, \mathcal{A}) = E_{shape}(\Phi, \mathcal{A}) + \mu E_{smooth}(\Phi) + \lambda E_{data}(\Phi) \qquad (1)$$

where $\Phi(\cdot)$, defined over the image domain Ω, is a signed distance function that encodes the segmentation boundary. μ and λ are constants manually tuned and

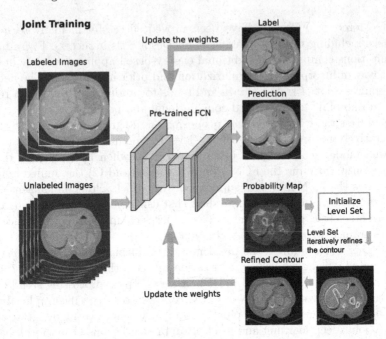

Fig. 1. Overview of the proposed FCN-levelset model. The pre-trained FCN is refined by further training with both labeled (top) and unlabeled data (bottom). The level set gets initialized with the probability map produced by the pre-trained FCN and provides a refined contour for fine-tuning the FCN. Green and blue arrows denote the forward pass and the back-propagation, respectively. (Color figure online)

kept fixed during experiments. The nonuniform smoothness term has the form of a weighted curve length:

$$E_{smooth}(\Phi) = \int\int_{\Omega} \delta_{\alpha}(\Phi(x))b(\cdot)|\nabla\Phi(x)|\, d\Omega, \tag{2}$$

where $\delta_{\alpha}(\cdot)$ is the regularized Dirac function. The weights $b(\cdot)$ are inversely proportional to the image gradients. The data term models the object/background intensity distribution as:

$$E_{data}(\Phi) = \int\int_{\Omega} H_{\alpha}(\Phi(x))\log(P_O(x))d\Omega + \int\int_{\Omega}(1 - H_{\alpha}(\Phi(x)))\log(P_B(x))d\Omega \tag{3}$$

where $H_{\alpha}(\cdot)$ is the Heaviside function. $P_O : \mathcal{R}^+ \to [0,1]$ and $P_B : \mathcal{R}^+ \to [0,1]$ are the probabilities belonging to object/background regions. In our model, the probabilities estimated by the FCN are used. The shape term is a critical component in knowledge-based segmentation. Based on the squared difference between the evolving level set and the shape prior level set, we choose [14]:

$$E_{shape}(\Phi) = \int\int_{\Omega} \delta_{\alpha}(\Phi(x))\big(S\Phi(x) - \Phi(x)_{\mathcal{M}}(\mathcal{A}(x))\big)^2 d\Omega \tag{4}$$

where $\Phi(\cdot)_{\mathcal{M}}$ denotes the shape model. $\mathcal{A}(x) = S\mathbf{R}x + \mathbf{T}$ is an affine transformation between the shape prior and the current segmentation.

To optimize the segmentation energy functional Eq. (1), calculus of variations is utilized. The position of the curve is updated by a geometric flow. Jointly with the evolution of Φ, the transformation parameters $\mathcal{A}$ are also estimated.

2.2 The Integrated FCN-Levelset Model

FCN. We create a shallow FCN to make the network less prone to over-fitting on a small training set. Fewer pooling layers are added to achieve a finer segmentation. We only have 4 convolutional layers with 23938 parameters and a total subsampling rate of 8 for liver segmentation, while 7 convolutional layers with 51238 parameters and a total subsampling rate of 6 are used for the left ventricle segmentation. Each convolution contains a $3 \times 3 \times 3$ filter followed by a rectified linear unit (ReLu), and then a max pooling with strides of two. An upconvolution is added right after a $1 \times 1 \times 1$ convolution to achieve a dense prediction. During training, we randomly initialize all new layers by drawing weights from a zero-mean Gaussian distribution with standard deviation 0.01 and ADADELTA [15] was used to optimize the cross entropy loss:

$$L(y, x; \theta) = -\frac{1}{N} \sum_{i=1}^{N} [y^{(i)} \ln P(x^{(i)}; \theta) + (1 - y^{(i)}) \ln(1 - P(x^{(i)}; \theta))] \quad (5)$$

where N is the number of pixels in one image, $(x^{(i)}, y^{(i)})$ denotes the pixel i and its label, θ is the network parameter and $P(x^{(i)}; \theta)$ denotes the network predicted probability of $x^{(i)}$ belonging to the object.

FCN-Levelset. The pre-trained FCN is further refined by the integrated level set module. Each unlabeled image is fed to the FCN and produces a probability map. It provides the level set module with a reliable initialization and foreground/background distribution for $E_{data}(\Phi)$ in Eq. (3). The level set further refines the output of the FCN. We compute the cross entropy loss between the FCN prediction and the level set output as the label. This loss is back propagated through the network to update the weights θ. In this manner, the FCN can implicitly learn the prior knowledge which is imposed on the level set especially from the unlabeled portion of dataset. Tuning the model weights only with the unlabeled data may cause drastic changes and corrupt the well learned weights θ from the labeled data. This is especially important at the beginning of the joint training when the performance of the system is not yet good. To make the learning progress smooth, the integrated FCN-levelset model is trained with both labeled and unlabeled data as illustrated in Fig. 1.

During the joint training, to ensure a stable improvement, the memory replay technique is used. It prevents some outliers from disrupting the training. In this technique, a dynamic buffer is used to cache the recently trained samples. Whenever the buffer is capped, a training on the buffer data is triggered which updates the network weights. Then the oldest sample is removed from the buffer and next training iteration is initiated.

Inference. To infer the segmentation, a forward pass through the FCN is required to get the probability map. The level set is initialized and the data term is set according to this probability map. Then the final segmentation is obtained by refining the output contour from the FCN. Different from the training where level set is mainly used to improve the performance of the FCN, inference is a post-processing step and the refined output is not back propagated to the FCN.

3   Experiments and Results

3.1   Data

Liver Segmentation contains a total of 20 CT scans [16]. All segmentations were created manually by radiology experts, working slice-by-slice in transversal view. The scans are randomly shuffled and divided into training (10 scans), validation (5 scans) and testing (5 scans). We select the middle 6 slices from each scan and form our data sets.

Left Ventricle Segmentation is a collection of 45 cine-MR sequences (about 20 slices each) taken during one breath hold cycle [17]. Manual segmentation is provided by an expert for images at end-diastolic (ED) and end-systolic (ES) phases. The original data division was used in our experiments where the 45 sequences are randomly divided into 3 subsets, each with 15 sequences, for training, validation, and testing. All slices are selected from the scans.

For both datasets, we divide the training set in two, labeled (30% of the liver data; 50% of the left ventricle data) and unlabeled part. The ground truth segmentation is not provided for the unlabeled part to simulate a scenario where only a percentage of data is available for supervised training.

3.2   Experiments

The networks used in the experiments are described in Sect. 2.2. As the level set method is sensitive to the initialization, the FCN needs to be pre-trained on the labeled part of data. During training, the batch size was 12, and the maximum number of epochs was set to 500. The best model on the validation set was stored and used for evaluations.

The level set models used for liver and left ventricle data are slightly different, depending on their properties. For liver data, we use a weighted curve length as the smoothness term with $b(\cdot) = \frac{1}{1+|\nabla I(x)|^2}$ in Eq. (2). The left ventricle data has no obvious edges, so $b(\cdot) = 1$ is used.

All experiments are performed on 2D slices. The shape prior is randomly taken from the ground truth segmentations. To show the superiority of the proposed integrated FCN-leveset model and its potential for semi-supervised learning we compared five models:

(i) *Pre-trained FCN:* trained with labeled images from the training set (30% of the liver data; 50% of the left ventricle data);

(ii) *Post-processing Level set:* Model (i) with post-processing level set;

(iii) *Jointly-trained FCN:* the FCN module of the integrated model jointly trained with the level set;

(iv) *FCN-levelset:* the proposed integrated model as described in Sect. 2.2.

(v) *Baseline FCN:* FCN trained on all images in the training set.

The performance of the above methods is evaluated using Dice score and Intersection over Union (IoU) score. Both scores measure the amount of overlap between the predicted region and the ground truth region.

3.3   Results

To illustrate the sensitivity of the level set method to initializations, we compared the output contours under different initializations, as shown in Fig. 2. The level set with manual initialization Fig. 2(a) converges to a proper shape (red) but with low precision. Initialization with the FCN predicted probability map Fig. 2(b) leads to a faster and more accurate convergence after a few iterations. Notably, when initialized with the probability map, the level set managed to eliminate the wrongly segmented parts (yellow) in the FCN output.

The performance of the five models described in Sect. 3.2 on the two datasets is summarized in Table 1 and illustrated in Fig. 3. Both datasets showed that the joint semi-supervised training improves the performance of the deep network. For liver data, the level set initialized with the probability map outperformed the pre-trained FCN by 4%. Through the joint training, FCN was fine-tuned and got an improvement of 3%. The performance of FCN-levelset was increased by 8% for Dice Score and 13% for IoU score compared to the pre-trained FCN. Notably, despite using only a few manually labeled images, the integrated model

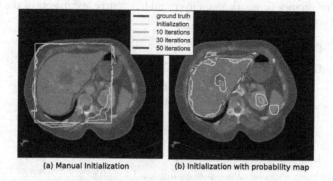

(a) Manual Initialization   (b) Initialization with probability map

Fig. 2. Impact of initialization to level set. Both (a) manual initialization and (b) initialized with probability map evolve the contour for 50 iterations. The blue contour is the ground truth; the yellow contour is the initialization and the red one is the segmentation results after 50 iterations. (Color figure online)

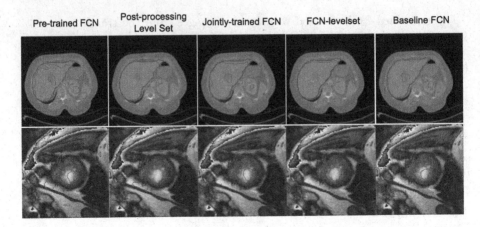

Fig. 3. Example segmentation of the five models. Output contour is in red while the ground truth is in blue. The top and bottom rows show the results on liver dataset and left ventricle dataset, respectively. The five columns correspond to the five models shown in Table 1.

performed even better than the baseline FCN, where the improvement was about 3% in Dice Score and 4% in IoU. The integrated FCN-levelset model has clear advantage in datasets where the segmented object presents a prominent shape.

Left ventricle data is more challenging for the level set model to distinguish the object from the background. Mainly due to the lack of a clear regional property. However, the joint model still improved the performance of the FCN by 2% during training. The FCN-levelset (trained with only 50% data) was able to improve the segmentation result by 3% compared to the pre-trained FCN and achieved a comparable performance to the baseline FCN (trained with all data).

Table 1. Performance comparison on liver and left ventricle datasets of the five models detailed in Sect. 3.2

Dataset	Model	Dice score	IoU
Liver	Pre-trained FCN	0.843	0.729
	Post-processing level set	0.885	0.794
	Jointly-trained FCN	0.879	0.784
	FCN-levelset	**0.923**	**0.857**
	Baseline FCN	0.896	0.811
Left Ventricle	Pre-trained FCN	0.754	0.605
	Post-processing level set	0.678	0.620
	Jointly-trained FCN	0.772	0.623
	FCN-levelset	**0.788**	**0.635**
	Baseline FCN	0.804	0.672

4 Discussion

In this paper a novel technique for integrating a level set with a fully convolutional network is presented. This technique combines the generality of convolutional networks with the precision of the level set method. Two advantages of this integration are shown in the paper. First, using the level set initialized with the FCN output can achieve better performance than the FCN alone. Second, as a training technique, by utilizing unlabeled data and jointly training the FCN with the level set, improves the FCN performance. While the proposed model is only 2D binary segmentation and has a simple shape prior, its extension to 3D and more complex probabilistic shape models is straightforward and we are currently working on it.

References

1. Cremers, D., Rousson, M., Deriche, R.: A review of statistical approaches to level set segmentation: integrating color, texture, motion and shape. Int. J. Comput. Vis. **72** (2007)
2. Chan, T., Vese, L.: Active contours without edges. IEEE Trans. Image Process. **10**(2), 266–277 (2001)
3. Salah, M.B., Mitiche, A., Ayed, I.B.: Effective level set image segmentation with a kernel induced data term. Trans. Img. Proc. **19**(1), 220–232 (2010)
4. Long, J., Shelhamer, E., Darrell, T.: Fully convolutional networks for semantic segmentation. In: CVPR, pp. 3431–3440 (2015)
5. Ronneberger, O., Fischer, P., Brox, T.: U-Net: convolutional networks for biomedical image segmentation. In: Navab, N., Hornegger, J., Wells, W.M., Frangi, A.F. (eds.) MICCAI 2015. LNCS, vol. 9351, pp. 234–241. Springer, Cham (2015). doi:10.1007/978-3-319-24574-4_28
6. Brosch, T., Yoo, Y., Tang, L.Y.W., Li, D.K.B., Traboulsee, A., Tam, R.: Deep convolutional encoder networks for multiple sclerosis lesion segmentation. In: Navab, N., Hornegger, J., Wells, W.M., Frangi, A.F. (eds.) MICCAI 2015. LNCS, vol. 9351, pp. 3–11. Springer, Cham (2015). doi:10.1007/978-3-319-24574-4_1
7. BenTaieb, A., Hamarneh, G.: Topology aware fully convolutional networks for histology gland segmentation. In: Ourselin, S., Joskowicz, L., Sabuncu, M.R., Unal, G., Wells, W. (eds.) MICCAI 2016. LNCS, vol. 9901, pp. 460–468. Springer, Cham (2016). doi:10.1007/978-3-319-46723-8_53
8. Kamnitsas, K., Ledig, C., Newcombe, V.F., Simpson, J.P., Kane, A.D., Menon, D.K., Glocker, B., Rueckert, D.: Efficient multi-scale 3D CNN with fully connected CRF for accurate brain lesion segmentation. Med. Image Anal. **36**, 61–78 (2017)
9. Cai, J., Lu, L., Zhang, Z., Xing, F., Yang, L., Yin, Q.: Pancreas segmentation in MRI using graph-based decision fusion on convolutional neural networks. In: Ourselin, S., Joskowicz, L., Sabuncu, M.R., Unal, G., Wells, W. (eds.) MICCAI 2016. LNCS, vol. 9901, pp. 442–450. Springer, Cham (2016). doi:10.1007/978-3-319-46723-8_51
10. Zheng, S., Jayasumana, S., Romera-Paredes, B., Vineet, V., Su, Z., Du, D., Huang, C., Tor, P.H.S.: Conditional random fields as recurrent neural network. In: ICCV, pp. 1529–1537 (2015)

11. Ngo, T.A., Lu, Z., Carneiro, G.: Combining deep learning and level set for the automated segmentation of the left ventricle of the heart from cardiac cine magnetic resonance. Med. Image Anal. **35**, 159–171 (2017)
12. Chen, F., Yu, H., Hu, R., Zeng, X.: Deep learning shape priors for object segmentation. In: Proceedings of the IEEE Conference on Computer Vision and Pattern Recognition, pp. 1870–1877 (2013)
13. Paragios, N., Deriche, R.: Geodesic active regions: a new paradigm to deal with frame partition problems in computer vision. Vis. Commun. Image Representation **13**, 249–268 (2002)
14. Cremers, D., Osher, S.J., Soatto, S.: Kernel density estimation and intrinsic alignment for shape priors in level set segmentation. Int. J. Comput. Vis. **69**(3), 335–351 (2006)
15. Zeiler, M.D.: Adadelta: an adaptive learning rate method. arXiv preprint arXiv:1212.5701 (2012)
16. Van Ginneken, B., Heimann, T., Styner, M.: 3D segmentation in the clinic: a grand challenge, pp. 7–15 (2007)
17. Radau, P.: Cardiac MR Left Ventricle Segmentation Challenge (2008). http://smial.sri.utoronto.ca/LV_Challenge/Home.html. Accessed 10 Dec 2016

Context-Based Normalization of Histological Stains Using Deep Convolutional Features

D. Bug[1(✉)], S. Schneider[1], A. Grote[2], E. Oswald[3], F. Feuerhake[2],
J. Schüler[3], and D. Merhof[1]

[1] Institute for Imaging and Computer Vision, RWTH-Aachen University,
Aachen, Germany
`daniel.bug@lfb.rwth-aachen.de`
[2] Hannover Medical School, Institute for Pathology, Hannover, Germany
[3] Oncotest GmbH, Freiburg im Breisgau, Germany

Abstract. While human observers are able to cope with variations in color and appearance of histological stains, digital pathology algorithms commonly require a well-normalized setting to achieve peak performance, especially when a limited amount of labeled data is available. This work provides a fully automated, end-to-end learning-based setup for normalizing histological stains, which considers the texture context of the tissue. We introduce *Feature Aware Normalization*, which extends the framework of batch normalization in combination with gating elements from Long Short-Term Memory units for normalization among different spatial regions of interest. By incorporating a pretrained deep neural network as a feature extractor steering a pixelwise processing pipeline, we achieve excellent normalization results and ensure a consistent representation of color and texture. The evaluation comprises a comparison of color histogram deviations, structural similarity and measures the color volume obtained by the different methods.

Keywords: Normalization · Machine learning · Deep learning · Histology

1 Introduction

In digital pathology, histological tissue samples undergo fixation and embedding, sectioning and staining, and are finally digitized via whole-slide scanners. Each individual processing step presents a potential source of variance. In addition to such inherent variation, different staining protocols typically exist between different institutions. While trained human observers can handle most of the resulting variability, algorithms typically require a unified representation of the data to run reliably. Essentially, the challenge in any normalization task is to transform the distribution of color values from an image acquired using an arbitrary staining protocol into a defined reference space, modeled by a different

---

D. Bug and S. Schneider—These authors contributed equally to this work.

© Springer International Publishing AG 2017
M.J. Cardoso et al. (Eds.): DLMIA/ML-CDS 2017, LNCS 10553, pp. 135–142, 2017.
DOI: 10.1007/978-3-319-67558-9_16

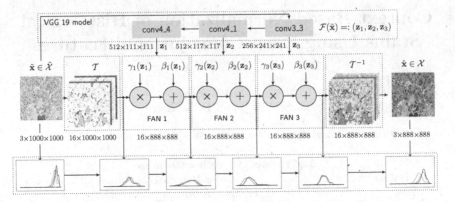

Fig. 1. Network architecture using Feature-Aware Normalization (FAN): Multi-scale deep network (here VGG-19) features are computed in a parallel path to the image processing. In the FAN units, the feature representations from different scales are upsampled to image size and perform pixelwise scaling and shifting of the feature maps computed by $\mathcal{T}$ in the main path. Below the network, changes in the RGB histograms are visualized along the path. Only the valid part of the convolution is used in the feature extractor, which is why the image gets cropped.

distribution. Chronologically, normalization algorithms applied in histopathology advanced from color matrix projections [8] to deconvolutional approaches [7,10], while many modern methods incorporate contextual information [1,6] to improve the normalization result. This trend is motivated by the amount of internal structures found in histological images, such as cell nuclei or blood-vessels, which may have similar color ranges under varying conditions and thus need to be normalized based on their context rather than individual pixel intensities. Particularly, the method "StaNoSa" [6], uses an autoencoder to provide features that steer a histogram-matching, which introduces the tissue context. In this work, we advance the idea by incorporating the entire normalization process into a deep neural network, where all algorithmic parameters are optimized via stochastic gradient descent. This requires several changes to the normalization setting, which we discuss in Sect. 2. We evaluate the approach using an extensive dataset dedicated to the normalization problem, which is described in Sect. 3. The dataset and an implementation of our approach are available at github.com/stes/fan.

2   Method

The goal of stain normalization is to adapt the color of each pixel based on the tissue context according to visually relevant features in the image. Similar requirements can be found in artistic style transfer [2,3], where the characteristics of a particular artistic style, e.g. van Gogh, are mapped onto an arbitrary image. However, while artistic style transfer may alter spatial relations between pixels, such as edges and corners, normalization in digital pathology requires changes that are limited to colors while preserving structures. This is similar to the problem of domain adaptation, where images from arbitrary source domains should be transferred to the target domain existing algorithms operate on.

Following this view, we consider manifolds $\mathcal{X}_i$ representing images from different color protocols. Furthermore, we assume that each image was generated from a shared underlying latent representation Ω, the image "content". In the process of image acquisition, we consider (unknown) functions $g_i : \Omega \to \mathcal{X}_i$ that map points from the latent representation to $\mathcal{X}_i$. In color normalization, we choose protocol k as a reference $\mathcal{X} := \mathcal{X}_k$, while all other datasets are considered to be part of a noise dataset $\tilde{\mathcal{X}} := \bigcup_{i \neq k} \mathcal{X}_i$. We propose a novel unsupervised learning algorithm capable of computing $f_\theta : \tilde{\mathcal{X}} \to \mathcal{X}$, only using samples from a fixed reference protocol k, such that for any latent feature representation $\mathbf{z} \in \Omega$, $f(\mathbf{x}_i) = \mathbf{x}_k$ given that $\mathbf{x}_i = g_i(\mathbf{z})$ and $\mathbf{x}_k = g_k(\mathbf{z})$.

2.1 Feature-Aware Normalization

From style transfer, we adapt the idea to use two parallel network paths, as depicted in Fig. 1. The main path is a transformer network $\mathcal{T}$ along with its inverse $\mathcal{T}^{-1}$ performing 1×1 convolutions to augment the color space by learning a meaningful latent representation, while a feature extractor $\mathcal{F}$ provides context information on the parallel path.

As a coupling mechanism we introduce Feature-Aware Normalization (FAN) units, which are inspired by Batch Normalization(BN) [5] and the gating mechanism in Long Short-Term Memory(LSTM) units [4]. As investigated by [2], different styles can be integrated into the same network by adapting the shifting and scaling parameters β and γ of the BN layers only. We treat color protocols as a particular simple form of style, restricted to only pixel-wise transformations based on the feature representation $\mathbf{z}$, which allows changes according to the pixel context, e.g. nucleus or plasma, while largely preserving the image structure. It is to note that minor structural changes may still occur since despite the restriction, each normalized pixel links back to a perceptive field via $\mathcal{F}$.

After transforming an input image $\tilde{\mathbf{x}}$ into the feature spaces $\mathbf{y} = \mathcal{T}(\tilde{\mathbf{x}})$ and $\mathbf{z} = \mathcal{F}(\tilde{\mathbf{x}})$, the normalization procedure for feature map k is given as

$$\overline{\mathbf{y}}^{(k)} = \frac{\mathbf{y}^{(k)} - \mu_k(\mathbf{y})}{\sqrt{\sigma_k^2(\mathbf{y}) + \epsilon}} \cdot \underbrace{\mathrm{sigm}(\sum_j W_M^{(k,j)} \cdot \mathbf{z}^{(j)})}_{=: \gamma^{(k)}(\mathbf{z})} + \underbrace{\max(\sum_j W_A^{(k,j)} \cdot \mathbf{z}^{(j)}, 0)}_{=: \beta^{(k)}(\mathbf{z})}, \quad (1)$$

where $\mu_k(\mathbf{y})$ and $\sigma_k^2(\mathbf{y})$ denote mean and variance computed over the spatial and batch dimension of feature map k. Think of this as a way of error correction, where sigm $\in (0, 1)$ suppresses incorrect color values and the additive ReLU activation contributes the correction. A small constant value ϵ provides numerical stability. The weight matrices for the multiplication gate W_M and addition gate W_A are used to map the representation from $\mathcal{F}$ onto the number of feature dimensions provided by $\mathcal{T}$. The result is upsampled to better match the input

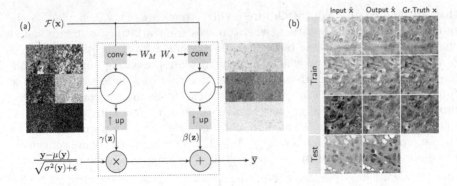

Fig. 2. (a) Block diagram of the FAN unit with examples of the pixel wise β and γ activations next to the nonlinearities. (b) Three examples of our PCA-based training augmentation and one application case.

size, however, some pixel information at the borders is lost due to the use of valid convolutions. A representation of the whole module along with example outputs of β and γ is depicted in Fig. 2. Together with Fig. 1, this illustrates the basic idea of gradually rescaling and shifting the internal representation.

We use a pretrained VGG-19 architecture with weights from the ILSVRC Challenge [9] as feature extractor $\mathcal{F}$. This network was chosen because of its success in style transfer and a variety of classification tasks, which indicates relevant internal features, computed on a dataset whose color variations exceed histopathological stains by far. However, we stress the fact that competing high scoring architectures, possibly finetuned to the image domain, could easily be deployed in this model as well. To incorporate context information from various scales, we deploy three FAN units, which is an empirical number. For $\mathcal{F}$ to provide meaningful information to the FAN units, we imply that there exists a transformation $\mathbf{z}' = W \cdot \mathbf{z}$ such that the result is independent of the staining protocol, i.e., the same tissue stained with two different protocols would yield identical representations in latent variable space, which is a feasible assumption. Throughout this work, the weights of $\mathcal{F}$ are fixed. Instead, the training process only adapts the parameters of all FAN units and the transformer network $\mathcal{T}$.

2.2 Normalization by Denoising

Training the network $f_\theta : \tilde{\mathcal{X}} \to \mathcal{X}$ is challenging since no samples $(\mathbf{x}, \tilde{\mathbf{x}})$ with matching latent representation $\mathbf{z}$ are available directly. To circumvent this issue, we propose the use of a noise model in which the reference image $\mathbf{x}$ is disturbed by a noise distribution $p(\tilde{\mathbf{x}}|\mathbf{x})$. During training, for each training example $\mathbf{x}_i \in \mathcal{X}$, we use the noise distribution to draw a corresponding sample $\tilde{\mathbf{x}}_i \sim p(\tilde{\mathbf{x}}|\mathbf{x}_i)$, $\tilde{\mathbf{x}}_i \in \tilde{\mathcal{X}}$ and minimize the mean-squared error between the normalized sample and the original one, yielding the objective

$$\min_{\theta} \sum_{i=0}^{N} \| f_{\theta}(\tilde{\mathbf{x}}_i) - \mathbf{x}_i \|^2, \quad \tilde{\mathbf{x}}_i \sim p(\tilde{\mathbf{x}}|\mathbf{x}_i) \; \forall i \in [N]. \tag{2}$$

A minimizer of this objective maps all disturbed images $\tilde{\mathbf{x}}$ from $\tilde{\mathcal{X}}$ onto $\mathcal{X}$ under the chosen noise model. For this training scheme to work, $p(\tilde{\mathbf{x}}|\mathbf{x})$ has to assign high probability to samples on $\tilde{\mathcal{X}}$. To ensure this property, we follow the commonly used principle of data augmentation according to the principal components of pixel values. The noise model is then given by a normal distribution $\mathcal{N}(\mathbf{x}, W\Sigma W^T \varepsilon)$. Herein, $W, \Sigma \in \mathbb{R}^{3\times3}$ denote the transformation from RGB color values to principal components and the component's explained variance, respectively. The noise magnitude is controlled by $\varepsilon \in (0, 1)$, determined empirically by visual inspection of the samples. An example of this training augmentation is given in Fig. 2b.

3 Experiments

In order to comprehensively evaluate the normalization performance, we introduce a dedicated dataset comprising five blocks of lung cancer tissue from patient-derived xenografts each of which was sliced consecutively into nine slides. In addition to variations of the Hematoxylin (H) and Eosin (E) component introduced in [6], we vary the slice thickness (T). For each slice, the parameters were iterated yielding the staining protocols 1. HET↑, 2. HET↓, 3. standard HET protocol, 4. HE↑ T, 5. HE↓ T, 6. H↑ ET, 7. H↓ ET, 8. H↑ E↑ T, 9. H↓ E↓ T, where '↑' denotes a doubled and '↓' a halved concentration or thickness, with the standard values being 1:10 for H (liquid component), 0.6 g/200 ml for E (powder) and 3 μm for T. From each digitized whole-slide image (Aperio AT2 Scanner, Leica Biosystems, Wetzlar, Germany), we extracted and manually registered five regions with 1000×1000 pixels, resulting in a test set of 225 images, wherein the tissue distribution is largely shared between the protocol sets of nine images due to consecutive slicing and registration. For each protocol, a separate normalization network was trained with the particular protocol chosen as the reference dataset $\mathcal{X}$. Approximately 15000 patches of size 192×192 were extracted separately from the slides and used for training. The validation was then done on tissue samples from the remaining protocols, excluding the training patches.

We define three experiments to evaluate the performance of our proposed algorithm in comparison to the previously presented methods of Reinhard (REI, [8]), Macenko (MAR, [7]), Vahadane (VAH, [10]) and Bejnordi (BEJ, [1]). For all methods, we repeat the experiments with each of the nine protocols as normalization target, with the exception of BEJ, where we use the provided reference protocol. Wherever possible, we show the metric computed on the unnormalized images as baseline (BAS).

1. Elimination of protocol deviations. As suggested by related publications, we measure distances between color distributions of registered patches in consecutive slices of tissue. As an estimator for the true distribution, we compute

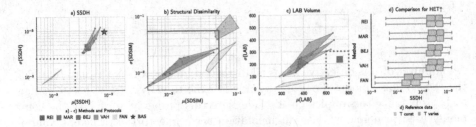

Fig. 3. Comparison of the methods from Reinhard (REI, [8]), Macenko (MAR, [7]), Vahadane (VAH, [10]), Bejnordi (BEJ, [1]) with the proposed FAN method. The metrics for unnormalized images are given as baseline (BAS) in case of SSDH and LAB. Plots (a)–(c) show the convex hulls of the metric distributions for each algorithm. Red dotted lines highlight preferred regions (best viewed in the digital version).

a kernel-density estimate (KDE) on a 256-bin histogram of each color channel. A binomial filter of length seven is used as KDE kernel. For clarity, we now slightly deviate from previous notation and regard $\mathbf{x}_i$ as a patch from protocol i and f_j as the normalization function to reference dataset j. This yields the metric $\ell_{i,j} = \mathrm{SSDH}(f_j(\mathbf{x}_i), \mathbf{x}_j)$, where SSDH denotes the sum of squared differences between the color histograms of $f_j(\mathbf{x}_i)$ and $\mathbf{x}_j$. In Fig. 3a the error distribution is visualized as a convex hull plot, where each point inside the hull represents the performance of a normalization f_j which is characterized by the mean and standard-deviation of all corresponding $\ell_{i,j}$.

2. Influence on texture. Whereas pixelwise methods do usually not interfere with pixel relations, introducing context into the normalization carries the risk of erroneous alterations in groups of pixels, since output pixels connect to a perceptive field via $\mathcal{F}$. Hence, we apply the structural dissimilarity index (SDSIM, [11]) as a metric, which quantifies perceptual dissimilarity by comparing first- and second-order statistics of the images before and after normalization. A value of zero indicates identical image structure, while higher values indicate dissimilarity. This yields the metric $\ell_{i,j} = \mathrm{SDSIM}(f_j(\mathbf{x}_i), \mathbf{x}_i)$ shown in Fig. 3b.

3. Color richness. Theoretically, a normalization effect with respect to a histogram deviation metric can be erroneously created by computing a lowly saturated subspace with correlated channels, which basically leads to a grayscale-like appearance. To reassure a valid normalization, we use the volume of colors in Lab color space as a measure for the amount of perceivable colors after normalization. Since all components of the Lab space are decorrelated, the approximate color volume is computed as product of the standard deviations in each channel $\mathrm{LAB} := \mathrm{std}(L) \cdot \mathrm{std}(a) \cdot \mathrm{std}(b)$. This yields the metric $\ell_{i,j} = \mathrm{LAB}(f_j(\mathbf{x}_i))$, shown in Fig. 3c. Ideally, a high color volume is achieved at a low variance.

4   Discussion

From quality metrics in Fig. 3 and the overview in Fig. 4, we conclude that in terms of SSDH and color volume our proposed method outperforms previous

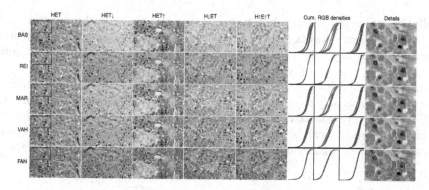

Fig. 4. Normalization of selected source protocols together with cumulated density per channel (all protocols) and detail view. Note that each algorithm is shown with the best performing target protocol according to SSDH and LAB.

approaches by a notable margin. As expected, the introduction of context information may result in minor texture changes, which originate from the different scales of the FAN units and the choice of $\mathcal{F}$ and display as faint 'tiles' in the output image. According to visual inspection, e.g. in the detailed views in Fig. 4, these effects can be considered negligible. In many cases, the volume of perceivable colors decreases in FAN compared to conventional methods. However, the low variance in the color volumes indicates a strong consistency of the resulting color spaces. In contrast, the high variance in some cases of the MAR method results from artifacts where a failed estimation of source or target color space leads to unrealistically intense and oversaturated colors. We may also note that the performance of both MAR and REI strongly depends on having similar tissue distributions in the source and reference patches, which is a property of the used dataset, but not given in general. Hence, the performance we report for these algorithms can be seen as an upper bound. VAH achieves a very reliable color normalization, but structures can be affected if the sparse decomposition does not find a good optimum. With the HET↑ protocol as target, we obtain a model which shows an exceptionally large color volume. In application, not every target protocol is equally important.

With the proposed approach, we overcome the challenge of providing a suitable reference image covering a variety of tissue statistics by finding an appropriate target protocol with high color volume and contrast. These properties are then encoded in the network parameters through training. As the low variance in all measures shows, the normalization characteristics encoded in FAN units exceed the capabilities of previous methods. We find further indication for this by comparing moderate stain variation (protocols 3–9) with deviations in thickness (protocols 1–3) in Fig. 3d. Herein, previous methods show a notable decay in the quality metrics, whereas FAN provides a very consistent normalization.

A beneficial advantage of the clear separation between the feature extractor and the transformer path is that the currently used VGG-19 architecture can easily be replaced by any other feature extractor. Particularly, computationally more efficient network structures with less parameters are good candidates. Additionally, it might be promising to train the feature extraction step jointly with the remaining network, which is left for further research.

In a greater perspective, the general approach to separate network functions into distinct parts may have application to other areas as well, such as the artistic style transfer networks that originally motivated our work. We assume that an artistic style transfer can be realized with this architecture simply by using larger filter sizes in the transformer network $\mathcal{T}$, allowing for spatial context aggregation on this path.

References

1. Bejnordi, B.E., Litjens, G., Timofeeva, N., Otte-Höller, I., Homeyer, A., Karssemeijer, N., van der Laak, J.A.: Stain specific standardization of whole-slide histopathological images. IEEE Trans. Med. Imaging **35**(2), 404–415 (2016)
2. Dumoulin, V., Shlens, J., Kudlur, M.: A learned representation for artistic style. CoRR abs/1610.07629 (2016). http://arxiv.org/abs/1610.07629
3. Gatys, L.A., Ecker, A.S., Bethge, M.: Image style transfer using convolutional neural networks. In: The IEEE Conference on Computer Vision and Pattern Recognition (CVPR) (2016)
4. Hochreiter, S., Schmidhuber, J.: Long short-term memory. Neural Comput. **9**(8), 1735–1780 (1997)
5. Ioffe, S., Szegedy, C.: Batch normalization: accelerating deep network training by reducing internal covariate shift. In: Proceedings of the 32nd International Conference on Machine Learning, ICML 2015, pp. 448–456 (2015)
6. Janowczyk, A., Basavanhally, A., Madabhushi, A.: Stain Normalization using Sparse AutoEncoders (StaNoSA): application to digital pathology. Comput. Med. Imaging Graph. **57**, 50–61 (2017)
7. Macenko, M., Niethammer, M., Marron, J.S., Borland, D., Woosley, J.T., Guan, X., Schmitt, C., Thomas, N.E.: A method for normalizing histology slides for quantitative analysis. In: 2009 IEEE International Symposium on Biomedical Imaging: from Nano to Macro, pp. 1107–1110, June 2009
8. Reinhard, E., Adhikhmin, M., Gooch, B., Shirley, P.: Color transfer between images. IEEE Comput. Graphics Appl. **21**(5), 34–41 (2001)
9. Simonyan, K., Zisserman, A.: Very deep convolutional networks for large-scale image recognition. CoRR abs/1409.1556 (2014)
10. Vahadane, A., Peng, T., Sethi, A., Albarqouni, S., Wang, L., Baust, M., Steiger, K., Schlitter, A.M., Esposito, I., Navab, N.: Structure-preserving color normalization and sparse stain separation for histological images. IEEE Trans. Med. Imaging **35**(8), 1962–1971 (2016)
11. Wang, Z., Bovik, A.C., Sheikh, H.R., Simoncelli, E.P.: Image quality assessment: from error visibility to structural similarity. IEEE Trans. Image Process. **13**(4), 600–612 (2004)

Transitioning Between Convolutional and Fully Connected Layers in Neural Networks

Shazia Akbar[1]([✉]), Mohammad Peikari[1], Sherine Salama[2],
Sharon Nofech-Mozes[2], and Anne Martel[1]

[1] Sunnybrook Research Institute, University of Toronto, Toronto, Canada
{sakbar,mpeikari,amartel}@sri.utoronto.ca
[2] Department of Pathology, Sunnybrook Health Sciences Centre, Toronto, Canada
{sherine.salama,sharon.nofech-mozes}@sunnybrook.ca

Abstract. Digital pathology has advanced substantially over the last decade however tumor localization continues to be a challenging problem due to highly complex patterns and textures in the underlying tissue bed. The use of convolutional neural networks (CNNs) to analyze such complex images has been well adopted in digital pathology. However in recent years, the architecture of CNNs have altered with the introduction of inception modules which have shown great promise for classification tasks. In this paper, we propose a modified "transition" module which learns global average pooling layers from filters of varying sizes to encourage class-specific filters at multiple spatial resolutions. We demonstrate the performance of the transition module in AlexNet and ZFNet, for classifying breast tumors in two independent datasets of scanned histology sections, of which the transition module was superior.

Keywords: Convolutional neural networks · Histology · Transition · Inception · Breast tumor

1 Introduction

There are growing demands for tumor identification in pathology for time consuming tasks such as measuring tumor burden, grading tissue samples, determining cell cellularity and many others. Recognizing tumor in histology images continues to be a challenging problem due to complex textural patterns and appearances in the tissue bed. With the addition of tumor, subtle changes which occur in the underlying morphology are difficult to distinguish from healthy structures and require expertise from a trained pathologist to interpret. An accurate automated solution for recognizing tumor in vastly heterogeneous pathology datasets would be of great benefit, enabling high-throughput experimentation, greater standardization and easing the burden of manual assessment of digital slides.

Deep convolutional neural networks (CNNs) are now a widely adopted architecture in machine learning. Indeed, CNNs have been adopted for tumor classification in applications such as analysis of whole slide images (WSI) of breast

© Springer International Publishing AG 2017
M.J. Cardoso et al. (Eds.): DLMIA/ML-CDS 2017, LNCS 10553, pp. 143–150, 2017.
DOI: 10.1007/978-3-319-67558-9_17

tissue using AlexNet [9] and voxel-level analysis for segmenting tumor in CT scans [16]. Such applications of CNNs continue to grow and the traditional architecture of a CNN has also evolved since its origination in 1998 [5]. A basic CNN architecture encompasses a combination of convolution and pooling operations. As we traverse deeper in the network, the network size decreases resulting in a series of outputs, whether that be classification scores or regression outcomes. In lower layers of a typical CNN, fully-connected (FC) layers are required to learn non-linear combinations of learned features. However the transition between a series of two-dimensional convolutional layers to a one-dimensional FC layer is abrupt, making the network susceptible to overfitting [7]. In this paper we propose a method for transitioning between convolutional layers and FC layers by introducing a framework which encourages generalization. Different from other regularizers [4,11], our method congregates high-dimensional data from features maps produced in convolutional layers in an efficient manner before flattening is performed.

To ease the dimensionality reduction between convolutional and FC layers we propose a transition module, inspired by the inception module [14]. Our method encompasses convolution layers of varying filter sizes, capturing learned feature properties at multiple scales, before collapsing them to a series of average pooling layers. We show that this configuration gives considerable performance gains for CNNs in a tumor classification problem in scanned images of breast cancer tissue. We also evaluate the performance of the transition module compared to other commonly used regularizers (Sect. 5.1).

2   Related Work

In the histopathology literature, CNN architectures tend to follow a linear trend with a series of layers sequentially arranged from an input layer to a softmax output layer [8,9]. Recently, however, there have been adaptations to this structure to encourage multiscale information at various stages in the CNN design. For example Xu *et al.* [17] proposed a multichannel side supervision CNN which

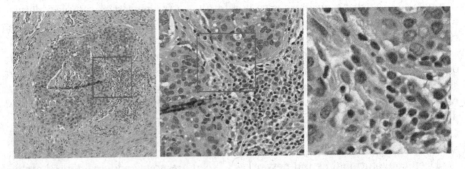

Fig. 1. Digital slide shown at multiple resolutions. Regions-of-interest outlined in red are shown at greater resolutions from left to right. (Color figure online)

merges edges and contours of glands at each convolution layer for gland segmentation. In cell nuclei classification, Buyssens *et al.* [2] learn multiple CNNs in parallel with input images at various resolutions before aggregating classification scores. These methods have shown to be particularly advantageous in histology images as it mimics pathologists' interpretation of digital slides when viewed at multiple objectives (Fig. 1).

However capturing multiscale information in a single layer has been a recent advancement after the introduction of the inception modules (Sect. 3.1). Since then, there have been some adaptations however these have been limited to convolution layers in a CNN. Liao and Carneiro [6] proposed a module which combines multiple convolution layers via a max-out operation as opposed to concatenation. Jin *et al.* [3] designed a CNN network in which independent FC layers are learned from the outputs of inception layers created at various levels of the network structure. In this paper, we focus on improving the network structure when changes in dimensionality occur between convolution and FC layers. Such changes occur when the network has already undergone substantial reduction and approaches the final output layer therefore generalization is key for optimal class separation.

3   Method

3.1   Inception Module

Inception modules, originally proposed by Szegedy *et al.* [14], are a method of gradually increasing feature map dimensionality thus increasing the depth of a CNN without adding extreme computational costs. In particular, the inception module enables filters to be learned at multiple scales simultaneously in a single layer, also known as a sub-network. Since its origination, there have been multiple inception networks incorporating various types of inception modules [13,15], including GoogleNet. The base representation of the original inception module is shown in Fig. 2 (left).

Each inception module is a parallel series of convolutional layers restricted to filter sizes 1 × 1, 3 × 3 and 5 × 5. By encompassing various convolution

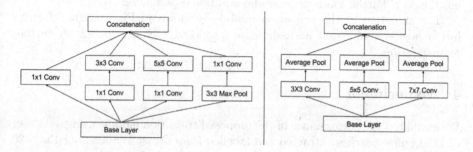

Fig. 2. Original inception module (left) and the proposed transition module (right).

sub-layers in a single deep layer, features can be explored at multiple scales simultaneously. During training, the combination of filter sizes which result in optimal performance are weighted accordingly. However on its own, this configuration results in a very large network with increased complexity. Therefore for practicality purposes, the inception module also encompasses 1 × 1 convolutions which act as dimensionality reduction mechanisms. The Inception *network* is defined as a stack of inception modules with occasional max-pooling operations. The original implementation of the Inception network encompasses nine inception modules [14].

3.2   Transition Module

In this paper, we propose a modified inception module, called the "transition" module, explicitly designed for the final stages of a CNN, in which learned features are mapped to FC layers. Whilst the inception module successfully captures multiscale from input data, the bridge between learned feature maps and classification scores is still treated as a black box. To ease this transition process we propose a method for enabling 2D feature maps to be downscaled substantially before tuning FC layers. In the transition module, instead of concatenating outcomes from each filter size, as in [14], independent global average pooling layers are configured after learning convolution layers which enable feature maps to be compressed via an average operation.

Originally proposed by Lin *et al.* [7], global average pooling layers were introduced as a method of enforcing correspondences between categories of the classification task (i.e. the softmax output layer) and filter maps. As the name suggests, in a global averaging pooling layer a single output is retrieved from each filter map in the preceeding convolutional layer by averaging it. For example, if given an input of 256 3 × 3 feature maps, a global average pool layer would form an output of size 256. In the transition module, we use global average pooling to sum out spatial information at multiple scales before collapsing each averaged filter to independent 1D output layers. This approach has the advantage of introducing generalizability and encourages a more gradual decrease in network size earlier in the network structure. As such, subsequent FC layers in the network are also smaller in size, making the task of delineating classification categories much easier. Furthermore there are no additional parameters to tune.

The structure of the transition module is shown in Fig. 2 (right). Convolution layers were batch normalized as proposed in Inception-v2 for further regularization.

4   Experiment

We evaluated the performance of the proposed transition module using a dataset of 1229 image patches extracted and labelled from breast WSIs scanned at ×20 magnification by a Scanscope XT (Aperio technologies, Leica Biosystems) scanner. Each RGB patch of size 512 × 512 was hand selected from 31 WSIs, each one from a single patient, by a trained pathologist. Biopsies were extracted from patients with invasive breast cancer and subsequently received neo-adjuvant therapy; post neoadjuvant tissue sections revealed invasive and/or ductal carcinoma *in situ*. 5-fold cross validation was used to evaluate the performance of this dataset. Each image patch was confirmed to contain either tumor or healthy tissue by an expert pathologist. "Healthy" refers to patches which are absent of cancer cells but may contain healthy epithelial cells amongst other tissue structures such as stroma, fat etc. Results are reported over 100 epochs.

We also validated our method on a public dataset, BreaKHis [10] (Sect. 5.3) which contains scanned images of benign (adenosis, fibroadenoma, phyllodes tumor, tubular adenoma) and malignant (ductal carcinoma, lobular carcinoma, mucinous carcinoma, papillary carcinoma) breast tumors at ×40 objective. Images were resampled into patches of dimensions 228 × 228, suitable for a CNN, resulting in 11, 800 image patches in total. BreaKHis was validated using 2-fold cross validation and across 30 epochs.

In Sect. 5.2, results are reported for three different CNN architectures (AlexNet, ZFNet, Inception-v3), of which transition modules were introduced in AlexNet and ZFNet. All CNNs were trained from scratch with no pretraining. Transition modules in both implementations encompassed 3 × 3, 5 × 5 and 7 × 7 convolutions, thus producing three average pooling layers. Each convolutional layer has a stride of 2, and 1024 and 2048 filter units for AlexNet and VFNet, respectively. Note, the number of filter units were adapted according to the size of the first FC layer proceeding the transition module.

CNNs were implemented using Lasagne 0.2 [1]. A softmax function was used to obtain classification predictions and convolutional layers encompassed ReLU activations. 10 training instances were used in each batch in both datasets. We used Nestorov Momentum [12] to perform updates with a learning rate of $1e^{-5}$.

5   Results

5.1   Experiment 1: Comparison with Regularizers

Our first experiment evaluated the performance of the transition model when compared to other commonly used regularizers including Dropout [11] and cross-channel local response normalization [4]. We evaluated the performance of each regularizer in AlexNet and report results for (a) a single transition module added before the first FC layer, (b) two Dropout layers, one added after each FC layer with $p = 0.5$, and lastly (c) normalization layers added after each max-pooling operation, similar to how it was utilized in [4].

The transition module achieved an overall accuracy rate of 91.5% which when compared to Dropout (86.8%) and local normalization (88.5%) showed considerable improvement, suggesting the transition module makes an effective regularizer compared to existing methods. When local response normalization was used in combination with the transition module in ZFNet (below), we achieved a slightly higher test accuracy of 91.9%.

5.2   Experiment 2: Comparing Architectures

Next we evaluated the performance of the transition module in two different CNN architectures: AlexNet and ZFNet. We also report the performance of Inception-v3 which already has built-in regularizers in the form of 1×1 convolutions [15], for comparative purposes. ROC curves are shown in Fig. 3.

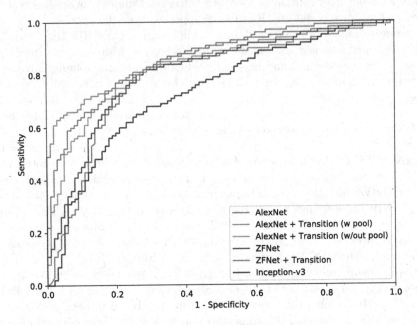

Fig. 3. ROC curves for AlexNet [4] and ZFNet [18], with and without the proposed transition module, and Inception-v3 [15]. ROC curves are also shown for the transition module with and without average pooling.

Both AlexNet and ZFNet benefited from the addition of a single transition module, improving test accuracy rates by an average of 4.3%, particularly at lower false positive rates. Smaller CNN architectures proved to be better for tumor classification in this case as overfitting was avoided, as shown by the comparison with Inception-v3. Surprisingly, the use of dimensionality reduction earlier in the architectural design does not prove to be effective for increasing classification accuracy. We also found that the incorporation of global average pooling in the transition module improved results slightly and resulted in 3.1% improvement in overall test accuracy.

5.3   Experiment 3: BreaKHis

We used the same AlexNet architecture used above to also validate BreaKHis. ROC curves are shown in Fig. 4. There was a noticeable improvement

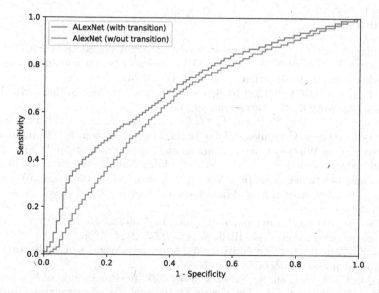

Fig. 4. ROC curves for BreaKHis [10] dataset with and without the proposed transition module.

$(AUC += 0.06)$ when the transition module was incorporated, suggesting that even when considerably more training data is available a smoother network reduction can be beneficial.

The transition module achieved an overall test accuracy of 82.7% which is comparable to 81.6% achieved with SVM in [10], however these results are not directly comparable and should be interpreted with caution.

6 Conclusion

In this paper we propose a novel regularization technique for CNNs called the transition module, which captures filters at multiple scales and then collapses them via average pooling in order to ease network size reduction from convolutional layers to FC layers. We showed that in two CNNs (AlexNet, ZFNet) this design proved to be beneficial for distinguishing tumor from healthy tissue in digital slides. We also showed an improvement in a larger publically available dataset, BreaKHis.

Acknowledgements. This work has been supported by grants from the Canadian Breast Cancer Foundation, Canadian Cancer Society (grant 703006) and the National Cancer Institute of the National Institutes of Health (grant number U24CA199374-01).

References

1. Lasagne. http://lasagne.readthedocs.io/en/latest/. Accessed 18 Feb 2017
2. Buyssens, P., Elmoataz, A., Lézoray, O.: Multiscale convolutional neural networks for vision–based classification of cells. In: Lee, K.M., Matsushita, Y., Rehg, J.M., Hu, Z. (eds.) ACCV 2012. LNCS, vol. 7725, pp. 342–352. Springer, Heidelberg (2013). doi:10.1007/978-3-642-37444-9_27
3. Jin, X., Chi, J., Peng, S., Tian, Y., Ye, C., Li, X.: Deep image aesthetics classification using inception modules and fine-tuning connected layer. In: 8th International Conference on Wireless Communications and Signal Processing (WCSP) (2016)
4. Krizhevsky, A., Sutskever, I., Hinton, G.E.: ImageNet classification with deep convolutional neural networks. In: Pereira, F., Burges, C.J.C., Bottou, L., Weinberger, K.Q. (eds.) Advances in Neural Information Processing Systems, vol. 25, pp. 1097–1105 (2012)
5. LeCun, Y., Bottou, L., Bengio, Y., Haffner, P.: Gradient-based learning applied to document recognition. Proc. IEEE **86**(11), 2278–2324 (1998)
6. Liao, Z., Carneiro, G.: Competitive multi-scale convolution. CoRR abs/1511.05635 (2015). http://arxiv.org/abs/1511.05635
7. Lin, M., Chen, Q., Yan, S.: Network in network. In: Proceedings of ICLR (2014)
8. Litjens, G., Sánchez, C.I., Timofeeva, N., Hermsen, M., Nagtegaal, I., Kovacs, I., Hulsbergen-van de Kaa, C., Bult, P., van Ginneken, B., van der Laak, J.: Deep learning as a tool for increased accuracy and efficiency of histopathological diagnosis. Sci. Rep. **6**, 26286 (2016)
9. Spanhol, F., Oliveira, L.S., Petitjean, C., Heutte, L.: Breast cancer histopathological image classification using convolutional neural network. In: International Joint Conference on Neural Networks, pp. 2560–2567
10. Spanhol, F.A., Oliveira, L.S., Petitjean, C., Heutte, L.: A dataset for breast cancer histopathological image classification. IEEE Trans. Biomed. Eng. **63**, 1455–1462 (2016)
11. Srivastava, N., Hinton, G., Krizhevsky, A., Sutskever, I., Salakhutdinov, R.: Dropout: a simple way to prevent neural networks from overfitting. J. Mach. Learn. Res. **15**(1), 1929–1958 (2014)
12. Sutskever, I.: Training recurrent neural networks. Thesis (2013)
13. Szegedy, C., Ioffe, S., Vanhoucke, V., Alemi, A.: Inception-v4, Inception-ResNet and the impact of residual connections on learning. arXiv preprint arXiv:1602.07261 (2016)
14. Szegedy, C., Liu, W., Jia, Y., Sermanet, P., Reed, S., Anguelov, D., Erhan, D., Vanhoucke, V., Rabinovich, A.: Going deeper with convolutions. In: Computer Vision and Pattern Recognition (CVPR) (2015)
15. Szegedy, C., Vanhoucke, V., Ioffe, S., Shlens, J., Wojna, Z.: Rethinking the inception architecture for computer vision. In: Proceedings of the IEEE Conference on Computer Vision and Pattern Recognition, pp. 2818–2826 (2016)
16. Vivanti, R., Ephrat, A., Joskowicz, L., Karaaslan, O.A., Lev-Cohain, N., Sosna, J.: Automatic liver tumor segmentation in follow up CT studies using convolutional neural networks. In: Proceedings of Patch-Based Methods in Medical Image Processing Workshop, MICCAI
17. Xu, Y., Li, Y., Wang, Y., Liu, M., Fan, Y., Lai, M., Chang, E.I.C.: Gland instance segmentation using deep multichannel neural networks. CoRR abs/1611.06661 (2016)
18. Zeiler, M.D., Fergus, R.: Visualizing and understanding convolutional networks. In: Computer Vision and Pattern Recognition (CVPR) (2013)

Quantifying the Impact of Type 2 Diabetes on Brain Perfusion Using Deep Neural Networks

Behrouz Saghafi[1], Prabhat Garg[1], Benjamin C. Wagner[1], S. Carrie Smith[2],
Jianzhao Xu[2], Ananth J. Madhuranthakam[1], Youngkyoo Jung[2],
Jasmin Divers[2], Barry I. Freedman[2], Joseph A. Maldjian[1],
and Albert Montillo[1]([✉])

[1] University of Texas Southwestern Medical Center, Dallas, TX, USA
Albert.Montillo@utsouthwestern.edu
[2] Wake Forest School of Medicine, Winston-Salem, NC, USA

Abstract. The effect of Type 2 Diabetes (T2D) on brain health is
poorly understood. This study aims to quantify the association between
T2D and perfusion in the brain. T2D is a very common metabolic disorder that can cause long term damage to the renal and cardiovascular
systems. Previous research has discovered the shape, volume and white
matter microstructures in the brain to be significantly impacted by T2D.
We propose a fully-connected deep neural network to classify the regional
Cerebral Blood Flow into low or high levels, given 16 clinical measures
as predictors. The clinical measures include diabetes, renal, cardiovascular and demographics measures. Our model enables us to discover any
nonlinear association which might exist between the input features and
target. Moreover, our end-to-end architecture automatically learns the
most relevant features and combines them without the need for applying
a feature selection method. We achieved promising classification performance. Furthermore, in comparison with six (6) classical machine learning algorithms and six (6) alternative deep neural networks similarly
tuned for the task, our proposed model outperformed all of them.

1 Introduction

More than 29 million people (9.3%) in the United States have diabetes [2]. In
adults, Type 2 Diabetes (T2D) accounts for 95% of all diagnosed cases of diabetes. T2D is a metabolic disorder characterized by high blood sugar caused by
insulin resistance or relative lack of insulin. It has been shown that prolonged
T2D can result in chronic kidney disease (CKD), cardiovascular disease (CVD)
and diabetic retinopathy. Since the brain consumes a disproportionately large
amount of the body's energy relative to its overall mass, it is reasonable to suspect that diabetes may impact brain health. Currently the effect of T2D on brain
health has been under-studied. Understanding these effects will help unravel this
complex disease and enable a more comprehensive evaluation of candidate new
therapies.

Initial studies on T2D patients have found that diabetes [3,11] and renal
[9,14] disease measures are associated with volumes of certain neuroanatomical

© Springer International Publishing AG 2017
M.J. Cardoso et al. (Eds.): DLMIA/ML-CDS 2017, LNCS 10553, pp. 151–159, 2017.
DOI: 10.1007/978-3-319-67558-9_18

structures, however the relationship between diabetes and the perfusion of the brain parenchyma has remained largely unknown. Additionally, according to the U.S. Center for Disease Control, African American adults are about twice as likely to be diagnosed with diabetes as European Americans [2]. Thus in this study we aim to identify the association between diabetes-related disease measures and regional brain perfusion in African Americans.

Several studies have linked diabetes to structural alterations in the brain. Sink et al. [14] found significant associations between renal measures and hippocampal white matter volume in African Americans with diabetic kidney disease, including the urine albumin to creatinine ratio (UACR) and the estimated glomerular filtration rate (GFR). Freedman et al. [3] reported an inverse association between aorta calcified plaque (a CVD measure) and the gray matter volume of hippocampus in African Americans with T2D. In a study using Diffusion Tensor Imaging (DTI), Hsu et al. [4] found that diabetes duration is significantly associated with white matter microstructure measures, such as mean diffusivity in several brain regions including bilateral cerebellum, temporal lobe, bilateral cingulate gyrus, pons, parahippocampal gyrus and right caudate.

Given that these studies have shown an association between diabetes and brain structure, we hypothesize that T2D also alters brain perfusion. This paper tests our hypothesis. Our main contributions are threefold. First, a massive univariate linear analysis approach is performed to identify candidate regions meeting the most stringent multiple comparisons correction criteria. Second, a fully-connected Deep Neural Network (DNN) architecture for predicting brain perfusion level is proposed which automatically learns optimal feature combinations and characterizes the T2D to perfusion association including any non-linearities. Third, permutation testing is conducted to access the reliability of proposed model's accuracy via the notion of statistical significance.

2   Materials

This cross-sectional study consisted of 152 African Americans with T2D. Laboratory tests were conducted to acquire measures of diabetes as well as related renal and cardiovascular disease measures. The diabetes measures included hemoglobin A_{1c} (HbA$_{1c}$) and diabetes duration. Renal disease measures included UACR, GFR, blood urea nitrogen (BUN), serum potassium, total serum protein, and urine microalbumin. Blood based measures of cardiovascular disease and inflammation included calcified atherosclerotic plaque in the coronary arteries (CAC) and C-reactive protein (CRP). Demographic measures obtained included gender (56.8% female), age (mean 59.2 years), body mass index, smoking status, and hypertension. These measures are summarized in Table 1.

Anatomical and perfusion MRI were acquired for every subject using a 3.0 Tesla Siemens Skyra MRI (Siemens Healthcare, Erlangen, Germany) with a high-resolution 20-channel head/neck coil. T1-weighted anatomic images were acquired using a 3D volumetric magnetization-prepared rapid acquisition gradient echo sequence (Repetition time [TR] 2,300 ms; echo time [TE] 2.02 ms;

Table 1. Diabetes and demographic measures included in the study.

Diabetes measures		Demographic measures	
UACR (mg/g)	83.7 (284.2)	Age (years)	59.2 (9.4)
CRP (mg/dL)	0.9 (1.5)	Female Sex (%)	56.8
HbA$_{1c}$ (%)	8.0 (1.8)	Education (%)	
Diabetes duration (years)	8.9 (7.5)	Less than High School	9.0
CAC (mass score, mg)	475.4 (1142.2)	High school diploma	23.9
GFR (mL/min/1.73 m^2)	90.6 (23.5)	Some college	39.4
Serum Potassium (mmol/L)	4.1 (0.4)	Associate degree	7.7
Total Serum Protein (g/dL)	7.2 (0.5)	College graduate	11.6
BUN (mg/dL)	15.6 (5.8)	After college	8.4
Urine Microalbumin	110.2 (344.0)	BMI (kg/m^2)	34.1 (7.2)
(mcg/mg creatinine)		Smoking	
		Never (%)	52.3
		Past smoker (%)	28.8
		Current smoker (%)	18.9
		Hypertension (%)	86.4

inversion time [TI] 900 ms; flip angle [FA] 9°; 192 slices; voxel dimensions $0.97 \times 0.97 \times 1\,\text{mm}^3$). Eight phase pseudo-Continuous Arterial Spin Labeling (pCASL) perfusion images were acquired with repetition time [TR] 4,000 ms; echo time [TE] 12 ms; inversion time [TI] 3000 ms; flip angle [FA] 90°; 26 slices/49 volumes; voxel dimensions $3.4 \times 3.4 \times 5\,\text{mm}^3$.

3   Methods

Our overall pipeline consists of these steps: (1) derivation of the mean Cerebral Blood Flow (CBF) per brain region, (2) identification of candidate brain regions via statistical analysis, and (3) fitting a DNN to quantify the association between the candidate region's CBF and diabetes measures. Each step is detailed below.

3.1   Compute Mean Gray Matter CBF per Anatomical Region

CBF volumes were computed from pCASL perfusion images in native space. To parcellate the CBF maps into regional measures, each subject's pCASL volume was co-registered to the same subject's T1-weighted image using affine transformation. Then each subject's T1-weighted image was spatially normalized to Montreal Neurological Institute (MNI) space using a non-linear transform [1] computed using the VBM8 toolbox[1]. These transforms were combined to spatially normalize the CBF maps into MNI space. The automated anatomical labeling (AAL) atlas [15], implemented in WFU PickAtlas [7] was used to parcellate the CBF map into 116 anatomical regions. A gray matter mask from VBM8 segmentation was applied to limit to the gray matter CBF voxels. Finally, the mean gray matter CBF of the voxels in each region was computed to form a vector containing the 116 mean regional CBF measures.

---

[1] http://dbm.neuro.uni-jena.de/vbm.html.

3.2   Identify candidate regions for further analysis

At this point the data consisted of 16 diabetes related predictors and 116 candidate regional CBF target measures. To prune the list of candidate regions a massive univariate approach was applied to the 152 subject cohort. In this approach 116 multiple linear regression models, each defined as $y = b_0 + b_1 x_1 + b_2 x_2 + ... + b_{16} x_{16}$ was fitted, where y is one of the regional CBF measures and x_i are the clinical measures. For each model, the coefficient of determination or R^2 score was computed to measure the goodness of fit while the probability of F-statistic, p(F-statistic), was computed to measure the significance of the regression model. Bonferroni multiple comparisons correction was applied, yielding a criterion for significance of $\alpha = 0.01/116 = 0.000086$.

Figure 1 shows the regions, sorted based on decreasing R^2 from the linear model fit to each region. The p(F-statistic) is also shown. 17 structures pass the significance test, p(F-statistic) < 0.000086. The most significant region is the right caudate with p(F-statistic) $= 1.16e - 07$ and $R^2 = 0.36$. This agrees with the finding in [4] of an association between diabetes duration and mean diffusivity in right caudate, discussed in Sect. 1. This analysis reveals the caudate as one of the structures significantly impacted by T2D, therefore in the following section we train a DNN to predict caudate CBF level from diabetes measures in order to quantify the association.

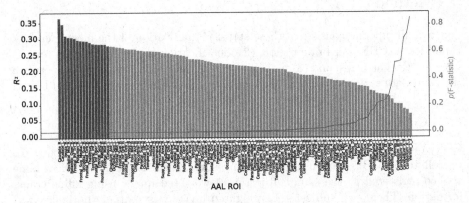

Fig. 1. Fitting 116 multivariable univariate linear models for prediction of perfusion in each AAL ROI from the 16 clinical features. The structures are ranked based on the model's R^2. Also the p(F-statistic) is shown in red. The green horizontal line indicates the significance threshold based on Bonferroni correction which has a height of $\alpha = 0.000086$. The significant regions are highlighted in green. (Color figure online)

3.3   Estimate Candidate Region Association Using a DNN

Subjects were ranked based on the perfusion of the CBF in the right caudate, then the top 30% and bottom 30% samples were considered for classification.

This resulted in 92 subjects: 46 with low and 46 with high CBF. Categorical features including education, sex, hypertension and amount of smoking were converted into numerical features. Each feature was scaled between zero and one. Several fully connected, DNN model architectures were evaluated to classify the caudate perfusion level. In each tested architecture, a strategy similar to the fully connected layers in AlexNet [6] and VGGNet [13], was chosen where the number of neurons is reduced in each successive hidden layer until the output layer. This allows a gradual build-up of a more and more abstract, high level features from lower level features. The rectified linear unit which is defined as $ReLU(z) = max(0, z)$ was applied as the activation function for each hidden layer neuron. A categorical output layer consisting of a single neuron per category was implemented via the softmax activation function defined as $S_j(z) = \frac{e^{z_j}}{\sum_{k=1}^{2} e^{z_k}}; j = 1, 2$. During training a batch size of 10 and learning rate of 0.001 was chosen based on empirical evidence. The ADAM optimization method [5] was used with $\beta_1 = 0.5$, $\beta_2 = 0.999$, and $\epsilon = 1e-08$ and weights were initialized to small random values near zero. In each validation test, early stopping with look ahead was employed, i.e. training was stopped when the network showed no improvement in validation accuracy for 15 epochs. To perform model selection, 72 subjects were randomly selected from the 92 to use as the training set while the remaining 20 subjects were held out as the test set and not used during model selection. Both training and test sets were balanced. The training set was further divided into training and validation via 5-fold cross-validation. The evaluated models and their average cross validation accuracy is shown in Table 2. These models include less deep architectures which underfit (model 16-8-2) and very deep architectures which overfit (model 16-16-8-8-8-4-4-4-2). The winning architecture and the DNN model that we propose is further illustrated in Fig. 2. The model contains 5 dense hidden layers, where 16 neurons were used in the first hidden layer, 8 neurons in the second and third and 4 neurons in the fourth and fifth layers. After selecting the proposed architecture, it was trained on the full training set and evaluated on the unseen held-out test set.

Table 2. Comparison of fully connected neural network architectures evaluated using 5-fold cross validation with 58 training subjects/fold and 14 test subjects/fold.

Model	5-fold cross-validation accuracy (%)
16-8-2	69.1 (±8.1)
16-16-8-2	75.9 (±14.9)
16-16-8-8-2	69.3 (±11.0)
16-16-8-8-4-2	74.8 (±8.9)
16-16-8-8-4-4-2	**76.3** (±9.9)
16-16-8-8-4-4-4-2	70.5 (±6.2)
16-16-8-8-8-4-4-4-2	65.0 (±9.7)

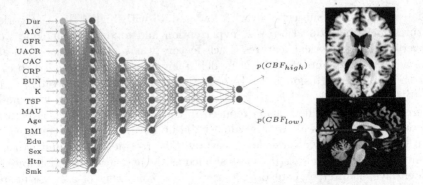

Fig. 2. Our proposed fully-connected deep neural network for the classification of caudate CBF into perfusion level based on clinical measures. Green neurons represent the input layer, while blue neurons constitute the hidden layers and red neurons are the output neurons which use softmax activation function to compute a categorical distribution. (Color figure online)

4 Results

4.1 Performance Comparison of the Learning Models

The proposed model achieves an accuracy of 90% with a sensitivity of 100% and specificity of 80%. Table 3 shows a comparison of the performance of the proposed model to widely used classical machine learning classifiers. The proposed DNN model outperforms the other algorithms in nearly all performance metrics. While the random forest had slightly higher specificity, it yielded inferior F1 score, AUC, accuracy, and sensitivity.

4.2 Statistical Significance of the Proposed Model

The null hypothesis was that the DNN cannot learn to predict the perfusion level based on the training set. The test statistic chosen was the accuracy on the unseen test set of 20 samples. The permutation testing procedure was as follows:

Table 3. Comparison of the performance of different classifiers on the held out test set. Each model is trained on 72 subjects and tested on 20 subjects.

Classifier	Accuracy (%)	Sensitivity (%)	Specificity (%)	AUC (%)	F1-score (%)
proposed DNN	**90.0**	**100.0**	80.0	**90.0**	**90.9**
Linear-SVM	80.0	90.0	70.0	80.0	81.8
RBF-SVM	80.0	90.0	70.0	80.0	81.8
Extra Trees	80.0	90.0	70.0	80.0	81.8
Random Forest	85.0	80.0	**90.0**	85.0	84.2
Adaboost	80.0	90.0	70.0	80.0	81.8
Gradboost	70.0	80.0	60.0	70.0	72.7

1. Repeat $R = 1000$ times:
   a. Randomly permute the N perfusion measures over the N diabetes feature vectors.
   b. Compute the value of the test statistic for the current permutation.
2. Construct an empirical probability distribution function (PDF) of the test statistic.
3. Compute the p-value of the test static without permutation.

The PDF for the accuracy test statistics are shown in Fig. 3. Upon evaluation the proposed model achieved statistically significant reliability; the probability of observing a classifier with higher accuracy than the proposed model is <1% ($p = 0.000999$). Thus with a significance level of $\alpha = 0.01$, we reject the null hypothesis in factor of the alternative hypothesis that the model has learned to predict the perfusion level with small expected error.

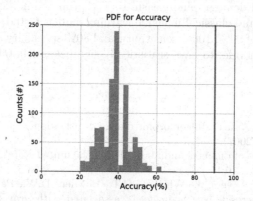

Fig. 3. Probability distribution function (PDF) from permutation analysis for the proposed DNN model. The red line indicates the classification accuracy obtained by the model.

5 Discussion

Our study found the caudate to be the structure whose blood perfusion is most impacted by diabetes. This is a noteworthy finding because the caudate is a structure vital for optimum brain health. The caudate is located within the dorsal striatum of the basal ganglia, and is associated with motor processes as well as cognitive functions including procedural learning and associative learning [8]. It is also one of the structures comprising the reward system [16].

Previous studies [4,10,12,17] have shown that the structure of the caudate nucleus, particularly the right caudate is significantly impacted by T2D. Peng et al. [10] reported a significant reduction of gray matter volume in the caudate in patients with T2D compared to normal controls. A similar study in pediatric population [12] showed caudate nucleus volume was significantly reduced in T2D patients compared to non-diabetic controls. Zhang et al. [17] found an association between higher plasma glucose (common in diabetics) and the shape of the caudate. Moreover, Hsu et al. [4] discovered a significant association between diabetes duration and white matter microstructural properties such as mean diffusivity in several brain regions including the right caudate. These complementary studies that associate caudate structural changes with T2D, corroborate our finding that T2D impacts blood perfusion in the caudate nucleus.

6   Conclusions

In this paper, we quantify the association between T2D-related measures and brain perfusion. We propose a fully connected deep neural network to classify the perfusion in caudate into low and high categories based on 16 diabetes, renal, cardiovascular and demographic measures. The proposed model outperforms all the deep learning and classical machine learning models tested, achieves a classification accuracy of 90%, 100% sensitivity, and 80% specificity, and permutation testing shows the model to have statistically significant reliability.

References

1. Ashburner, J.: A fast diffeomorphic image registration algorithm. Neuroimage **38**(1), 95–113 (2007)
2. CDC: Estimates of diabetes and its burden in the united states. National diabetes statistics report (2014)
3. Freedman, B.I., Divers, J., Whitlow, C.T., Bowden, D.W., Palmer, N.D., et al.: Subclinical atherosclerosis is inversely associated with gray matter volume in African Americans with type 2 diabetes. Diab. Care **38**(11), 2158–2165 (2015)
4. Hsu, J.L., Chen, Y.L., Leu, J.G., Jaw, F.S., Lee, C.H., Tsai, Y.F., Hsu, C.Y., Bai, C.H., Leemans, A.: Microstructural white matter abnormalities in type 2 diabetes mellitus: a diffusion tensor imaging study. NeuroImage **59**(2), 1098–1105 (2012)
5. Kingma, D., Ba, J.: Adam: a method for stochastic optimization. arXiv preprint arXiv:1412.6980 (2014)
6. Krizhevsky, A., Sutskever, I., Hinton, G.E.: Imagenet classification with deep convolutional neural networks. In: NIPS (2012)
7. Maldjian, J.A., Laurienti, P.J., Kraft, R.A., Burdette, J.H.: An automated method for neuroanatomic and cytoarchitectonic atlas-based interrogation of FMRI data sets. Neuroimage **19**(3), 1233–1239 (2003)
8. Malenka, R.C., Nestler, E., Hyman, S., Sydor, A., Brown, R., et al.: Molecular Neuropharmacology: A Foundation for Clinical Neuroscience. McGrawHill Medical, New York (2009)
9. Murea, M., Hsu, F.C., Cox, A.J., Hugenschmidt, C.E., Xu, J., Adams, J.N., et al.: Structural and functional assessment of the brain in European Americans with mild-to-moderate kidney disease: diabetes heart study-mind. Nephrol. Dial. Transplant. **30**(8), 1322–1329 (2015)
10. Peng, B., Chen, Z., Ma, L., Dai, Y.: Cerebral alterations of type 2 diabetes mellitus on MRI: a pilot study. Neurosci. Lett. **606**, 100–105 (2015)

11. Raffield, L.M., Cox, A.J., Freedman, B.I., Hugenschmidt, C.E., Hsu, F.C., et al.: Analysis of the relationships between type 2 diabetes status, glycemic control, and neuroimaging measures in the diabetes heart study mind. Acta Diabetologica **53**(3), 439–447 (2016)

12. Rofey, D.L., Arslanian, S.A., El Nokali, N.E., Verstynen, T., Watt, J.C., Black, J.J., et al.: Brain volume and white matter in youth with type 2 diabetes compared to obese and normal weight, non-diabetic peers: a pilot study. Int. J. Dev. Neurosci. **46**, 88–91 (2015)

13. Simonyan, K., Zisserman, A.: Very deep convolutional networks for large-scale image recognition. In: ICLR (2015)

14. Sink, K.M., Divers, J., Whitlow, C.T., Palmer, N.D., Smith, S.C., Xu, J., et al.: Cerebral structural changes in diabetic kidney disease: African American-diabetes heart study mind. Diab. Care **38**(2), 206–212 (2015)

15. Tzourio-Mazoyer, N., Landeau, B., Papathanassiou, D., Crivello, F., Etard, O., et al.: Automated anatomical labeling of activations in SPM using a macroscopic anatomical parcellation of the MNI MRI single-subject brain. Neuroimage **15**(1), 273–289 (2002)

16. Yager, L., Garcia, A., Wunsch, A., Ferguson, S.: The ins and outs of the striatum: role in drug addiction. Neuroscience **301**, 529–541 (2015)

17. Zhang, T., Shaw, M., Humphries, J., Sachdev, P., Anstey, K.J., Cherbuin, N.: Higher fasting plasma glucose is associated with striatal and hippocampal shape differences: the 2sweet project. BMJ Open Diab. Res. Care **4**(1), 1–8 (2016)

Multi-stage Diagnosis of Alzheimer's Disease with Incomplete Multimodal Data via Multi-task Deep Learning

Kim-Han Thung, Pew-Thian Yap, and Dinggang Shen[✉]

Department of Radiology and BRIC,
University of North Carolina, Chapel Hill, USA
dgshen@med.unc.edu

Abstract. Utilization of biomedical data from multiple modalities improves the diagnostic accuracy of neurodegenerative diseases. However, multi-modality data are often incomplete because not all data can be collected for every individual. When using such incomplete data for diagnosis, current approaches for addressing the problem of missing data, such as imputation, matrix completion and multi-task learning, implicitly assume linear data-to-label relationship, therefore limiting their performances. We thus propose multi-task deep learning for incomplete data, where prediction tasks that are associated with different modality combinations are learnt jointly to improve the performance of each task. Specifically, we devise a multi-input multi-output deep learning framework, and train our deep network subnet-wise, partially updating its weights based on the availability of modality data. The experimental results using the ADNI dataset show that our method outperforms the state-of-the-art methods.

1 Introduction

Neuroimaging technologies such as Magnetic Resonance Imaging (MRI) and Positron Emission Topography (PET) afford different views of the human brain, both in terms of structure and function. Utilizing neuroimaging data jointly with other non-imaging data can significantly improve diagnostic accuracy [13]. However, multimodal or longitudinal datasets are quite often incomplete because not all data can be collected for every individual. We will show in this work how the Alzheimer's Disease Neuroimaging Initiative (ADNI) dataset, which is incomplete, can be used effectively for diagnosis.

Though incomplete data can still be used by discarding incomplete samples, this will reduce the sample size. Doing so with the ADNI data will cause a loss of about 50% of the data. To use the data without discarding samples, one can impute the missing values using methods like k-nearest neighbor, expectation maximization, low-rank matrix completion (LRMC), etc. [2,11]. However, these

This work was supported in part by NIH grants NS093842, EB006733, EB008374, EB009634, EB022880, AG041721, MH100217, and AG042599.

M.J. Cardoso et al. (Eds.): DLMIA/ML-CDS 2017, LNCS 10553, pp. 160–168, 2017.
DOI: 10.1007/978-3-319-67558-9_19

imputation methods only work well if data are missing in random. This is seldom the case in clinical study design, e.g., data in ADNI are missing in blocks (e.g., an entire modality is missing). To tackle this problem, Goldberg et al. [4] propose to simultaneously impute the missing data and the diagnostic labels using low rank matrix completion. This approach is improved in [7] by selecting discriminative features and samples before matrix completion.

Another approach to tackle this problem is by using multi-task learning [16]. Specifically, this method first divides the incomplete dataset into several subsets of complete data with different combinations of modalities. Then, a classification model is trained for each subset, with all models trained jointly across subsets. This avoids the need to perform imputation and allows utilization of all available data. However, like the matrix completion method, this method still assumes *linear* relationship between the predictors (multimodal data) and the targets (class labels), therefore imposing a strong assumption on their relationships [13].

Deep learning can learn complex nonlinear mappings, but most of the deep learning models assume the data are complete. Though some models are applicable to incomplete data, almost all of them are based on data synthesis/imputation [5,6], which may not work well for large amount of missing data [7]. In this work, we show how one can use deep learning on incomplete multimodal data via multi-task learning, without involving data synthesis. Similar to [16], we associate a learning task with each data subset and learn these tasks jointly. More specifically, we design a multi-input multi-output deep learning framework, i.e., one input for one modality and one output for one learning task, and train the network iteratively and partially, based on data availability. The proposed method allows multi-class classification, in contrast to most current methods that focus on binary classification [1,7–10,15,16].

2   Method

2.1   Multi-task Learning

Figure 1 shows a typical multi-task learning (MTL) framework for dealing with classification problems involving incomplete multimodal data [16]. Without loss of generality, here we assume that there are two modalities, denoted by M1 and M2, which could be incomplete (i.e., some subjects do not have data from M1 or M2). Due to the missing data, we cannot build a prediction model directly. However, by grouping the data into subsets of complete data with different combinations of modalities, we can define a classification task for each of these subsets (right in Fig. 2). Yuan et al. [16] then learned these tasks jointly, by constraining all the tasks to select common features for each modality. However, there are several limitations in Yuan's method. First, it implicitly assumes linear data-to-label relationship. Second, it did not utilize the maximum number of samples for each task. For example, for Task 1 where the relationship between modality M1 and target Y is to be learned, only a portion of M1 samples is used (left in Fig. 1). Third, it uses original features of all the the modalities for classification, ignoring heterogeneity among modalities. Fourth, its current

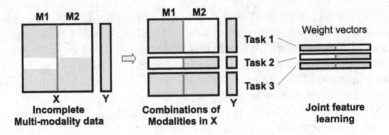

Fig. 1. Left: Original classification problem with incomplete multimodal data. Right: Multi-task learning for incomplete multimodal data. (X: input data; Y: target data; M1, M2: multimodal data; White: Missing data; Grey: Available data)

formulation only supports binary classification. Many of these weaknesses can be readily overcome by using deep learning.

2.2 Multi-task Deep Learning for Incomplete Multimodal Data

Figure 2 shows an overview of the proposed Multi-Task Deep Learning (MTDL) framework for two incomplete modalities. This is a multi-input multi-output deep learning framework, as we have two types of input (i.e., one for each modality), and three types of output (i.e., one for each classification task). The architecture consists of mainly three layers, i.e., input layer, hidden layers, and output layer. Each blue rectangular box denotes a set of neurons and the arrows denote the connections between the neurons. Note that there are two sets of neurons in the input layer, each of which receives input data from one modality. In addition, there are three sets of neurons in the output layer, which correspond to three different classification tasks, i.e., one for each combination of modalities. Each classification task is learned by using the maximum number of available samples, as shown on the right side of Fig. 2. There are two types of hidden layers in our framework, i.e., "modality-specific layers" and "task-specific layers", focusing on learning the latent data representation for specific modality and also learning the data-to-label relationship for specific classification task, respectively. Details on these two layers are described next.

Modality-Specific Layers. Heterogeneity is inevitable in multimodal data. In our case, we use MRI and PET data for information on structure and functional. The number of samples for each modality is different. PET data have far less number of samples compared with MRI data. Thus, to use the maximum number of samples to effectively learn the latent representation, i.e., high-level features for each modality, we adopt modality-specific hidden layers in our framework. The number of neurons and the number of layers can be different for different modalities.

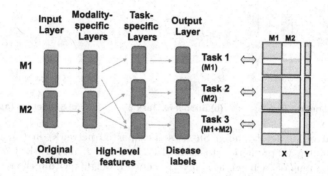

Fig. 2. The proposed multi-task deep learning (MTDL) via multi-input multi-output neural network.

Task-Specific Layers. A task is a classification problem associated with each modality combination. Using the example in Fig. 2, we have a total of three tasks, i.e., (Task 1) classification using only M1 data, (Task 2) classification using only M2 data, and (Task 3) classification using M1+M2 data. In "task-specific layers", we first concatenate the high-level features learned from each modality (via "modality-specific layers") based on the combination mentioned above for all the three classification tasks (Fig. 2). Then we learn the data-to-label mapping for each task in this layer.

Iterative Subnet Training. Training the network in Fig. 2 is not straightforward, as the number of samples differs across modalities. To maximally utilize all samples during training, we train the MTDL network in Fig. 2 iteratively and subnet-wise, as shown in Fig. 3. Again, using the example of Fig. 2, we first divide the data into three subsets, i.e., M1, M2, and M1+M2, each consists of maximum number of samples for each data combination of modalities. Then, in each training batch, only a certain part of the network (i.e., subnet) is updated, using the corresponding subset of data.

Implementation. We used Keras [3] to implement our framework. To regularize the network weights, we use $l_{2,1}$-norm plus l_2-norm for the first hidden layer, and use l_2-norm for other hidden layers. We use $l_{2,1}$-norm to regularize the network weights as it is useful for feature selection by imposing row sparsity on the network weight matrix. We use softmax activation function for the output layer, and use categorical crossentropy as the loss function for the output of each task. We also divided our data into ten parts according to [14], and then use eight parts for training, one part for validation, and the remaining one part for testing. For each subnet, we train it using all the available training samples that contain the needed modality combination for 200 epochs, and stop the training if the validation loss stops improving for 20 epochs. We then update the network using the network weights that is obtained when the validation loss is minimum, for each subnet training. We iteratively trained all the subnets for several iterations

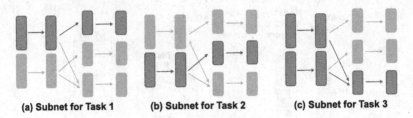

(a) Subnet for Task 1   **(b) Subnet for Task 2**   **(c) Subnet for Task 3**

Fig. 3. Iterative subnet training. Subnets (i.e., partial network) in (a), (b) and (c) are corresponding to prediction tasks of Task 1, Task 2, and Task 3, respectively. During the training of each subnet, only the network weights within the red blocks are updated, while the remaining network weights (in gray blocks) are frozen. The network in Fig. 2 is learned iteratively in the sequence of $\{(a) \to (b) \to (c) \to (a) \to \cdots\}$ until convergence.

(Fig. 3) until convergence condition (e.g., no improvement of loss minimization) or maximum number of iterations was met.

3   Materials, Preprocessing and Feature Extraction

In this study, we used MRI and PET data from the ADNI dataset[1] (MRI data are complete, while PET data are incomplete, at baseline) for multi-stage AD diagnosis, i.e., to differentiate between Normal Control (NC), Mild Impairment Cognitive (MCI - prodromal stage of AD) and Alzheimer's Disease (AD). In addition, to improve the diagnostic performance, we also included demographic information (i.e., age, gender, education) and genetic information (i.e., Apoe4). Moreover, we also used samples from all the time points for each training subject to significantly increase the number of samples for training. We summarize the demographic information of the subjects used in this study in Table 1. From the Table 1, we can see that 805 samples (here we define a sample as the data collected from one subject at a certain time point), which corresponded to 805 subjects used in this study, were collected at the baseline. After included the samples from all the time points for these 805 subjects, we ended up with a total of 4830 samples in this study.

We extracted the region-of-interest (ROI)-based features from the MRI and PET images by using the following processing pipeline. Each MR image was AC-PC aligned, N3 intensity inhomogeneity corrected, skull stripped, tissue segmented, and registered to a template with 93 ROIs [12]. We then used the normalized Gray matter (GM) volumes from the 93 ROIs as MRI features. We also linearly aligned each PET image to its corresponding MRI image, and used the mean intensity values of each ROI as a PET feature.

---

[1] http://adni.loni.ucla.edu.

Table 1. Demographic information of the subjects involved in this study. (Edu.: Education)

	# Subjects	Gender (M/F)	Age (years)	Edu. (years)
NC	226	118/108	75.8 ± 5.0	16.0 ± 2.9
MCI	393	253/140	74.9 ± 7.3	15.6 ± 3.0
AD	186	99/87	75.3 ± 7.6	14.7 ± 3.1
Total	805	470/335	-	-

4   Results and Discussions

We compared our proposed method with a Baseline method (i.e., sparse multi-label feature selection followed by a multi-class support vector classifier) that uses only the complete portion of the data, and two other popular methods that use incomplete multimodal data, i.e., low-rank matrix completion with sparse feature selection (LRMC) [7], and incomplete multi-source joint feature learning (iMSF) [16]. We used the standardized cross-validation indices in [14] to perform 10-fold cross validation experiment for all the comparison methods, and reported the mean accuracy as the evaluation metric. The accuracy of the classification is defined as the total number of correctly classified samples divided by the total number of samples.

We performed two experiments. The first experiment (i.e., Exp.1) used incomplete multimodal data for multi-stage AD diagnosis via the proposed multi-task deep learning method. The second experiment (i.e., Exp.2) used extra demographic information (i.e., age, gender, education and Apoe4). We appended this extra information to the MRI and PET data to increase their feature dimension to 97 (93 + 4) and use the same framework to solve the problem. Figure 4 shows the results of these two experiments using the proposed framework. The two bar groups on the left are for the first experiment, and the other two on the right are for the second experiment. There are 4 types of results for each group of bars, namely "MRI", "MRI+PET", "Overall-bl" and "Overall", which respectively denote the results for prediction task that uses only MRI data, prediction task that uses MRI and PET data, overall classification result (for all the prediction tasks) using all the baseline testing samples, and overall classification result (for all the prediction tasks) using all the testing samples at different time points. As can be seen, "MRI+PET" in general has higher classification accuracy than others, as it uses more information. To show that all the tasks in our framework are complementary to each other, and training them jointly and iteratively will improve each individual task, we also present the results of the first and last iterations, respectively denoted by "Initial" and "Final" in Fig. 4. As can be seen from Fig. 4, the final network in general has higher classification accuracy than the initial network, e.g., for the first experiment, the initial classification accuracy using MRI data is 55.5%, but after several iterations, it is improved to 58.0%. Similarly for the second experiment, the classification accuracy using

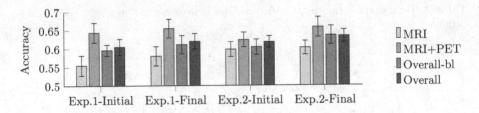

Fig. 4. Accuracy of multi-stage AD diagnosis using MRI and PET data for the proposed multi-task deep learning method. Exp.1 and Exp.2 denote the experiments that ignored and used the demographic information, respectively. Error bars: standard deviations.

MRI+PET data has improved from 62.4% to 65.8%. Also note that, the multi-stage classification results reported here is lower than the binary classification results reported in the literature [7], as multi-stage classification is much more challenging as it needs to differentiate the subtle differences among NC, MCI and early AD samples simultaneously.

In addition, we also compared our results with the state-of-the-art methods in Table 2, i.e., iMSF [16] and LRMC [7]. We modified the iMSF code to one-vs-others setting for multi-class classification. The baseline method can only work with complete data and therefore used a smaller sample size of data (marked with '*'). Its results are not comparable with those that use all the samples (i.e., column 3, 4 and 5 of Table 2). For comparison, we also included our results for these subset of samples in the last column of Table 2, and marked them with '*' as well. From the Table, it can be seen that our methods is better than the compared methods that using either incomplete or complete data. It is interesting that the iMSF algorithm, which also uses multi-task learning strategy, did not perform well for multi-stage classification. This is probably due to its use of linear regression model, which is unable to learn the complex data-to-label relationship in this multi-stage AD study. LRMC is the best performing method after MTDL. It is noteworthy that MTDL can be easily extended to use more complicated neuroimaging data, e.g., the raw images, by simply replacing the fully connected modality-specific layers with the convolutional layers.

Table 2. Comparison with the baseline, iMSF [16] and LRMC [7]. **Bold**: Best results, *: Complete data results (less number of samples)

Data	Baseline	iMSF	LRMC	MTDL	
MRI	55.1	54.4	56.9	**58.0**	
MRI+PET	56.9*	54.0	58.5	**61.1**	65.6*
MRI+PET+DEM	62.0*	58.3	61.9	**63.6**	65.8*

5   Conclusion

We have demonstrated how a deep neural network can be applied on incomplete multimodal data by using a multi-task learning strategy, so that no data synthesis is needed. To this end, we have devised a multi-input multi-output deep neural network, which can be trained using an iterative subnet training strategy. The experiment results show that different classification tasks can benefit from each other during training. In addition, the results also show that the proposed method (MTDL) outperforms other compared methods, such as LRMC and iMSF that were designed for incomplete multimodal data. Moreover, our framework can be easily extended to use more complicated imaging data by replacing the full-connectivity network in the modality-specific layers with the convolutional layers.

References

1. Adeli-Mosabbeb, E., et al.: Robust feature-sample linear discriminant analysis for brain disorders diagnosis. In: Advances in Neural Information Processing Systems, pp. 658–666 (2015)
2. Candès, E.J., et al.: Exact matrix completion via convex optimization. Found. Comput. Math. **9**(6), 717–772 (2009)
3. Chollet, F.: Keras (2015). https://github.com/fchollet/keras
4. Goldberg, A., et al.: Transduction with matrix completion: three birds with one stone. Adv. Neural Inf. Process. Syst. **23**, 757–765 (2010)
5. Li, F., et al.: A robust deep model for improved classification of AD/MCI patients. IEEE J. Biomed. Health Inf. **19**(5), 1610–1616 (2015)
6. Liu, S., et al.: Multimodal neuroimaging feature learning for multiclass diagnosis of Alzheimer's disease. IEEE Trans. Biomed. Eng. **62**(4), 1132–1140 (2015)
7. Thung, K.H., et al.: Neurodegenerative disease diagnosis using incomplete multi-modality data via matrix shrinkage and completion. Neuroimage **91**, 386–400 (2014)
8. Thung, K.H., et al.: Identification of progressive mild cognitive impairment patients using incomplete longitudinal MRI scans. Brain Struct. Funct. **221**(8), 3979–3995 (2016)
9. Thung, K.-H., Yap, P.-T., Adeli-M, E., Shen, D.: Joint diagnosis and conversion time prediction of progressive mild cognitive impairment (pMCI) using low-rank subspace clustering and matrix completion. In: Navab, N., Hornegger, J., Wells, W.M., Frangi, A.F. (eds.) MICCAI 2015. LNCS, vol. 9351, pp. 527–534. Springer, Cham (2015). doi:10.1007/978-3-319-24574-4_63
10. Thung, K.-H., Adeli, E., Yap, P.-T., Shen, D.: Stability-weighted matrix completion of incomplete multi-modal data for disease diagnosis. In: Ourselin, S., Joskowicz, L., Sabuncu, M.R., Unal, G., Wells, W. (eds.) MICCAI 2016. LNCS, vol. 9901, pp. 88–96. Springer, Cham (2016). doi:10.1007/978-3-319-46723-8_11
11. Troyanskaya, O., et al.: Missing value estimation methods for DNA microarrays. Bioinformatics **17**(6), 520–525 (2001)
12. Wang, Y., Nie, J., Yap, P.-T., Shi, F., Guo, L., Shen, D.: Robust deformable-surface-based skull-stripping for large-scale studies. In: Fichtinger, G., Martel, A., Peters, T. (eds.) MICCAI 2011. LNCS, vol. 6893, pp. 635–642. Springer, Heidelberg (2011). doi:10.1007/978-3-642-23626-6_78

13. Weiner, M.W., et al.: The Alzheimer's disease neuroimaging initiative: a review of papers published since its inception. Alzheimer's Dementia **9**(5), e111–e194 (2013)
14. Wyman, B.T., et al.: Standardization of analysis sets for reporting results from ADNI MRI data. Alzheimer's Dementia **9**(3), 332–337 (2013)
15. Yu, G., et al.: Multi-task linear programming discriminant analysis for the identification of progressive MCI individuals. PLoS ONE **9**(5), e96458 (2014)
16. Yuan, L., et al.: Multi-source feature learning for joint analysis of incomplete multiple heterogeneous neuroimaging data. NeuroImage **61**(3), 622–632 (2012)

A Multi-scale CNN and Curriculum Learning Strategy for Mammogram Classification

William Lotter[1,2(✉)], Greg Sorensen[2], and David Cox[1,2]

[1] Harvard University, Cambridge, MA, USA
[2] DeepHealth Inc., Cambridge, MA, USA
{lotter,sorensen,davidcox}@deephealth.io

Abstract. Screening mammography is an important front-line tool for the early detection of breast cancer, and some 39 million exams are conducted each year in the United States alone. Here, we describe a multi-scale convolutional neural network (CNN) trained with a curriculum learning strategy that achieves high levels of accuracy in classifying mammograms. Specifically, we first train CNN-based patch classifiers on segmentation masks of lesions in mammograms, and then use the learned features to initialize a scanning-based model that renders a decision on the whole image, trained end-to-end on outcome data. We demonstrate that our approach effectively handles the "needle in a haystack" nature of full-image mammogram classification, achieving 0.92 AUROC on the DDSM dataset.

1 Introduction

Roughly one eighth of women in the United States will develop breast cancer during their lifetimes [26]. Early intervention is critical—five-year relative survival rates can be up to 3–4 times higher for cancers detected at an early stage versus at a later stage [5]. An important tool for early detection is screening mammography, which has been attributed with a significant reduction in breast cancer mortality [3]. However, the overall value of screening mammography is limited by several factors: reading mammograms is a tedious and error-prone process, and not all radiologists achieve uniformly high levels of accuracy [9,15]. In particular, empirically high false positive rates in screening mammography can lead to significant unnecessary cost and patient stress [4,19]. For these reasons, effective machine vision-based solutions for reading mammograms hold significant potential to improve patient outcomes.

Traditional computer-aided diagnosis (CAD) systems for mammography have typically relied on hand-engineered features [20]. With the recent success of deep learning in other fields, there have been several promising attempts to apply these techniques to mammography [1,6–8,10,14,16,17,27,29]. Many of these approaches have been designed for specific tasks or subtasks of a full evaluation pipeline, for instance, mass segmentation [7,30] or region-of-interest (ROI) microcalcification classification [17]. Here, we address the full problem of binary cancer status classification: given an entire mammogram image, we seek

M.J. Cardoso et al. (Eds.): DLMIA/ML-CDS 2017, LNCS 10553, pp. 169–177, 2017.
DOI: 10.1007/978-3-319-67558-9_20

to classify whether cancer is present [6,14,29]. As recent efforts have shown [10], creating an effective end-to-end differentiable model, the cornerstone of supervised deep learning, is challenging given the "needle in a haystack" nature of the problem. To address this challenge, we have developed a two-stage, curriculum learning-based [2] approach. We first train patch-level CNN classifiers at multiple scales, which are then used as feature extractors in a sliding-window fashion to build an image-level model. Initialized with the patch-trained weights, the image-level model can then effectively be trained end-to-end. We demonstrate the efficacy of our approach on the largest public mammography database, the Digital Database for Screening Mammography (DDSM) [12]. Evaluated against the final pathology outcomes, we achieve an AUROC of 0.92.

2   Multi-scale CNN with Curriculum Learning Strategy

Figure 1 shows typical examples of the two most common classes of lesions found in mammograms, masses and calcifications. Segmentation masks drawn by radiologists are shown in red [12]. Even though the masks often encompass the surrounding tissue, the median size is only around 0.5% of the entire image area for calcifications, and 1.2% for masses. The insets shown in Fig. 1 illustrate the high level of fine detail required for detection. As noted in [10], the requirement to find small, subtle features in high resolution images (e.g. ~5500 × 3000 pixels in the DDSM dataset) means that the standard practice of downsampling images, which has proven effective in working with many standard natural image

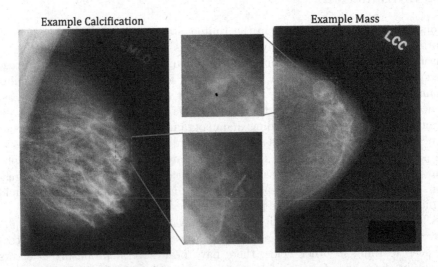

Fig. 1. Examples of the two most common categories of lesions in mammograms, calcifications and masses, from the DDSM dataset [12]. Radiologist-annotated segmentation masks are shown in red. The examples were chosen such that the mask sizes approximately match the median sizes in the dataset. (Color figure online)

datasets [22], is unlikely to be successful for mammograms. It is for these reasons that traditional mammogram classification pipelines typically consist of a sequence of steps, such as candidate ROI proposals, followed by feature extraction, perhaps segmentation, and finally classification [14]. While deep learning could in principle be used for any or all of these individual pieces, a variety of studies have suggested that the greatest gains from deep learning are seen when the system is trained "end-to-end" such that errors are backpropagated uninterrupted through the entire pipeline.

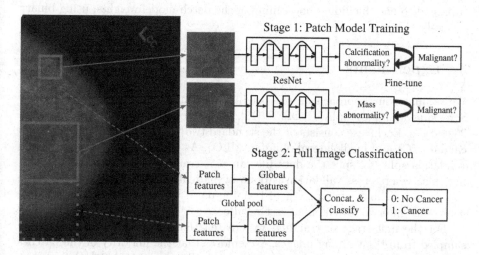

Fig. 2. Schematic of our approach, which consists of first training a patch classifier, followed by image-level training using a scanning window scheme. We train separate patch classifiers for calcifications and masses, at different scales, using a form of a ResNet CNN [11,28]. For image-level training, we globally pool the last ResNet feature layer at each scale, followed by concatenation and classification, with end-to-end training on binary cancer labels.

Our training strategy is illustrated in Fig. 2. The first stage consists of training a classifier to estimate the probability of the presence of a lesion in a given image patch. For training, we randomly sample patches from a set of training images, relying on (noisy) segmentation maps to create labels for each patch. Given the different typical scales of calcifications and masses, we train a separate classifier for each. We use ResNet CNNs [11] for the classifiers, specifically using the "Wide ResNet" formulation [28]. We first train for abnormality detection (i.e. is there a lesion present), followed by fine-tuning for pathology-determined malignancy.

The second stage of our approach consists of image-level training. Given the high level of scrutiny that is needed to detect lesions, and because of its compatibility with end-to-end training, we use a scanning-window (e.g. convolutional) scheme. We partition the image into a set of patches such that each patch is contained entirely within the image, and the image is completely tiled, but there is as minimal overlap and number of patches as possible. Features are extracted for each patch using the last layer before classification of the patch model. Final classification involves aggregation across patches and the two scales, for which we tried various pooling methods and number of fully-connected layers. The strategy that ultimately performed best, as assessed on the validation data, was using global average pooling across patch features at each scale, followed by concatenation and a single softmax classification layer. Using a model of this form, we train end-to-end, including fine-tuning of the patch model weights, using binary image-level labels.

3   Experiments

We evaluate our approach on the original version of the Digital Database for Screening Mammography (DDSM) [12], which consists of 10480 images from 2620 cases. Each case consists of the standard two views for each breast, craniocaudal (CC) and mediolateral-oblique (MLO). As there is not a standard cross validation split, we split the data into an 87% training/5% validation/8% testing split, where cross validation is done by patient. The split percentages were chosen to maximize the amount of training data, while ensuring an acceptable confidence interval for the final test results.

For the first stage of training, we create a large dataset of image patches sampled from the training images. We enforce that the majority of the patches come from the breast, by first segmenting using Otsu's method [21]. Before sampling, we resize the original images with different factors for calcification and mass patches. Instead of resizing to a fixed size, which would cause distortions because the aspect ratio varies over the images in the dataset, or cropping, which could cause a loss of information, we resize such that the resulting image falls within a particular range. We set the target size to 2750×1500 and 1100×600 pixels, for the calcification and mass scales, respectively. Given an input image, we calculate a range of allowable resize factors as the min and max resize factors over the two dimensions. That is, given an example of size, say 3000×2000, the range of resize factors for the calcification scale would be $[1500/2000 = 0.75, 2750/3000 = 0.92]$, from which we sample uniformly. For other sources of data augmentation, we use horizontal flipping, rotation of up to $30°$, and an additional rescaling by a factor chosen between 0.75 and 1.25. We then sample patches of size 256×256. In the first stage of patch classification training, lesion detection without malignancy classification, we create 800 K patches for each lesion category, split equally between positive and negative samples. In the second stage, we create 900 K patches split equally between normal, benign, and malignant.

As mentioned above, we use ResNets [11] for the patch classifiers. We specifically use the "Wide ResNet" formulation [28], although, for the sake of training speed and to avoid overfitting, our networks are not particularly wide. The Wide ResNet consists of groups of convolutional blocks with residual connections, and

2×2 average pooling between the groups. Each convolution in a block is proceeded by batch normalization [13] followed by ReLU activation. After the final group, features are globally average pooled, followed by a single classification layer. The main hyperparameters of the model are the number of filters per layer and the number of residual blocks per group, N. For our models, we use five groups with the number of filters per group of (16, 32, 48, 64, 96) and an N of 2 and 4 for the calcification and mass models, respectively. For more details of the architecture, the reader is referred to [28]. The only deviation we make is using 5×5 convolutions with a 3×3 stride for the initial convolutional layer, accounting for the relatively large input size we use of 256×256.

For training the patch models, we use RMSProp [25] with a learning rate of 2×10^{-4} and batch size of 32. In the initial abnormality detection stage, we train for 50 epochs with 10 K patch samples per epoch and an equal proportion of positive and negative samples, followed by 125 epochs with 15 K per epoch and a positive sample rate of 25%. We then fine-tune for malignancy. For the calcification model, we use a normal/benign/malignant labeling scheme. We train for 225 epochs with 15 K samples per epoch, and an equal proportion of the three classes. The three-way labeling scheme caused overfitting with the mass model, so we instead use binary malignant/non-malignant labels. The model is trained for 150 epochs with a sampling ratio of 20% normal/40% benign/40% malignant. To illustrate the information learned by the patch classifiers, we show several of the highest scoring patches for malignancy in the test set in Fig. 3.

For the image-level training, we initialize the model with the final patch weights and follow a similar image resizing scheme. We again augment using

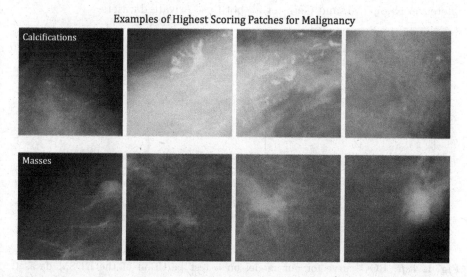

Fig. 3. Examples of the highest scoring patches for malignancy in the test set. The top row corresponds to the calcification model and the bottom row corresponds to the mass model.

horizontal reflections and an additional rescaling by a factor chosen between 0.8 to 1.2. At each scale, we divide the image into 256×256 patches, using the stride strategy explained earlier. We also keep track of the regions of overlap between patches, and normalize these areas when global pooling, since otherwise the final features would be biased towards these locations. For image-level labels, we categorize according to if there is a malignant lesion in either view of the breast. Due to the possible different number of patches per image and because of the high memory footprint, we use a batch size of 1 during training. We train for $100\,\mathrm{K}$ iterations using RMSProp [25] with a learning rate of 2×10^{-4}, followed by 4×10^{-5} for $50\,\mathrm{K}$ iterations. Final weights were chosen by monitoring the area-under-the-curve (AUC) for a receiver operating characteristic (ROC) plot on the validation set. While we train on a per image basis, we report final results on a (patient, laterality) basis by averaging final scores across the CC and MLO views of the breast. For final test results, we average predictions across both horizontal orientations and five resize factors, chosen equally spaced between the allowable factors per image.

Figure 4 contains the ROC curve on the test set for our proposed model. We obtain standard deviation error bars using a bootstrapping estimate. Our model achieves an AUC of 0.92 ± 0.02. As full image mammogram classification lacks a standardized evaluation framework, it is somewhat difficult to directly compare our results to other work. Related papers include the work of Carneiro et al. [6], Zhu et al. [29], Kooi et al. [14], and Geras et al. [10]. Carneiro et al. use the InBreast [18] dataset and a different version of DDSM, and use radiologist segmentation masks as input into their final model. Zhu et al. use InBreast for mass classification, and do not rely on segmentation masks for training or inference. Kooi et al. and Geras et al. both use private datasets.

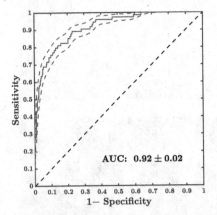

	Pre-Training	Data Augment.	AUC
Multi-scale CNN	DDSM lesions	size, flips	0.92 ± 0.02
Multi-scale CNN	DDSM lesions	flips	0.89 ± 0.02
Multi-scale CNN	none	flips	0.65 ± 0.04
InceptionV3	ImageNet	flips	0.77 ± 0.03
InceptionV3	none	flips	0.59 ± 0.04

Fig. 4. *Left*: ROC curve for our model on a test partition of the DDSM dataset. Predictions and ground-truth are compared at a breast-level basis. *Below*: ROC AUC by pre-training and data augmentation. InceptionV3 assumes a fixed input size, so "size" augmentation, i.e. random input image resizing, isn't directly feasible. ROC curve on left corresponds to the top row.

To provide some form of a meaningful comparison to our model, we report results here using a CNN designed for ImageNet classification. We use the InceptionV3 model [23,24], choosing this model over alternatives because its input size is relatively large at 299×299. Because InceptionV3 is designed for a fixed input size, training with resizing augmentation isn't feasible, but we do train with horizontal flip augmentation. Consistent with many other results in the literature [6,29], we find that ImageNet pre-training of InceptionV3 helps for eventual training on the DDSM dataset. However, the InceptionV3 model still underperforms our model, achieving an AUC of 0.77 ± 0.03. Without ImageNet pre-training, the InceptionV3 model achieves an AUC of 0.59 ± 0.04. In both cases, results are still reported on a (patient, laterality) basis with averaging across views and possible horizontal orientations. A summary of the results is contained in the table by Fig. 4. To make a more controlled comparison, we also report the results for our model without size augmentation training. The performance drops slightly to 0.89 ± 0.02, but is still significantly higher than the InceptionV3 model. The third row of the table also contains results for our model without the DDSM lesion pre-training, which decreases performance to 0.65 ± 0.04, however the model still performs slightly better than the InceptionV3 model without pre-training (last row). Altogether, these results suggest that all elements of our approach — including model formulation and pre-training scheme — are important for accurate full image mammogram classification performance.

4   Conclusions

Computer-aided diagnosis for mammography is a heavily studied problem given its potential for large real-world impact. This field, like many others, is transitioning from hand-engineered features to features learned in a deep learning framework. While there have been many efforts to apply deep learning to subcomponents of the mammography pipeline, here we are concerned with full image classification. Given the high resolution and relatively small ROIs, effectively designing an end-to-end solution is challenging. We have presented a multi-scale CNN scanning window scheme with a lesion-specific curriculum learning strategy that achieves promising results. Our approach performs significantly better than standard "out of the box" CNN models, and our experiments show that both the choice of architecture and training scheme play an important role in achieving this performance. Future work includes learning interest points in a more unsupervised way, to reduce reliance on hand-drawn segmentation masks. Overall, we argue that mammogram classification is a task that is well matched to the capabilities of modern deep learning approaches, and that it can serve as a natural testbed for the development of new deep learning techniques in the context of an application with clear potential for societal impact.

References

1. Arevalo, J., Gonzalez, F., Ramos-Pollan, R., et al.: Representation learning for mammography mass lesion classification with convolutional neural networks. Comput. Methods Programs Biomed. **127**, 248–257 (2016)
2. Bengio, Y., Louradour, J., Collobert, R., et al.: Curriculum learning. In: ICML (2009)
3. Berry, D.A., Cronin, K.A., Plevritis, S.K., et al.: Effect of screening and adjuvant therapy on mortality from breast cancer. In: NEJM (2005)
4. Brewer, N.T., Salz, T., Lillie, S.E.: Systematic review: the long-term effects of false-positive mammograms. Ann. Internal Med. **146**(7), 502–510 (2007)
5. Cancer Stat Facts: Female breast cancer. https://seer.cancer.gov/
6. Carneiro, G., Nascimento, J., Bradley, A.P.: Unregistered multiview mammogram analysis with pre-trained deep learning models. In: Navab, N., Hornegger, J., Wells, W.M., Frangi, A.F. (eds.) MICCAI 2015. LNCS, vol. 9351, pp. 652–660. Springer, Cham (2015). doi:10.1007/978-3-319-24574-4_78
7. Dhungel, N., Carneiro, G., Bradley, A.P.: Deep structured learning for mass segmentation from mammograms. arXiv:1410.7454 (2014)
8. Dhungel, N., Carneiro, G., Bradley, A.P.: Automated mass detection in mammograms using cascaded deep learning and random forests. In: DICTA (2015)
9. Elmore, J.G., Jackson, S.L., Abraham, L., et al.: Variability in interpretive performance at screening mammography and radiologists characteristics associated with accuracy. Radiology **253**(3), 587–589 (2009)
10. Geras, K.J., Wolfson, S., Kim, S.G., et al.: High-resolution breast cancer screening with multi-view deep convolutional neural networks. arXiv:1703.07047 (2017)
11. He, K., Zhang, X., Ren, X., Sun, J.: Deep residual learning for image recognition. arXiv:1512.03385 (2012)
12. Heath, M., Bowyer, K., Kopans, D., et al.: The digital database for screening mammography. In: Proceedings of the Fifth International Workshop on Digital Mammography (2001)
13. Ioffe, S., Szegedy, C.: Batch normalization: accelerating deep network training by reducing internal covariate shift. arXiv:1502.03167 (2015)
14. Kooi, T., Litjens, G., van Ginneken, B., et al.: Large scale deep learning for computer aided detection of mammographic lesions. Med. Image Anal. **35**, 303–312 (2017)
15. Lehman, C., Arao, R., Sprague, B., et al.: National performance benchmarks for modern screening digital mammography: update from the breast cancer surveillance consortium. Radiology **283**(1), 49–58 (2017)
16. Levy, D., Jain, A.: Breast mass classification from mammograms using deep convolutional neural networks. arXiv:1612.00542 (2016)
17. Mordang, J.-J., Janssen, T., Bria, A., Kooi, T., Gubern-Mérida, A., Karssemeijer, N.: Automatic microcalcification detection in multi-vendor mammography using convolutional neural networks. In: Tingberg, A., Lång, K., Timberg, P. (eds.) IWDM 2016. LNCS, vol. 9699, pp. 35–42. Springer, Cham (2016). doi:10.1007/978-3-319-41546-8_5
18. Moreira, I.C., Amaral, I., Domingues, I., et al.: INbreast: toward a full-field digital mammographic database. Acad. Radiol. **19**(2), 236–248 (2012)
19. Myers, E.R., Moorman, P., Gierisch, J.M., et al.: Benefits and harms of breast cancer screening: a systematic review. JAMA **314**(15), 1615–1634 (2015)

20. Nishikawa, R.M.: Current status and future directions of computer-aided diagnosis in mammography. Comput. Med. Imaging Graph. **31**, 224–235 (2007)
21. Otsu, N.: A threshold selection method from gray-level histograms. IEEE Trans. Syst. Man Cybern. **9**(1), 62–66 (1979)
22. Russakovsky, O., Deng, J., Su, H., et al.: ImageNet large scale visual recognition challenge. arXiv:1409.0575 (2014)
23. Szegedy, C., Liu, W., et al.: Going deeper with convolutions. In: CVPR (2015)
24. Szegedy, C., Vanhoucke, V., Ioffe, S., et al.: Rethinking the inception architecture for computer vision. arXiv:1512.00567 (2015)
25. Tieleman, T., Hinton, G.: Lecture 6.5 - rmsprop, coursera. Neural networks for machine learning (2012). http://www.cs.toronto.edu/~tijmen/csc321/slides/lecture_slides_lec6.pdf
26. U.S. Breast Cancer Statistics. http://www.breastcancer.org/
27. Yi, D., Sawyer, R.L., Cohn III., D., et al.: Optimizing and visualizing deep learning for benign/malignant classification in breast tumors. arXiv:1705.06362 (2017)
28. Zagoruyko, S., Komodakis, N.: Wide residual networks. arXiv:1605.07146 (2016)
29. Zhu, W., Lou, Q., Vang, Y.S., Xie, X.: Deep multi-instance networks with sparse label assignment for whole mammogram classification. arXiv:1612.05968 (2016)
30. Zhu, W., Xie, X.: Adversarial deep structural networks for mammographic mass segmentation. arXiv:1612.05970 (2016)

Analyzing Microscopic Images of Peripheral Blood Smear Using Deep Learning

Dheeraj Mundhra[✉], Bharath Cheluvaraju, Jaiprasad Rampure,
and Tathagato Rai Dastidar

SigTuple Technologies Pvt. Ltd., Bangalore, India
{dmundhra,bharath,jaiprasad,trd}@sigtuple.com
https://www.sigtuple.com

Abstract. This paper presents a new automated peripheral blood smear analysis system, *Shonit™* [1]. It consists of an automated microscope for capturing microscopic images of a blood sample, and a software component for analysis of the images. The software component employs an ensemble of deep learning models to analyze peripheral blood smear images for localization and classification of the three major blood cell types (red blood cells, white blood cells and platelets) and their subtypes [2]. We present the results of the classification and segmentation on a large variety of blood samples. The specificity and sensitivity of identification for the common cell types were above 98% and 91% respectively. The primary advantage of *Shonit™* over other automated blood smear analysis systems [3–5] is its robustness to quality variation in the blood smears, and the low cost of its image capture device.

1 Introduction

Manual microscopic review of peripheral blood smear (PBS) is still considered as a gold standard for detecting several haematological disorders [2]. The process involves counting different types of cells, observing morphological abnormalities, etc. The manual classification technique is error prone and labourious. Consequently, automating the process, which enhances reproducability of the results and reduces cost, is desirable. *Shonit™* is inspired by such a vision.

Peripheral blood smear (PBS) consists primarily of three cell types – RBC (red blood cell or erythrocyte), WBC (white blood cell or leukocyte) and platelet (or thrombocyte). Each of these primary classes have sub-classes. The interested reader is directed to [2] for more details on PBS analysis. Since manual analysis is laborious and error prone, attempts have been made to automate the process. Existing automatic systems [3–6] use different form of image based or flow cytometry techniques, not all of which are published. [3] uses Artificial Neural Networks (ANN) for classification of blood cells.

Shonit™ aims to automate the process of PBS analysis and provide quantitative metrics which are difficult to calculate manually. It consists of a hardware and a software component. The hardware component – a low cost automated microscope – is used to capture images of the PBS. The software component analyzes these images. It performs the following functions:

© Springer International Publishing AG 2017
M.J. Cardoso et al. (Eds.): DLMIA/ML-CDS 2017, LNCS 10553, pp. 178–185, 2017.
DOI: 10.1007/978-3-319-67558-9_21

- Localizes and classifies *all* WBCs visible in the images and computes a *differential count* of the WBC subtypes.
- Localizes and classifies thousands of RBCs and platelets.
- It also computes a differential count of the RBC subtypes. This is typically not done in the manual process as RBCs are too high in number for manual counting. Yet, this metrics has medical significance for certain types of diseases like anemia [2].

Shonit^{TM} uses an ensemble of deep learning techniques for the localization and classification tasks. To the best of our knowledge, this is the first attempt to use a deep learning network (U-net [7]) towards object localization in PBS images. The advantages of *Shonit*^{TM} over the existing PBS analysis systems are:

- It is able to analyze smears prepared both manually and through a machine. Existing systems either rely on automated smears only [3], or provide their own system for creating the blood smears [4].
- The cost of its hardware component – used to create the digital images of PBS – is extremely low compared to the existing systems.

The paper is organized as follows: Sect. 2 describes the overall functioning of the system. Section 3 gives details of the methods used. Experimental results are presented in Sect. 4. Finally, Sect. 5 concludes the paper.

2   The *Shonit*^{TM} System for Analysis of Peripheral Blood Smears

Shonit^{TM} has a hardware and a software component. The hardware part consists of an automated microscope. It is built from a standard light microscope (currently a Labomed LX500 [8]), fitted with robotic attachments which automate the movement of the stage (the platform on which the slide is placed). Images are captured through a cell phone camera (currently an iPhone-6s [9]) attached to the binocular eyepiece of the microscope. The cell phone doubles up as the controller of the robotic components. It controls the movement and focus.

Multiple images (currently, in excess of 120) are captured from different parts of the smear at a magnification of 400X. The scanning software automatically recognizes suitable (manually analyzable) areas of the smear and avoids others. Typically, images are captured only from those areas of the smear where the RBCs are just touching each other, and there are limited overlapping cells [2].

The captured images are transferred to a compute cloud hosting the software component – an artificial intelligence (AI) based platform which analyses these images. All WBCs visible in the captured images are classified. Thousands of RBCs (approximately 30,000) and platelets (approximately 5,000) are also extracted and classified into different categories.

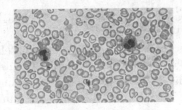

Fig. 1. A portion of a field of view (FOV) showing all three types of cells (Color figure online)

3 Deep Learning Techniques for Analyzing PBS Images

An example field of view (FOV) captured through the microscope is shown in Fig. 1. The large blue cells are WBCs, the small blue dot-like objects are platelets, and the remaining are RBCs. As can be seen, a typical image has a multitude of cells. Thus, different cell types need to be first *localized* before they can be classified. The analysis proceeds in two separate steps: an extraction step where cells of the three major types are separately localized, followed by a classification step where they are classified into subtypes.

3.1 Cell Extraction

In recent times, convolution neural network (CNN) based simultaneous localization and classification techniques have gained prominence [10,11]. However, such techniques are difficult to apply for this problem as most of them employ a pretrained classification model, typically one trained on the ImageNet dataset [12].

Extraction of WBCs and Platelets. As can be seen in Fig. 1, WBCs and platelets have a characteristic dark blue or purple color, which is caused by the *stain* – a chemical applied on the raw blood to get the coloration [2]. Naive methods based on thresholds – either fixed or adaptive – are inaccurate due to variation in the colouration of manual stains.

To overcome this problem, we employ the U-net deep learning architecture [7], which has shown good results for cell segmentation, for WBC and platelet segmentation. A deep learning model understands the "context" in which an object occurs, and thus is expected to perform better than the naive approach.

U-net Training. We created a small training set of 300 images where the WBCs and platelets were segmented manually by an expert. Image size was kept small (128×128 pixels for WBC and 32×32 for platelets), with typically only one object of interest at the center of the image. Using this training data, we trained a U-net like model with 4 convolutional layers in each arm of the 'U'. We obtained a intersection-over-union score of 0.93 on the held out validation data. Though U-net provides exact segmentation masks, we use it as a technique to localize WBC.

An example of detected WBCs in a very lightly stained image is shown in Fig. 2. Notice the faded out color of the WBCs compared to Fig. 1. The model was not confused by the varying coloration of the WBCs. Threshold based techniques typically fail on such images.

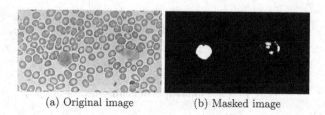

(a) Original image   (b) Masked image

Fig. 2. Example of WBC extraction through U-net model on a lightly stained image (Color figure online)

RBC Extraction. RBC extraction, on the other hand, is more effectively done through image processing alone. RBCs are much more numerous than WBCs, and it is not necessary to localize each one of them – a random sample suffices.

We use a method similar to [13] to extract RBCs from the images. It works on the green channel of the RGB image, and relies on an Otsu's thresholding [14] based method to separate the white background from the pinkish foreground. We then apply thresholds on the minimum and maximum diameter of an RBC to reject small artifacts and clumped RBCs respectively.

3.2   Cell Classification

As described in the previous section, three different extractors are employed to extract cell candidates for each of the major cell types. The candidates are extracted as small patches (128 × 128 for WBC, 64 × 64 for RBC and 48 × 48 for platelet) with the object of interest at the center of the patch. Separate extractors are used for each major cell type. Each extractor can potentially extract objects from other classes too. We next employ an ensemble of CNNs to classify each of the major classes into subtypes and reject the artifacts which were extracted. Artifacts include objects of other classes and other things like stain smudges, foreign objects, etc.

The subtypes of each major cell type are as follows:

– **WBC:** The subtypes are neutrophil, lymphocyte, monocyte, eosinophil, basophil, and a class encompassing all atypical cells like immature granulocytes, blasts, etc [2]. Examples are shown in Fig. 3.
– **RBC:** The subtypes are normocyte, microcyte, macrocyte, elliptocyte, teardrop, target, echinocyte, fragmented and an 'invalid' class which is used to

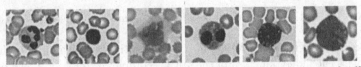

Neutrophil Lymphocyte Monocyte Eosinophil Basophil Atypical cell

Fig. 3. Different WBC subtypes

Round Elliptocyte Echinocyte Target Teardrop Fragmented Invalid

Fig. 4. Different RBC subtypes

reject artifacts, clumped or overlapped cells, degenerated cells, etc. The first three types are round in shape and are differentiated by size alone. Examples are shown in Fig. 4.

- **Platelet:** We do not subclassify platelets, but use a model to distinguish between true platelets and artifacts which look like platelets.

While CNNs have shown great efficacy in classification of natural images [15], there are several challenges in applying the state-of-the-art deep architectures for classification of blood cell images.

- Annotation of blood cell images requires certified medical expert. Thus, generating training data on the scale of ImageNet is nearly impossible. Small training sets are inevitable.
- The differentiation between cell subtypes are often vague at best. Experts thus vary in their annotation for such cells. Our experience shows that this inter-observer variability can be as high as 25% with 3 experts, going higher with the increasing number of experts.
- The natural imbalance in the frequency of the cell types makes it difficult to build larger data sets for the rarer types. For example, neutrophil and lymphocyte together constitute more than 85% of the WBC population, while round cells constitute close to 90% of the population for RBCs.

All the above can lead to significant overfitting in the training process. The state-of-the-art deep architectures, such as ResNet [15] can be difficult to apply on this problem. To counter overfitting, we use the following techniques:

- Shallower architecture with a maximum of 6 convolutional layers and a relatively small fully connected layer (maximum 256 units with 2 layers).
- Aggressive data augmentation through rotation, reflection and translation. For example, we use 12 rotations (at 30° each) on the cell images to create 12 rotated copies. Reflection is performed along the vertical and horizontal axes. Small translations (<30 pixels) are added either in the horizontal or vertical direction on random images.

- Using a large L2 regularization (0.005 for WBC, 0.001 for RBC and Platelet model) on the weights and increasing it throughout the training process.
- Using a large dropout [16] of 0.5 in the fully connected layers.
- Stopping early as soon as the training and validation errors start diverging.

We use a hierarchy of models for WBC classification. The extraction process for WBCs yields, apart from valid WBCs, giant platelets, clumps of platelets, nucleated RBCs and other artifacts. All of these can be similar in size and appearance to certain types of WBCs. The first model in the hierarchy differentiates the extracted patches between 5 classes: WBC, large platelet, clump of platelets, nucleated RBC and artifact. The valid WBCs identified in the first model is then further sub-classified by a second model into the WBC subtypes.

RBC and platelet subclassification is done using a single model, one for each cell type. The RBC model uses a single class "round" for normocytes, microcytes and macrocytes. The sub-classification between these classes is done based on the size of the cell, which is obtained from the mask generated in the extraction process. Both models have an 'artifact' output class used for rejecting patches not belonging to the respective cell class.

In each model, we set a probability cut-off for each class. If the predicted probability of the class with maximum probability is less than the cut-off, we put the cell into an "unclassified" bucket (similar to [3]). For example, the probability thresholds for WBC subtypes are – Neutrophil:0.6, Lymphocyte:0.4, Monocyte:0.7, Eosinophil:0.6, Basophil:0.5, Atypical cells:0.5. The thresholds are chosen using grid search to strike a balance between specificity and sensitivity.

4   Experimental Results

Existing peripheral blood smear analyzers typically work with machine prepared slides only [3], or have their own internal slide creation process [4]. $Shonit^{TM}$, on the other hand, was trained to work with both automated and manually prepared smears, and also over multiple stain types – May-Grünwald-Giemsa (MGG) and Leishman, with results comparable to that of [3,4]. This ensures that it finds applicability even in smaller laboratories which may not have access to advanced equipments. The manual smears were prepared based on our standard operation procedure (SOP), which resembles the procedure to prepare a blood smear under normal conditions. Details of the SOP are beyond the scope of this paper.

We did a validation study of $Shonit^{TM}$ on 40 anonymised abnormal blood samples, 10 each of the four stain and smear combination – automated MGG, automated Leishman, manual MGG and manual Leishman. Samples were collected from the normal workload of three major laboratories. An Institutional Ethics Committee approved the study. All WBCs visible in each of the samples were classified by the machine. Approximately 2.8% of WBCs were tagged as "unclassified" by our system, which is acceptable by medical standards [17,18]. The classification results were verified independently by three medical experts. A random sample of around 12,000 RBCs and 20,000 platelets were also verified. Cells where the experts did not agree with each other were rejected, to

Table 1. Specificity and sensitivity of WBC subtype classification

Cell type	Specificity (%)	Sensitivity (%)	Cell count
Neutrophil	99.16	98.28	13,640
Lymphocyte	98.85	98.97	2,838
Monocyte	99.93	99.00	482
Eosinophil	99.86	91.49	380

Table 2. Specificity and sensitivity of RBC subtype classification

Cell type	Specificity (%)	Sensitivity (%)	Cell count
Round	100	97.17	6719
Elliptocyte	99.11	100	962
Target	99.91	100	236
Teardrop	99.87	100	120
Echinocyte	99.57	99.88	2606
Fragmented	99.62	100	141

Table 3. Specificity and sensitivity of platelet classification

Cell type	Specificity (%)	Sensitivity (%)	Cell count
Platelet	93.4	99.7	10,355
Artifact	99.7	93.4	9,240

avoid ambiguity. The specificity and sensitivity of each class are listed in the Tables 1, 2 and 3. The sensitivity for WBC extraction was 99.5%. There were extremely few samples of Basophil and Atypical cells (4 and 12, respectively), hence their specificity and sensitivity are not reported.

5   Conclusion

Application of deep learning techniques on microscopic images of peripheral blood smear has its unique set of challenges, distinct from those for natural images. In this paper, we have described *Shonit*^{TM}, a system for analysis of PBS images using deep learning methods, which aims to address these challenges effectively. It consists of a low cost automated microscope and a software component. An ensemble of deep learning and image processing techniques are used for cell localization and classification. Experimental results are shown for a variety of smear and stain types. The specificity and sensitivity compare favourably with those of [3] reported in [17]. Due to its robustness to input variation and the low cost of its hardware, *Shonit*^{TM} can find wide applicability.

References

1. SigTuple, Shonit™: The complete peripheral blood smear analysis solution. https://www.sigtuple.com. Accessed 1 May 2017
2. Bain, B.J.: A Beginner's Guide to Blood Cells. Wiley, Chichester (2008)
3. Cellavision: Introducing CellaVision DM9600. http://www.cellavision.com/en/our-products/products/cellavision-dm9600. Accessed 20 Oct 2016
4. Roche Diagnostics Limited, cobas m511 integrated hematology analyser. http://www.cobas.com/home/product/hematology-testing/cobas-m-511.html. Accessed 15 Oct 2016
5. West Medica, Vision Hema Ultimate. http://visionhemaultimate.com/. Accessed 1 Nov 2016
6. Fujimoto, K.: Principles of measurement in hematology analyzers manufactured by Sysmex corporation. Sysmex J. Int. **9**(1), 31–44 (1999). SEAS SUM
7. Ronneberger, O., Fischer, P., Brox, T.: U-net: convolutional networks for biomedical image segmentation. In: Navab, N., Hornegger, J., Wells, W.M., Frangi, A.F. (eds.) MICCAI 2015. LNCS, vol. 9351, pp. 234–241. Springer, Cham (2015). doi:10.1007/978-3-319-24574-4_28
8. Labo America, Lx 500. http://www.laboamerica.com/products/compound/lx500. Accessed 27 Sept 2016
9. Apple, iPhone 6s. http://www.apple.com/iphone-6s/. Accessed 1 Nov 2016
10. Ren, S., He, K., Girshick, R., Sun, J.: Faster R-CNN: towards real-time object detection with region proposal networks. In: Advances in neural information processing systems, pp. 91–99 (2015)
11. He, K., Gkioxari, G., Dollár, P., Girshick, R.: Mask R-CNN (2017). arXiv preprint arXiv:1703.06870
12. Deng, J., Dong, W., Socher, R., Li, L. J., Li, K., Fei-Fei, L.: Imagenet: a large-scale hierarchical image database. In: IEEE Conference on Computer Vision and Pattern Recognition, CVPR 2009, pp. 248–255. IEEE (2009)
13. Angulo, J., Flandrin, G.: Automated detection of working area of peripheral blood smears using mathematical morphology. Anal. Cell. Pathol. **25**(1), 37–49 (2003)
14. Otsu, N.: A threshold selection method from gray-level histograms. IEEE Trans. Syst. Man Cybern. **9**(1), 62–66 (1979)
15. He, K., Zhang, X., Ren, S., Sun, J.: Deep residual learning for image recognition. In: Proceedings of the IEEE Conference on Computer Vision and Pattern Recognition, pp. 770–778 (2016)
16. Srivastava, N., Hinton, G.E., Krizhevsky, A., Sutskever, I., Salakhutdinov, R.: Dropout: a simple way to prevent neural networks from overfitting. J. Mach. Learn. Res. **15**(1), 1929–1958 (2014)
17. Rollins-Raval, M.A., Raval, J.S., Contis, L.: Experience with CellaVision DM96 for peripheral blood differentials in a large multi-center academic hospital system. J. Pathol. Inf. **3**(1), 29 (2012)
18. Kratz, A., et al.: Performance evaluation of the Cellavision DM96 system. Am. J. Clin. Pathol. **124**(5), 770–781 (2005)

AGNet: Attention-Guided Network for Surgical Tool Presence Detection

Xiaowei Hu[1]($\boxtimes$), Lequan Yu[1], Hao Chen[1], Jing Qin[2], and Pheng-Ann Heng[1]

[1] Department of Computer Science and Engineering,
The Chinese University of Hong Kong, Hong Kong, The People's Republic of China
xwhu@cse.cuhk.edu.hk

[2] Centre for Smart Health, School of Nursing, The Hong Kong Polytechnic University, Hong Kong, People's Republic of China

Abstract. We propose a novel approach to automatically recognize the presence of surgical tools in surgical videos, which is quite challenging due to the large variation and partially appearance of surgical tools, the complicated surgical scenes, and the co-occurrence of some tools in the same frame. Inspired by human visual attention mechanism, which first orients and selects some important visual cues and then carefully analyzes these focuses of attention, we propose to first leverage a global prediction network to obtain a set of visual attention maps and a global prediction for each tool, and then harness a local prediction network to predict the presence of tools based on these attention maps. We apply a gate function to obtain the final prediction results by balancing the global and the local predictions. The proposed attention-guided network (AGNet) achieves state-of-the-art performance on *m2cai16-tool* dataset and surpasses the winner in *2016* by a significant margin.

Keywords: Surgical tool recognition · Attention-guided network · Laparoscopic videos · Cholecystectomy · Deep learning

1 Introduction

Automatically annotating surgical tools in surgical videos is essential for intelligent surgical workflow analysis and can find a lot of relevant applications including operation room scheduling optimization, surgical report generation and real-time decision support. However, this task is quite challenging for the following reasons. First, the complicated surgical scenes make it difficult to accurately distinguish various surgical tools from the background. Second, the large variation of surgical tools makes this task even harder. Particularly, the partial appearances of some surgical tools in some frames make it complex to recognize them based on the features designed or learned according to the full view of the surgical tools. Third, when two or more surgical tools appear in the same scene (or frame) and are even overlapped with each other, it is difficult to pick out all of them from the background. In addition, there is only an image-level label

© Springer International Publishing AG 2017
M.J. Cardoso et al. (Eds.): DLMIA/ML-CDS 2017, LNCS 10553, pp. 186–194, 2017.
DOI: 10.1007/978-3-319-67558-9_22

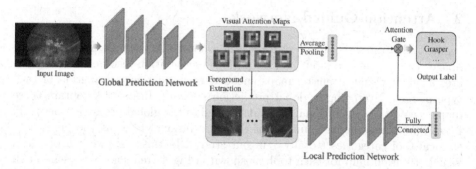

Fig. 1. The schematic illustration of proposed AGNet. (Color figure online)

(no bounding boxes) provided for each frame of surgical videos. In principle, this task is a multi-label classification problem as we have to apply multiple labels to describe one frame in case that different kinds of tools may appear simultaneously.

Existing approaches address this task in two ways. One is to directly recognize each kind of tools individually based on hand-crafted features or learned features [2,3], which ignores the underlying relationships among different kinds of tools. The other regards this task as a multi-label classification problem and learns the features of the tools in a data-driven manner [5,8,10,12]. Although it is effective to use the inherent visual features for tool present detection, the accuracy is still not satisfactory especially when just a small part of tool appears in the image or the tool is occluded by other kinds of tools. This is because the response of the attended tool is suppressed by other tools or complex background, and hence the attended one is not spatially dominant.

In order to meet above-mentioned challenges and solve this task better, we propose a novel deep learning based approach inspired by human visual attention mechanism. Similar to human visual attention, which first orients and selects some important visual cues and then carefully analyzes these focuses of attention [6], our method aims at first extracting some regions with relatively high probability of containing surgical tools by a deep neural network and then carefully analyzing these regions by another deep neural network (as shown in Fig. 1). We call the first neural network as global prediction network, which takes the original frame as input and outputs a visual attention map and a global prediction for each tool. The visual attention map is generated based on the activation regions of the feature maps produced by the neural network. We call the second neural network as local prediction network, as it takes the visual attention maps encoded rich local information as inputs and outputs another prediction for each tool, which is produced more based on the local information. We apply a gate function to obtain the final prediction results by balancing the global and the local predictions. The proposed attention-guided network (AGNet) achieves state-of-the-art performance on *m2cai16-tool* dataset [11] and surpasses the winner in *2016* by a significant margin.

2   Attention-Guided Network

Figure 1 illustrates the architecture of the proposed AGNet. It takes the whole images (frames) as the input and outputs the recognition results. It first generates a set of visual attention maps encoded with location information of each type of tools leveraging a global prediction network. In order to obtain these maps, we add a convolutional layer at the end of the global prediction network, by assigning the channel number of the convolutional layer as the number of surgical tool categories (totally 7 in our study). In this case, we can obtain a visual attention map for each tool, as shown in Fig. 1 (red color represents high responses while blue represents low responses). Then the presence confidence of each tool is predicted by averaging the feature values on each visual attention map.

After that, we extract the regions in the input image, which contain feature values exceeding a threshold in the visual attention maps, and then resize these regions to images with the same size of the input image. We call these generated images as foregrounds of surgical tools, as they include the regions with high probability of containing tools. Each foreground is fed into another deep neural network, i.e. the local prediction network, to obtain another set of predictions. Finally, these predictions are applied as gates to adjust the predictions generated from the global prediction network to achieve the final results. As the local prediction network focuses on the attention regions, it is much easier for it to learn more discriminative local features to more accurately recognize the occluded or partially appeared surgical tools and distinguish attended tools from complex backgrounds or other different kinds of tools.

2.1   Global Prediction Network

In order to obtain the visual attention maps based on the image-level supervision, we employ a residual network with 101 layers [1] as the basic network to generate a set of feature maps from the input image and add a new convolution layer following the last convolution layer of original residual network. We assign the channel number of the newly added convolution layer as the number of surgical tool categories so that we can obtain a visual attention map for each tool. After getting these maps, we can calculate the global prediction of each tool by figuring out the mean value of each map.

During the training process, cross-entropy loss is used for each kind of tools and the total loss L is defined as the summation of loss on all kinds of tools:

$$L = -\frac{1}{N} \sum_{i=1}^{N} \sum_{j=1}^{c} (T_{i,j} log(p_{i,j}) + (1 - T_{i,j}) log(1 - p_{i,j})), \tag{1}$$

where N and c denote the number of images and the number of tool categories respectively; the $T_{i,j}$ is equal to 1 when the j-th surgical tool is in the i-th image and it is 0 otherwise; the confidence of tool j in image i is represented as $p_{i,j}$ which is normalized by sigmoid function into $[0, 1]$. Note that the prediction generated from this global network is based on the whole image, which is easy to neglect the information of occluded, partially visible tools and/or tools with small size since the activations of these tools are usually much lower than large fully visible tools.

However, as mentioned, the visual attention maps generated by the last convolutional network of the global network are able to represent each kind of surgical tools independently and contain rich local information. After all, the tools in hard cases, albeit having lower activations than easily recognized tools, still contains higher activations compared with the background (as shown in Fig. 1). On each visual attention map, we employ Otsu algorithm [4] to dynamically separate the pixels into two categories by maximizing the inter-class variance. The category with large feature values on each visual attention map has relatively high probability to contain the surgical tools. Then the minimum bounding box of the region belonging to this category is extracted and resized into an image with the same size of the input image as the foreground. And we extract the foregrounds for every type of tools as illustrated in Fig. 1. The labels of these foregrounds are the corresponding types of tools.

2.2   Local Prediction Network

The local prediction network takes the foregrounds inferred by the global prediction network as inputs and predicts another confidence of each type of tools based on these foregrounds. This network has the same structure with the global prediction network except using fully connected layer to produce a fixed-length vector for surgical tool prediction instead of just averaging the feature maps. Moreover, the same loss function as shown in Eq. 1 is used in this network.

The predictions generated by the local network are used as gates to control the confidences predicted by the global prediction network. The final prediction confidence P_{fi} of each kind of tools i is:

$$P_{fi} = G(\sigma(P_{gi}), \sigma(P_{li})), \qquad (2)$$

where G is the gate function and it calculates the inner product of the confidences P_{gi} and P_{li} of the global and local networks for i-th tool; σ is the sigmoid function used for normalization. Here, we use the sigmoid instead of softmax since the categorical distribution of the tools is independent in this multi-label problem.

According to Eq. 2, in our AGNet, if the "attention" really contains surgical tool(s), regardless of its size or if occluded by other tools or not, it is likely to get a high confidence since the attention map mainly contains one type of tool and includes more local context information. On the other hand, if the "attention" map is the background or other kinds of tools, it is likely to get a low confidence even if this part of the image is similar with the real type of tool because these hard instances are subsampled and retrained by focusing on them.

In summary, the proposed AGNet can meet the challenges of surgical tool presence detection in three aspects. First, the occluded or partial visible tools can be more easily recognized since there is only one kind of tool in one attention map and hence we can obtain more discriminative features of this kind of tool from the local network. Second, the small tools are easy to be recognized since the regions with high attention are extracted from the input image and enlarged for more contextual information. Finally, the tools or backgrounds which are hard to be recognized are retrained by this network and hence it has the same function of hard negative mining [9], which increases the accuracy by focusing on hard examples.

3   Experiments

3.1   Datasets and Preprocessing

Dataset. We evaluate the proposed AGNet on m2cai16-tool dataset [11] which includes 10 cholecystectomy videos for training and 5 videos for testing. These videos are captured at 25 fps, but only one frame has the label in every 25 frames. We ignore the frames (images) without labels. Finally, we have around 23k training images and 12k testing images. And there are in total seven kinds of surgical tools in these videos (as shown in Fig. 2). The performance is evaluated by average precision and the overall performance is the mean value of the average precisions on these seven types of tools.

Grasper   Bipolar   Hook   Scissors   Clipper   Irrigator   Specimen Bag

Fig. 2. The seven kinds of tools on m2cai16-tool dataset.

Image Alignment. The surgical videos captured by laparoscope always contains "black borders" with zero gradient which are useless for understanding images (as shown in Fig. 1). In order to reduce space and time complexity, we cut off the border automatically by finding the minimum bounding box of the maximum connected area with non-zero gradient. Only the image region with context information is sent to the deep neural network, so that we can keep more detail context of the images, which is essential to recognize small or partially visible surgical tools.

3.2   Training Procedure

Fine-tuning the network based on the weights pre-trained on other task related datasets with numerous training images is beneficial to accelerate the training process, as the knowledge of images is transferred from large datasets to the new dataset. In our AGNet, we first use the weights of residual network with 101 layers pre-trained on ImageNet [7] to initialize the weights of global prediction network. Then we fine-tune the global prediction network on the m2cai16 dataset and take this well-trained weights of the global prediction network as the pre-trained weights for the local prediction network, since these two networks have similar structures and training images. Stochastic gradient descent (SGD) is used to optimize these two networks. We set the learning rate as 0.001 and it reduced by a factor of 0.1 at 4,000 and 8,000 iterations. Learning stops after 10,000 iterations. We use a batch size of 16 for the global prediction network and 14 for the local prediction network. In addition, images are flipped up-down and left-right randomly for data argumentation.

3.3 Ablation Analysis

We perform the ablation analysis on the m2cai16-tool training dataset including ten videos. We split these videos into training (last eight videos) and validation (first two videos) sets, and all video frames are resized to 224 × 224 as the inputs.

Image Alignment and Attention Guidance. Table 1 shows how different steps affect the performance. Here, the prediction results of global prediction network are denoted as GPN. After applying the image alignment to select the image parts with rich context information (GPN-A), the recognition accuracy improves a lot. And there are huge improvements on "Scissors", "Irrigator" and "Clipper", which indicates the detail features are critical to distinguish them with other types of tools. Moreover, our Attention-guided Network (AGNet) further enhances the accuracy, especially for "Hook" and "Clipper" that get more than 99% average precision. It shows that the discriminative local features are very useful to recognize the hardly visible tools and distinguish tools with similar appearance. Note that the accuracy of "Irrigator" is much lower than other categories, because the appearance of this tool is not obvious (a cylinder with metallic luster as shown in Fig. 2) and the training samples are limited.

Table 1. Average precision (%) on m2cai16-tool validation dataset.

Method	Overall	Grasper	Bipolar	Hook	Scissors	Clipper	Irrigator	Specimen bag
GPN	81.295	95.091	88.526	96.381	63.827	87.659	50.458	87.120
GPN-A	88.540	**95.200**	91.057	96.528	88.486	90.964	69.334	88.214
AGNet	**91.102**	94.963	**91.556**	**99.616**	**90.409**	**99.312**	**72.645**	**89.213**

Importance of Attention Guidance. Model ensemble is a common technique to enhance the image recognition accuracy by combining the prediction results of multiple models, as the diversity among the models is able to reduce the risk of overfitting. In our AGNet, we also use two sub-networks for surgical tool presence detection. However, our two networks not only take the advantage of

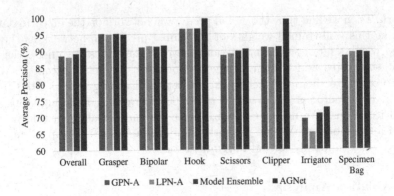

Fig. 3. Comparison with model ensemble.

diversity but also leverage different levels of information of images. In these two sub-networks, one recognizes the surgical tools based on the global images while another looks into the local regions deduced from visual attention maps. In order to prove the effectiveness of the attention guidance mechanism, we firstly train the global prediction network and the local prediction network (GPN-A and LPN-A) with the whole images and make the predictions individually. The training parameters are same as described in Subsect. 3.2 for these two networks. Then, after getting the prediction results of these two networks, we average them as the model ensemble result (Model Ensemble). As shown in Fig. 3, averaging these two models is slight better than each individual network. However, it is much worse than our AGNet, especially for "Hook" and "Clipper" which are easy to be occluded by other tools. It indicates that our AGNet is not simply assembling multiple networks, but leveraging different levels of information.

3.4   Comparison with the State-of-the-Arts

We compare the proposed AGNet with the state-of-the-arts on m2cai16-tool dataset. In this experiment, the entire training set is used to train our model. Table 2 lists all methods submitted to this challenge. We can find that the algorithm based on hand-crafted features [2] has the lowest accuracy. For the algorithms based on deep convolution neural network, Luo et al. [3] use seven convolutional neural networks to recognize the seven kinds of tools individually while other methods [5,8,10,12] use a single deep neural network to extract features based on the whole image. However, none of them use enough deep network and local information to obtain more discriminative features. It is clear that our AGNet which combines the global and local information achieves the highest precision and exceeds the second-place algorithm over 23%.

Table 2. Tool presence detection results on m2cai16-tool dataset.

Method	Mean average precision (%)
Ours	**86.9**
Raju et al. [5]	63.8
Sahu et al. [8]	61.5
Winanda et al. [10]	52.5
Zia et al. [12]	37.8
Luo et al. [3]	27.9
Letouzey et al. [2]	21.1

4   Conclusion

In this paper, we proposed an attention-guided network, denoted as AGNet, for surgical tool presence detection. Global prediction network is proposed to obtain the regions containing surgical tools based on image-level labels as well as make the predictions for the whole images while local prediction network makes the predictions for the attended regions extracted by global prediction network. By considering global and local features, AGNet can recognize the partially visible or small tools and discriminate tools from complex backgrounds. Our algorithm achieves the state-of-the-art performance on m2cai16-tool dataset and surpasses all other methods by a significant margin.

Acknowledgements. The work described in this paper was supported by the following grants from the Research Grants Council of the Hong Kong Special Administrative Region, China (Project No. CUHK 14202514 and CUHK 14203115).

References

1. He, K., Zhang, X., Ren, S., Sun, J.: Deep residual learning for image recognition. In: CVPR, pp. 770–778 (2016)
2. Letouzey, A., Decrouez, M., Agustinos, A., Voros, S.: Instruments localisation and identification for laparoscopic surgeries (2016). http://camma.u-strasbg.fr/m2cai2016/reports/Letouzey-Tool.pdf
3. Luo, H., Hu, Q., Jia, F.: Surgical tool detection via multiple convolutional neural networks (2016). http://camma.u-strasbg.fr/m2cai2016/reports/Luo-Tool.pdf
4. Otsu, N.: A threshold selection method from gray-level histograms. IEEE Trans. Syst. Man Cybern. **9**(1), 62–66 (1979)
5. Raju, A., Wang, S., Huang, J.: M2CAI surgical tool detection challenge report (2016). http://camma.u-strasbg.fr/m2cai2016/reports/Raju-Tool.pdf
6. Rosen, M.L., Stern, C.E., Michalka, S.W., Devaney, K.J., Somers, D.C.: Cognitive Control Network Contributions to Memory-Guided Visual Attention. Cerebral Cortex, New York (2015). bhv028
7. Russakovsky, O., Deng, J., Su, H., Krause, J., Satheesh, S., Ma, S., Huang, Z., Karpathy, A., Khosla, A., Bernstein, M., et al.: Imagenet large scale visual recognition challenge. IJCV **115**(3), 211–252 (2015)

8. Sahu, M., Mukhopadhyay, A., Szengel, A., Zachow, S.: Tool and phase recognition using contextual CNN features. arXiv preprint arXiv:1610.08854 (2016)
9. Shrivastava, A., Gupta, A., Girshick, R.: Training region-based object detectors with online hard example mining. In: CVPR, pp. 761–769 (2016)
10. Twinanda, A.P., Mutter, D., Marescaux, J., de Mathelin, M., Padoy, N.: Single- and multi-task architectures for tool presence detection challenge at M2CAI 2016. arXiv preprint arXiv:1610.08851 (2016)
11. Twinanda, A.P., Shehata, S., Mutter, D., Marescaux, J., de Mathelin, M., Padoy, N.: EndoNet: a deep architecture for recognition tasks on laparoscopic videos. IEEE Trans. Med. Imaging **36**(1), 86–97 (2017)
12. Zia, A., Castro, D., Essa, I.: Fine-tuning deep architectures for surgical tool detection (2016). http://camma.u-strasbg.fr/m2cai2016/reports/Zia-Tool.pdf

Pathological Pulmonary Lobe Segmentation from CT Images Using Progressive Holistically Nested Neural Networks and Random Walker

Kevin George, Adam P. Harrison, Dakai Jin, Ziyue Xu[(⊠)], and Daniel J. Mollura

National Institutes of Health, Bethesda, MD, USA
ziyue.xu@nih.gov

Abstract. Automatic pathological pulmonary lobe segmentation(PPLS) enables regional analyses of lung disease, a clinically important capability. Due to often incomplete lobe boundaries, PPLS is difficult even for experts, and most prior art requires inference from contextual information. To address this, we propose a novel PPLS method that couples deep learning with the random walker (RW) algorithm. We first employ the recent progressive holistically-nested network (P-HNN) model to identify potential lobar boundaries, then generate final segmentations using a RW that is seeded and weighted by the P-HNN output. We are the first to apply deep learning to PPLS. The advantages are independence from prior airway/vessel segmentations, increased robustness in diseased lungs, and methodological simplicity that does not sacrifice accuracy. Our method posts a high mean Jaccard score of 0.888 ± 0.164 on a held-out set of 154 CT scans from lung-disease patients, while also significantly ($p < 0.001$) outperforming a state-of-the-art method.

Keywords: Lung lobe segmentation · CT · Holistically nested neural network · Fissure · Random walker

1 Introduction

Due to the prevalence of lung disease worldwide, automated lung analysis tools are a major aim within medical imaging. Human lungs divide into five relatively functionally independent lobes, each with separate bronchial and vascular systems. This causes several lung diseases to preferentially affect specific lobes [1].

This work is supported by the Intramural Research Program of the National Institutes of Health, Clinical Center and the National Institute of Allergy and Infectious Diseases. Support also received by the Natural Sciences and Engineering Research Council of Canada. We also thank Nvidia for the donation of a Tesla K40 GPU. The rights of this work are transferred to the extent transferable according to title 17 § 105 U.S.C.

M.J. Cardoso et al. (Eds.): DLMIA/ML-CDS 2017, LNCS 10553, pp. 195–203, 2017.
DOI: 10.1007/978-3-319-67558-9_23

As a result, pathological pulmonary lobe segmentation (PPLS) is an important step toward computerized disease analysis tools.

There are several anatomical challenges to PPLS. In healthy lungs, lobar borders are typically defined by visible fissures. However, fissures are often incomplete, *i.e.*, they fail to extend across the full lobe boundary [2]. Thus, fissure detection and lobar boundary detection are not always equivalent problems. Even when fissures are visible, they can vary in thickness, location, and shape due to respiratory and cardiac motion [3]. Furthermore, accessory fissures, structures visually identical to major fissures but that do not actually divide lobe regions, are common [4]. Finally, although PPLS provides the greatest clinical utility in pathological lungs, lung diseases often completely obscure fissures.

For these reasons, most of prior art relies on pulmonary vessel/airway segmentations or on semi-automated approaches [1]. Bragman et al.'s work is an exemplar of the former category [5], achieving good accuracy, but with a relatively complex scheme involving pulmonary vessel and airway segmentations, in addition to population-based models of fissure location and surface fitting. Apart from the complexity involved, pulmonary airway and vessel segmentation are not always reliable, especially in the presence of pathologies. This is why Doel et al. [6], for example, attempt to minimize dependence on the quality of these priors by using them mainly to initialize fissure detection in their method.

Only two lobe segmentation methods have been proposed that are both automatic and non-reliant on prior airway/vessel segmentations [7,8]. However, both methods have noticeable limitations. Ross et al. [7] use a manually defined atlas, which is laborious to create, prone to poor results when tested on highly variable pathological lungs, and increases execution time [1]. Pu et al. [8] also achieve lobe segmentation without reliance on prior airway/vessel segmentations, but they report less success in cases with incomplete fissures, large accessory fissures, and where lobe boundaries have abnormal orientations. Finally their method is only validated on healthy volumes and only qualitative results were reported. So far, no deep learning based algorithm has been proposed.

To fill this gap, we develop a PPLS approach that combines a state-of-art deep fully convolutional network (FCN) with the random walker (RW) algorithm. Observing that FCNs have proven highly powerful for segmentation, we use the recently proposed progressive holistically-nested network (P-HNN) architecture [9] to segment lobar boundaries within the lung. Importantly, P-HNN is able to perform well even when fissures are incomplete. The P-HNN probability map then is used to generate seeds and edge probabilities for the RW algorithm [10], which produces a final lobe segmentation mask. We call our method P-HNN+RW, which is the first to apply deep learning to PPLS. Importantly, P-HNN+RW only requires a lung mask to function and remains simple, fast, and efficient, yet robust. We test P-HNN+RW on the highly challenging Lung Tissue Research Consortium (LTRC) dataset [11], which includes CT scans of patients suffering from chronic obstructive pulmonary disease (COPD) and interstitial lung disease (ILD), and demonstrate improvement over a state-of-the-art method [6] despite being much simpler in execution. On a held-out test set of

154 volumes, the largest analysis of lobe segmentation to date, we achieve a high Jaccard score (JS) of 0.888 ± 0.164.

2   Method

Figure 1 illustrates a flowchart of our PPLS pipeline. To avoid relying on prior vessel/airway segmentation, we instead employ a principled combination of the P-HNN deep-learning architecture [9] with the celebrated RW algorithm [10]. The only pre-requisite is a lung mask, where reliable methods do exist [9].

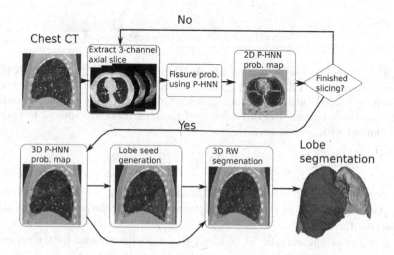

Fig. 1. Flowchart of P-HNN+RW's operation.

2.1   Lobar Boundary Segmentation

Given a volume and a lung mask, if a reliable boundary map between lobes were calculated, segmenting the five lobes would be relatively straightforward. Unfortunately, lobar boundaries are difficult to detect, given the common occurrence of incomplete features and other confounding factors, causing much of the prior art to rely on supplemental information given by airway or vessel segmentations. Reliably segmenting lobar boundaries directly from attenuation values remains attractive as it would avoid this considerable complexity. The recent prominence and success of FCNs in highly challenging segmentation tasks has opened up the possibility of more reliable lobar boundary segmentation.

Along those lines, our method's first step is to apply an FCN model to lobar boundary segmentation. Without loss of generality, we adopt the recent P-HNN model [9], which has been effectively applied to pathological lung segmentation. Figure 2 illustrates the P-HNN model, which is an adaptation of the highly successful HNN model [12] that introduced the concept of deeply supervised side-outputs to FCNs. Importantly, the model avoids the complicated upsampling

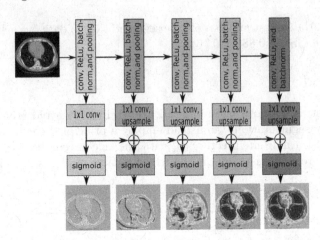

Fig. 2. Lobar boundary segmentation with P-HNN. Like HNN [12], the P-HNN model computes side-outputs, which are all connected to a loss function. Unlike HNN, P-HNN does not have a fused output, but constrains the side-outputs to progressively improve upon the previous output. Regions outside of the lung mask are ignored in training. Figure adapted with permission [9]

pathways used in recent FCN architectures, instead using progressive multipath connections to help address the well-known coarsening resolution problem of FCNs. As such, the model remains straightforward and relatively simple, making it attractive for PPLS.

Regardless of the specific FCN used, the model functions by training on a set of data $S = \{(X_n, Y_n), n = 1 \ldots, N\}$, where $X_n = \{x_j^{(n)}, j = 1 \ldots, |X_n|\}$ and $Y_n = \{y_j^{(n)}, j = 1 \ldots, |X_n|\}$, $y_j^{(n)} \in \{0, 1\}$ represent the input and binary ground-truth images, respectively. The latter denotes which pixels are located on a boundary region, $i.e.$, $x_j \in \Omega_{\text{bound}}$, where we do not discriminate based on which exact lobes are being separated. Given an image and a set of parameters, $\mathbf{W}$, the network then estimates

$$p_j = P\left(x_j \in \Omega_{\text{bound}}|X_n; \mathbf{W}\right). \tag{1}$$

We perform training and inference on 2D axial slices, choosing the 2D domain for two main reasons. For one, current 3D FCN solutions require a sliding box, limiting the spatial context, which we posit is important for delineating lobar boundaries, especially when fissures are incomplete or obscured. In addition, sliding boxes add to the complexity and inference time, which we aim to minimize. Second, due to memory constraints, 3D FCNs are less wide and deep than their 2D counterparts. Given that excellent performance has been achieved in segmenting anatomical structures in 2D [9,13,14], we opt for a 2D solution that requires no compromise in network depth and width and remains simpler in application. Regions outside the lung mask are ignored in training.

2.2   3D Random Walker

Although the P-HNN model can provide a relatively reliable lobar boundary map, the output is still noisy, often resulting in gaps. Thus, the final segmentation cannot be obtained from the boundary maps alone, meaning a subsequent processing step, one that is robust to noise, is required. Conventionally, this can be done by simply fitting a surface to potential fissure locations [8], but this can be inefficient and error-prone. Instead, we combine all axial slices from the P-HNN output and execute 3D RW segmentation [10]. Typically used in interactive settings, the RW algorithm can be also used whenever there are clearly defined seeds, *i.e.*, regions constrained to fall in a specific segmentation mask. We choose RW over other graph-based methods, *e.g.*, dense CRFs, due to its globally optimal solution, efficiency, and simplicity.

Formally, assume we have five sets of seed points, R_k, $k = 1 \ldots 5$, corresponding to regions that definitively belong to one of five lobes. If we partition the seed points into $S_f = R_1$ and $S_b = \{R_m, m = 2 \ldots 5\}$, then the pseudo-probability that the unseeded points belong to lobe 1 is calculated by minimizing the following energy functional:

$$E(\mathbf{y}) = \sum_{e_{ij}} w_{ij}(y_i - y_j)^2 \text{ s.t., } \begin{matrix} y_i = 1, \ x_i \in S_f \\ y_i = 0, \ x_i \in S_b \end{matrix}, \tag{2}$$

where y_j are the resulting probabilities and w_{ij} denotes the weights penalizing differences across the edges, e_{ij}, connecting each pixel to its neighbors. We use a 6-connected neighborhood and ignore regions outside the lung mask. Minimizing (2) can be done linearly [10]. For all other lobes, the process is repeated, except that different lobes must be treated as foreground, *i.e.*, S_f, with the rest as background, *i.e.*, S_b. Because all pseudo-probabilities sum to 1 [10], the last lobe's segmentation can be calculated from the first four, saving one RW execution.

To minimize (2), seed regions and edge weights must be generated. To compute seed regions, we first invert the P-HNN boundary mask and then iteratively erode it until we obtain five fully connected regions, one for each lobe. Region centroids are used to identify specific lobes and a stopping threshold is used in cases where five regions cannot be determined. This iterative erosion approach breaks connections between lobes caused by incomplete or weak P-HNN boundaries, and is stopped at the point of the largest possible seeds for the RW. These enlarged seed regions reduce the size of the linear system, keeping a fully 3D RW implementation within memory constraints.

With the seeds generated, we calculate edge weights based on the P-HNN output:

$$w_{ij} = \exp(-\beta(p_i - p_j)^2), \tag{3}$$

where $p_{(.)}$ is obtained from (1) and β is a global parameter analogous to a diffusion constant in physical systems, whose value is chosen based on validation subsets. Thus, the RW penalizes lobar mask leakage across the P-HNN boundaries, while producing a result robust to gaps or noise.

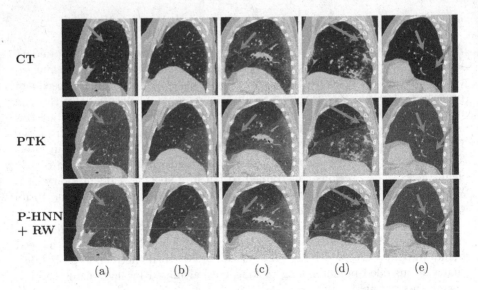

Fig. 3. Example PPLSs with ground-truth boundaries rendered as red lines. (a) P-HNN+RW correctly segments the left lung lobes, while PTK follows an erroneous boundary. (b) P-HNN+RW successfully handles an incomplete fissure. (c) PTK over-segments one lobe. (d) P-HNN+RW, unlike PTK, successfully differentiates fibrosis from a lobe boundary. (e) Despite no visual fissure, P-HNN+RW is able to use context to infer a more reasonable lobe boundary. Note, accessory fissure is ignored. (Color figure online)

3   Experiments and Results

We train on axial slices from 540 annotated volumes from the LTRC dataset [11], testing on a held-out test set of 154 volumes. Importantly, every volume exhibits challenging attenuation patterns from ILD and/or COPD; thus, we test on pathological cases where PPLS could provide clinical utility. We rescale each slice to a 3-channel 8-bit image using three windows of $[-1000, 200]$, $[-160, 240]$, and $[-1000, -775]$ Hounsfield units. We fine-tune from the ImageNet VGG-16 model [15] and stop training after roughly 2 epochs. Parameters for the erosion step and RW were selected based on a validation set of 60 volumes. During inference, the full pipeline takes on average 4–8 min for a typical volume using a Tesla K40 GPU and 12-core processor. To our knowledge, this represents the largest evaluation of a PPLS technique published to date.

To measure P-HNN+RW's performance, without the lung mask as a confounding factor, we first test using the ground-truth lung mask. Table 1 depicts the mean JSs and their standard deviation for each lobe over the full test set. P-HNN+RW is strongest in the left lung and weakest in the right middle lobe, a consistent trend among current methods [5, 16]. Importantly, we obtain a high overall Jaccard score of 0.888 ± 0.164, which compares very favorably with the highest performing published methods tested on similarly challenging pathological lungs [5, 16, 17],

Table 1. Mean JSs and their standard deviation for each lobe on the our test set. Starred results only use the 114 volumes the PTK could segment, otherwise results include the entire 154 volume test set, which may include more difficult cases. Overall scores average each lobe score, weighted by their relative number of pixels.

Lobe	LU	LL	RU	RM	RL	Overall
P-HNN+RW (GT Mask)	0.925 ± 0.175	0.945 ± 0.116	0.929 ± 0.057	0.824 ± 0.142	0.887 ± 0.239	0.888 ± 0.164
P-HNN+RW*	0.930 ± 0.144	0.925 ± 0.130	0.927 ± 0.044	0.848 ± 0.091	0.873 ± 0.244	0.888 ± 0.146
PTK*	0.884 ± 0.171	0.853 ± 0.202	0.835 ± 0.148	0.683 ± 0.222	0.865 ± 0.170	0.838 ± 0.140

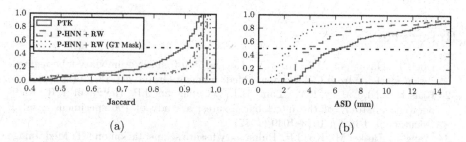

(a)   (b)

Fig. 4. Cumulative histograms of our method's performance on 114 CT scans, *i.e.*, excluding those that the PTK could not segment. (a) and (b) depict JS and ASD results, respectively.

but without requiring vessel/airway segmentations, groupwise priors, and other complex processes.

We also directly compare against prior art. Since no other deep-learning method has been published, we instead compare against the pulmonary toolkit (PTK)'s [6] automatic and non-deep-learning scheme. For fair comparison, we regenerated our results using the PTK lung masks. The PTK failed to produce a final lobe segmentation for 40 volumes. For the remaining 114 volumes, Fig. 4 depicts cumulative histograms of both JSs and average surface distances (ASDs) for P-HNN+RW, with and without the ground-truth mask, and for the PTK. The last two rows of Table 1 provide lobe-specific results for comparison with PTK. As can be seen, when using the same mask, our method (0.89 JS) convincingly outperforms the PTK's (0.84 JS), producing statistically significant improvements ($p < 0.001$ Wilcoxon signed-rank test) for both metrics, despite excluding volumes that the PTK could not segment. Figure 3 provides some qualitative examples, demonstrating the impact of the P-HNN+RW's improvements in terms of visual quality and segmentation usability.

4   Conclusion

This work introduces a simple, yet highly powerful, approach to PPLS that uses deep FCNs combined with a RW refinement. We achieve high accuracy

202 K. George et al.

without reliance on prior airway or vessel segmentations. Furthermore, the proposed method is developed and validated specifically for pathological lungs. When tested on 154 held-out pathological computed tomography (CT) scans, the largest evaluation of PPLS to date, the proposed method produces strong lobe-specific and overall JSs. Our proposed method also consistently outperforms a state-of-art airway and vessel based PPLS method [6] in both ASD and JS ($p < 0.001$). Importantly, these results are obtained with less complexity relative to the state of the art, increasing our method's generalizability. Thus, our proposed method represents an important step forward toward clinically useful PPLS tools.

References

1. Doel, T., Gavaghan, D.J., Grau, V.: Review of automatic pulmonary lobe segmentation methods from CT. Comput. Med. Imaging Graph. **40**, 13–29 (2015)
2. Raasch, B.N., Carsky, E.W., Lane, E.J., O'Callaghan, J.P., Heitzman, E.R.: Radiographic anatomy of the interlobar fissures: a study of 100 specimens. Am. J. Roentgenol. **138**(6), 1043–1049 (1982)
3. Wang, J., Betke, M., Ko, J.P.: Pulmonary fissure segmentation on CT. Med. Image Anal. **10**(4), 530–547 (2006)
4. Cronin, P., Gross, B.H., Kelly, A.M., Patel, S., Kazerooni, E.A., Carlos, R.C.: Normal and accessory fissures of the lung: evaluation with contiguous volumetric thin-section multidetector CT. Eur. J. Radiol. **75**(2), e1–e8 (2010)
5. Bragman, F., McClelland, J., Jacob, J., Hurst, J., Hawkes, D.: Pulmonary lobe segmentation with probabilistic segmentation of the fissures and a groupwise fissure prior. IEEE Trans. Med. Imaging **36**(8), 1650–1663 (2017)
6. Doel, T., Matin, T.N., Gleeson, F.V., Gavaghan, D.J., Grau, V.: Pulmonary lobe segmentation from CT images using fissureness, airways, vessels and multilevel b-splines. In: 2012 9th IEEE Internationl Symposium on Biomedical Imaging (ISBI), pp. 1491–1494, May 2012
7. Ross, J.C., San José Estépar, R., Kindlmann, G., Díaz, A., Westin, C.-F., Silverman, E.K., Washko, G.R.: Automatic lung lobe segmentation using particles, thin plate splines, and maximum a posteriori estimation. In: Jiang, T., Navab, N., Pluim, J.P.W., Viergever, M.A. (eds.) MICCAI 2010. LNCS, vol. 6363, pp. 163–171. Springer, Heidelberg (2010). doi:10.1007/978-3-642-15711-0_21
8. Pu, J., Zheng, B., Leader, J.K., Fuhrman, C., Knollmann, F., Klym, A., Gur, D.: Pulmonary lobe segmentation in CT examinations using implicit surface fitting. IEEE Trans. Med. Imaging **28**(12), 1986–1996 (2009)
9. Harrison, A.P., Xu, Z., George, K., Lu, L., Summers, R.M., Mollura, D.J.: Progressive and multi-path holistically nested neural networks for pathological lung segmentation from CT images. In: MICCAI 2017, Proceedings (2017)
10. Grady, L.: Random walks for image segmentation. IEEE Trans. Pattern Anal. Mach. Intell. **28**(11), 1768–1783 (2006)
11. Karwoski, R.A., Bartholmai, B., Zavaletta, V.A., Holmes, D., Robb, R.A.: Processing of CT images for analysis of diffuse lung disease in the lung tissue research consortium. In: Proceedings of SPIE 6916, Medical Imaging 2008: Physiology, Function, and Structure from Medical Images (2008)
12. Xie, S., Tu, Z.: Holistically-nested edge detection. In: The IEEE International Conference on Computer Vision (ICCV), December 2015

13. Roth, H.R., Lu, L., Farag, A., Sohn, A., Summers, R.M.: Spatial aggregation of holistically-nested networks for automated pancreas segmentation. In: Ourselin, S., Joskowicz, L., Sabuncu, M.R., Unal, G., Wells, W. (eds.) MICCAI 2016. LNCS, vol. 9901, pp. 451–459. Springer, Cham (2016). doi:10.1007/978-3-319-46723-8_52
14. Zhou, Y., Xie, L., Shen, W., Fishman, E., Yuille, A.: Pancreas segmentation in abdominal CT scan: a coarse-to-fine approach. CoRR/abs/1612.08230 (2016)
15. Simonyan, K., Zisserman, A.: Very deep convolutional networks for large-scale visual recognition. In: ICLR (2015)
16. Lassen, B., van Rikxoort, E.M.: Automatic segmentation of the pulmonary lobes from chest CT scans based on fissures, vessels, and bronchi. IEEE Trans. Med. Imaging 32(2), 210–222 (2013)
17. van Rikxoort, E.M., Prokop, M., de Hoop, B., Viergever, M.A., Pluim, J.P.W., van Ginneken, B.: Automatic segmentation of pulmonary lobes robust against incomplete fissures. IEEE Trans. Med. Imaging 29(6), 1286–1296 (2010)

End-to-End Unsupervised Deformable Image Registration with a Convolutional Neural Network

Bob D. de Vos[1]($\boxtimes$), Floris F. Berendsen[2], Max A. Viergever[1], Marius Staring[2], and Ivana Išgum[1]

[1] Image Sciences Institute, University Medical Center Utrecht, Utrecht, The Netherlands
B.D.deVos-2@umcutrecht.nl
[2] Division of Image Processing, Leiden University Medical Center, Leiden, The Netherlands

Abstract. In this work we propose a deep learning network for deformable image registration (DIRNet). The DIRNet consists of a convolutional neural network (ConvNet) regressor, a spatial transformer, and a resampler. The ConvNet analyzes a pair of fixed and moving images and outputs parameters for the spatial transformer, which generates the displacement vector field that enables the resampler to warp the moving image to the fixed image. The DIRNet is trained end-to-end by unsupervised optimization of a similarity metric between input image pairs. A trained DIRNet can be applied to perform registration on unseen image pairs in one pass, thus non-iteratively. Evaluation was performed with registration of images of handwritten digits (MNIST) and cardiac cine MR scans (Sunnybrook Cardiac Data). The results demonstrate that registration with DIRNet is as accurate as a conventional deformable image registration method with short execution times.

Keywords: Deep learning · Deformable image registration · Convolution neural network · Spatial transformer · Cardiac cine MRI

1 Introduction

Image registration is a fundamental step in many medical image analysis tasks. Traditionally, image registration is performed by exploiting intensity information between pairs of fixed and moving images. Since recently, deep learning approaches are used to aid image registration. Wu et al. [11] used a convolutional stacked auto-encoder (CAE) to extract features from fixed and moving images that are subsequently used in conventional deformable image registration algorithms. However, the CAE is decoupled from the image registration task and hence, it does not necessarily extract the features most descriptive for image registration. The training of the CAE was unsupervised, but the registration task was not learned end-to-end. Miao et al. [8] and Liao et al. [6] have used deep

© Springer International Publishing AG 2017
M.J. Cardoso et al. (Eds.): DLMIA/ML-CDS 2017, LNCS 10553, pp. 204–212, 2017.
DOI: 10.1007/978-3-319-67558-9_24

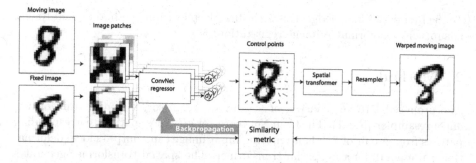

Fig. 1. Schematics of the DIRNet with two input images from the MNIST data. The DIRNet takes one or more pairs of moving and fixed images as its inputs. The fully convolutional ConvNet regressor analyzes spatially corresponding image patches from the moving and fixed images and generates a grid of control points for a B-spline transformer. The B-spline transformer generates a full displacement vector field to warp a moving image to a fixed image. Training of the DIRNet is unsupervised and end-to-end by backpropagating an image similarity metric as a loss.

learning to learn rigid registration with predefined registration examples. Miao et al. [8] used a convolutional neural network (ConvNet) to predict a transformation matrix for rigid registration of synthetic 2D to 3D images. Liao et al. [6] used a ConvNet for intra-patient rigid registration of CT to cone-beam CT applied to either cardiac or abdominal images. This ConvNet learned to predict iterative updates of registration using reinforcement learning. Both methods are end-to-end but use supervised techniques, i.e. registration examples are necessary for training, which are often task specific and highly challenging to obtain.

Jaderberg et al. [3] introduced the spatial transformer network (STN) that can be used as a building block that aligns input images in a larger network that performs a particular task. By training the entire network end-to-end, the embedded STN deduces optimal alignment for solving that specific task. However, alignment is not guaranteed, and it is only performed when required for the task of the entire network. The STNs were used for affine transformations, as well as deformable transformations using thin-plate splines. However, an STN needs many labeled training examples, and to the best of our knowledge, have not yet been used in medical imaging.

In this work, we present the deformable image registration network (DIRNet). The DIRNet takes pairs of fixed and moving images as inputs, and it outputs moving images warped to the fixed images. Training of the DIRNet is unsupervised. Unlike previous methods, the DIRNet is not trained with known registration examples, but learns to register images by directly optimizing a similarity metric between the fixed and the moving image. Hence, similar to conventional intensity-based image registration, it directly learns the registration task end-to-end and it is truly unsupervised. In addition, a trained DIRNet is able to perform deformable image registration non-iteratively on unseen data.

To the best of our knowledge, this is the first deep learning method for end-to-end unsupervised deformable image registration.

2 Method

The proposed DIRNet consists of a ConvNet regressor, a spatial transformer, and a resampler (Fig. 1). The ConvNet regressor analyzes spatially corresponding patches from a pair of fixed and moving input images and outputs local deformation parameters for the spatial transformer. The spatial transformer generates a dense displacement vector field (DVF) that enables the resampler to warp the moving image to the fixed image. The DIRNet learns the registration task end-to-end by unsupervised training with an image similarity metric. Since the training phase involves simultaneous optimization of registration of many image pairs, the ConvNet implicitly learns a representation of the features in images that are important for predictions of local displacement. Unlike regular image registration methods that typically perform iterative optimization for each image pair at hand, a trained DIRNet registers images in one pass.

The ConvNet regressor expects concatenated pairs of moving and fixed images as its input, and applies four alternating layers of 3×3 convolutions with 0-padding and 2×2 downsampling layers. Downsampling reduces the number of the ConvNet parameters, but it is associated with translational invariance. We postulate that this effect should be minimal in a ConvNet used for image registration, thus we use average pooling which should retain the most information during downsampling. Subsequently, 3×3 convolutional layers are added to increase the receptive field of the ConvNet to coincide with the capture range of the control points of the spatial transformer. Finally, three 1×1 convolutional layers are applied. Batch normalization [2] and exponential linear units [1] are used throughout the network, except in the final layer. The number of kernels per layer can be of arbitrary size, but the number of kernels of the final layer is determined by the dimensionality of the input images (e.g. 2 kernels for 2D images that require 2D displacement). The fully convolutional design in combination with downsampling allows fast analysis of separate spatially corresponding patch pairs from the fixed and moving images. The input image sizes and the number of downsampling layers jointly define the number of output parameters, i.e. the size and spacing of the control point grid. This way, for images of different sizes, similar grid spacing is ensured. Using the control point displacements, the spatial transformer generates a DVF used to warp the moving image to the fixed image. Like in [3], a thin-plate spline could be used as a spatial transformer, but due to its global support it is deemed less suitable for a patched-based approach. Therefore, we implemented a cubic B-spline [10] transformer which has local support. Thereafter, a resampler is used to generate warped moving images with linear interpolation.

The DIRNet is trained by optimizing an image similarity metric (i.e. by backpropagating dissimilarity) between pairs of moving and fixed images from a training set using mini-batch stochastic gradient descent (Adam [4]). Any similarity metric used in conventional image registration could be used. In this work normalized cross correlation is employed.

After training, the DIRNet can be applied for registration of unseen image pairs from a separate dataset.

3   Data

The DIRNet was evaluated with handwritten digits from the MNIST database [5] and clinical MRI scans from the Sunnybrook Cardiac Data (SCD) [9].

The MNIST database contains 28 × 28 pixel grayscale images of handwritten digits that were centered by computing the center of mass of the pixels. The test images (10,000 digits) were kept separate from the training images (60,000 digits). One sixth of the training data was used for validation to monitor overfitting during training.

The SCD contains 45 cardiac cine MRI scans that were acquired on a single MRI-scanner. The scans consist of short-axis cardiac image slices each containing 20 timepoints that encompass the entire cardiac cycle. Slice thickness and spacing is 8 mm, and slice dimensions are 256×256 with a pixel size of 1.25 mm×1.25 mm. The SCD is equally divided in 15 training scans (183 slices), 15 validation scans (168 slices), and 15 test scans (176 slices). An expert annotated the left ventricle myocardium at end-diastolic (ED) and end-systolic (ES) time points following the annotation protocol of the SCD. Annotations were made in the test scans and only used for final quantitative evaluation. In total, 129 image slices were annotated, i.e. 258 annotated timepoints.

4   Experiments and Results

DIRNet was implemented with Theano[1] and Lasagne[2], and conventional registration was performed with SimpleElastix [7].

4.1   Registration of Handwritten Digits

To demonstrate feasibility of the method we first applied it to registration of handwritten digits from the MNIST data. Separate DIRNet instances were trained for image registration of a specific class: one for each digit. The DIRNets were designed with 16 kernels per convolution layer, the third and fourth downsampling layers were removed. This resulted in a control point grid of 7 × 7 (grid spacing of 4 pixels). Each DIRNet was trained separately with random combinations of digits from its class with mini-batches of 32 random fixed and moving image pairs in 5,000 iterations (i.e. backpropagations). See Fig. 2 (left) for the learning curves.

---

[1] http://deeplearning.net/software/theano/ *(version 0.8.2)*.
[2] https://lasagne.readthedocs.io/en/latest/ *(version 0.2.dev1)*.

Registration performance of the trained DIRNets was qualitatively assessed on the test data. For each digit, one sample was randomly chosen to be the fixed image. Thereafter, all remaining digits (approximately 1,000 per class) were registered to the corresponding fixed image. Figure 2 (right) shows the registration results.

Fig. 2. Left: Learning curves showing the negative normalized cross correlation loss (L_{NCC}) on the validation set of DIRNets trained in 5,000 iterations for registration of MNIST digits. Right: Registration results of the trained DIRNets on a separate test set. The top row shows an average of all moving images per class (about 1,000 digits), the middle row shows one randomly chosen fixed image per class, and the bottom row shows an average of the registration results of independent registrations of the moving images to the chosen fixed image. Averages of the moving images after registration are much sharper than before registration indicating a good registration result. The blurry areas in the resulting images indicate where registration is challenging.

4.2   Registration of Cardiac MRI

Next, to demonstrate feasibility of the method on real-world medical data, we register cine cardiac MR images from the SCD. The DIRNet was trained by randomly selecting pairs of fixed and moving image slices from cardiac cine MRI scans (4D data). The pairs of fixed and moving images were anatomically corresponding slices from the same 4D scan of a single patient but acquired at different time points in the cardiac cycle. This resulted in 69,540 image pairs for training, and 63,840 pairs for validation.

A baseline DIRNet, as described in Sect. 2, was designed with 16 kernels per convolution layer. This resulted in a grid of 16 × 16 control points, i.e. a grid spacing of 16 pixels (20 mm). To evaluate effect of various DIRNet parameters, additional experiments were performed. First, to evaluate the effect of the downsampling method, DIRNet-A1 was designed with max-pooling layers, and DIRNet-A2 was designed with 2×2 strided convolutions. Second, to evaluate the effect of the spatial transformer, DIRNet-B1 was designed with a quadratic B-spline transformer, and DIRNet-B2 with a thin-plate spline transformer. Finally, to show the effect of the size of the receptive field (i.e. patch size), DIRNet-C1 was designed with neighbouring (i.e. non-overlapping) patches, by leaving out the last two 3 × 3 convolutional layers. In addition, DIRNet-C2 analyzed full image slices for each control point by replacing the 1 × 1 convolution layers with a 3 × 3 convolution layer, followed by a downsampling layer, two fully connected layers of 1,024 nodes, and a final output layer of 16 × 16 2D control points.

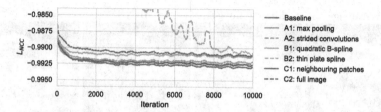

Fig. 3. Learning curves showing the negative normalized cross correlation loss (L_{NCC}) on the validation set of DIRNets The Net loss over 10,000 iteration for the baseline DIRNet, DIRNets with different downsampling techniques (A1, A2), DIRNets with different spatial transformers (B1, B2), and DIRNets with different receptive fields (C1, C2).

Table 1. Registration performance was quantified in cardiac MRI by registration of image slices and subsequent transformation of corresponding left ventricle annotations. Mean and standard deviations of the Dice coefficients between the reference and warped segmentations were computed. Additionally, 95th percentiles of the surface distance (95thSD), and mean absolute surface distance (MAD) were calculated. The rows list results before registration, with conventional iterative image registration using SimpleElastix, and results obtained using the DIRNet. The rightmost column shows the runtime at inference for the conventional methods and the best performing DIRNet.

		Iterations	Dice	95th SD (mm)	MAD (mm)	Time (s)
No registration			0.62 ± 0.15	7.79 ± 2.92	2.89 ± 1.07	-
SimpleElastix		2×100	0.79 ± 0.08	5.09 ± 2.36	1.91 ± 0.94	0.51 ± 0.07
SimpleElastix		2×2000	$\mathbf{0.81 \pm 0.08}$	5.09 ± 7.25	$\mathbf{1.75 \pm 1.29}$	7.38 ± 0.94
DIRNet	BL		0.80 ± 0.08	$\mathbf{5.03 \pm 2.30}$	1.83 ± 0.89	0.049 ± 0.004
	A1		0.78 ± 0.08	5.26 ± 2.16	1.95 ± 0.85	-
	A2		0.78 ± 0.08	5.30 ± 2.28	1.97 ± 0.87	-
	B1		0.72 ± 0.11	6.41 ± 2.61	2.40 ± 0.96	-
	B2		0.78 ± 0.09	5.48 ± 2.36	2.01 ± 0.89	-
	C1		0.79 ± 0.08	5.20 ± 2.30	1.92 ± 0.89	-
	C2		0.76 ± 0.09	5.55 ± 2.24	2.10 ± 0.90	-

Each DIRNet was trained until convergence in mini-batches of 32 image pairs in at least 10,000 iterations. The training loss closely followed the validation loss in each experiment, and no signs of overfitting were apparent. Figure 3 shows the validation loss of 10,000 iterations during training for all experiments. The DIRNets converged quickly in each experiment, except DIRNet-B2, where convergence was reached after approximately 30,000 iterations. The final loss was lowest for baseline DIRNet.

Quantitative evaluation was performed on the test set by registering image slices at ED to ES, and vice versa, which resulted in 258 independent registration experiments. The obtained transformation parameters were used to warp

Fixed image   Moving image   DIRNet   SimpleElastix

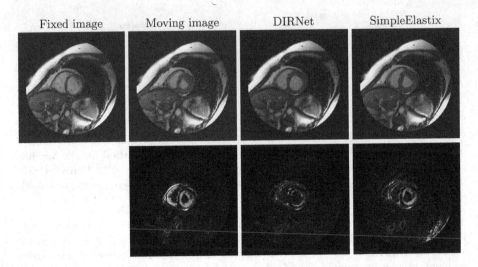

Fig. 4. Top, from left to right: The fixed (ED), the moving (ES), the DIRNet warped, and the SimpleElastix warped images. Bottom: Heatmaps showing absolute difference images between the fixed image and (from left to right) the original, the DIRNet warped, and the SimpleElastix warped moving images.

the left ventricle annotations of the moving image to the fixed image. The transformed annotations were compared with the reference annotations in the fixed images. The results are listed in Table 1. For comparison, the table also lists conventional iterative intensity-based registrations (SimpleElastix), with parameters specifically tuned for this task. A grid spacing was used similar to the DIRNet but in a multi-resolution approach, downsampling first with a factor of 2 and thereafter using the original resolution. Two conventional image registration experiments were performed, one for optimal speed (2 times 100 iterations), and one for optimal registration accuracy (2 times 2000 iterations), but at the cost of longer computation time. Experiments with the DIRNets were performed on an NVIDIA Titan X Maxwell GPU and experiments with SimpleElastix were performed on an Intel Xeon 1620-v3 3.5 GHz CPU using 8 threads. Figure 4 shows registration results for a randomly chosen image pair.

5   Discussion and Conclusion

A deep learning method for unsupervised end-to-end learning of deformable image registration has been presented. The method has been evaluated with registration of images with handwritten digits and image slices from cine cardiac MRI scans. The presented DIRNet achieves a performance that is as accurate as a conventional deformable image registration method with substantially shorter execution times. The method does not require training data, which is often difficult to obtain for medical images. To the best of our knowledge this is the first deep learning method that uses unsupervised end-to-end training for deformable image registration.

Even though registration of images with handwritten digits is an easy task, the performed experiments demonstrate that a single DIRNet architecture can be used to perform registration in different image domains given domain specific training. It would be interesting to further investigate whether a single DIRNet instance can be trained for registration across different image domains.

Registration of slices from cardiac cine MRI scans was quantitatively evaluated between the ES and ED timepoints, so at maximum cardiac compression and maximum dilation. The conventional registration method (SimpleElastix) was specifically tuned for this task and the DIRNet was not, because it was trained for registration of slices from any timepoint of the cardiac cycle. Nevertheless, the results of the DIRNet were comparable to the conventional approach.

The data used in this work did not require pre-alignment of images. However, to extend the applicability of the proposed method in future work, performing affine registration will be investigated. Furthermore, proposed method is designed for registration of 2D images. In future work the method will be extended for registration of 3D images. Moreover, experiments were performed using only normalized cross correlation as a similarity metric, but any differentiable metric could be used.

To conclude, the DIRNet is able to learn image registration tasks in an unsupervised end-to-end fashion using an image similarity metric for optimization. Image registration is performed in one pass, thus non-iteratively. The results demonstrate that the network achieves a performance that is as accurate as a conventional deformable image registration method within shorter execution times.

Acknowledgments. This study was funded by the Netherlands Organization for Scientific Research (NWO): project 12726.

References

1. Clevert, D.A., Unterthiner, T., Hochreiter, S.: Fast and accurate deep network learning by exponential linear units (ELUs). In: ICLR (2016)
2. Ioffe, S., Szegedy, C.: Batch normalization: accelerating deep network training by reducing internal covariate shift. In: PMLR, vol. 37, pp. 448–456 (2015)
3. Jaderberg, M., Simonyan, K., Zisserman, A., Kavukcuoglu, K.: Spatial transformer networks. In: Cortes, C., Lawrence, N.D., Lee, D.D., Sugiyama, M., Garnett, R. (eds.) Advances in Neural Information Processing Systems 28, pp. 2017–2025. Curran Associates, Inc., Red Hook (2015)
4. Kingma, D., Ba, J.: Adam: a method for stochastic optimization. In: ICLR (2015)
5. LeCun, Y., Cortes, C.: The MNIST database of handwritten digits (1998)
6. Liao, R., Miao, S., de Tournemire, P., Grbic, S., Kamen, A., Mansi, T., Comaniciu, D.: An artificial agent for robust image registration. arXiv preprint arXiv:1611.10336 (2016)

7. Marstal, K., Berendsen, F., Staring, M., Klein, S.: SimpleElastix: a user-friendly, multi-lingual library for medical image registration. In: Proceedings of the IEEE Conference on Computer Vision and Pattern Recognition Workshops (2016)
8. Miao, S., Wang, Z.J., Liao, R.: A CNN regression approach for real-time 2D/3D registration. IEEE Trans. Med. Imaging 35(5), 1352–1363 (2016)
9. Radau, P., Lu, Y., Connelly, K., Paul, G., Dick, A., Wright, G.: Evaluation framework for algorithms segmenting short axis cardiac MRI (2009)
10. Rueckert, D., Sonoda, L.I., Hayes, C., Hill, D.L.G., Leach, M.O., Hawkes, D.J.: Nonrigid registration using free-form deformations: application to breast MR images. IEEE Trans. Med. Imaging 18(8), 712–721 (1999)
11. Wu, G., Kim, M., Wang, Q., Munsell, B.C., Shen, D.: Scalable high performance image registration framework by unsupervised deep feature representations learning. IEEE Trans. Biomed. Eng. 63(7), 1505–1516 (2016)

Stain Colour Normalisation to Improve Mitosis Detection on Breast Histology Images

Azam Hamidinekoo[(⊠)] and Reyer Zwiggelaar

Department of Computer Science, Aberystwyth University, Aberystwyth, UK
azh2@aber.ac.uk

Abstract. The mitosis count, one of the main components considered for grading breast cancer on histology images, is used to assess tumour proliferation. On breast histology sections, different coloured stains are used to highlight existing cellular components. To keep the wealth of information in the stain colour representation and decrease the sensitivity of detection and classification models to stain variations, the RGB Histogram Specification method is used as a preprocessing step in the training of a modified deep convolutional neural network. Different models are trained on raw and stain normalised images using different databases. Evaluation results show more stable detection performance for various imaging conditions. Combining different data sources and employing stain colour normalisation and transfer learning, a network is trained that can be used for the general mitosis detection task and dealing with staining and scanner variations.

1 Introduction

Breast cancer is the most prevalent type of cancer diagnosed and accounts for 25.2% of cancer related death worldwide [10]. In breast histological imaging, different tissue components are visualised by using staining. The standard staining protocol for breast tissue is Hematoxylin and Eosin (H&E) that selectively stains nuclei structures blue and cytoplasm pink [12]. Different H&E stained breast tissue image frames acquired from two different Whole Slide Imaging (WSI) scanners are shown in Fig. 1a and c. Tumour differentiation in breast cancer histopathology is analysed via the Bloom-Richardson and Nottingham grading scheme, recommended by the World Health Organisation [3]. It is derived from the assessment of three morphological features: tubule formation, nuclear pleomorphism and mitotic count. A numerical score from 1–3 is assigned to each component and when combined, it provides the overall breast cancer grade.

Recently, deep learning has been used for the automatic analysis of breast histopathology slides and several competitions have been held [2,7]. Among the three main components considered for grading breast cancer, mitosis detection for assessing tumour proliferation is prognostically critical. Several appropriately annotated databases have become publicly available for mitosis detection. In deep learning based approaches, a network, that is trained on one specific dataset, is only useful for data gathered at that centre. Such model often needs

© Springer International Publishing AG 2017
M.J. Cardoso et al. (Eds.): DLMIA/ML-CDS 2017, LNCS 10553, pp. 213–221, 2017.
DOI: 10.1007/978-3-319-67558-9_25

to be readjusted when used for slides from other centres. Nevertheless, transfer learning is one of the solutions where knowledge gained during training on one type of problem is used to train on another similar problem [11]. However, the dependency issue of deep learning based methods on specific training sources gets worse if that training source does not provide a sufficient amount of data required by deep learning. For mitosis detection in breast tissue histology images, one can pre-train a deep convolutional neural network (CNN) on a large annotated dataset, but subsequently fine-tuning using a new but smaller dataset, the performance of the resulting network might be decreased. This data dependency issue does exist in both deep learning and conventional machine learning based methods. The motivation behind this study is to propose an approach that can reduce such data dependency issue and provide a more general model for classifying mitosis and non-mitosis samples, detecting mitosis on breast histopathology sections and estimating the score of mitotic count considered in the breast grading system.

In this work, a stain colour normalisation method is proposed as a preprocessing stage along with a modified deep CNN for tackling the issue of stain colour variations among different data sources. The robustness of several trained deep networks is investigated with regards to sensitivity to different clinical conditions in the task of detecting mitoses and estimating the number of mitosis occurrences in a region of interest (RoI). We show how stain colour normalisation and transfer learning concepts can be employed to improve mitosis detection results and provide a general framework to be used for external test data.

2 Related Work

Estimating the mitosis count on breast histological images is challenging due to their small size with various shape configurations. Promising results using deep learning based methods were reported recently. Cireşan et al. [6] used a deep max-pooling CNN directly on raw data. Using the same approach with a multi-column CNN, they won the mitotic detection competition [2]. They averaged the outputs of multiple classifiers with different architectures to reduce the CNN variance and bias on a specific dataset. Wang et al. [13] combined a lightweight CNN with hand-crafted features and a cascade of two random forest classifiers. Chen et al. [5] suggested a deep cascade neural network for firstly retrieving probable mitosis candidates and secondly detecting mitotic cells in all positive samples determined by the first CNN. To overcome the lack of sufficient annotated training samples, the concept of learning from crowds and generating a ground-truth labelling from non-expert crowd sourced annotation was presented in [1]. However, these methods are trained and tested on a specific dataset and can not be generalised to be used for additional/different image data. Moreover, the issue of robustness to various data sources with different clinical/protocols has not been investigated which is covered in this study.

3   Dataset

A WSI for one case is examined by a pathologist to identify the RoI and then image frames are created by subdividing RoIs. We used image frames from two datasets (AMIDA13 [2] and ICPR14 [7]) with x40 magnification acquired from two different scanners in various labs and split them by patient into training, validation and test sets. Detailed information can be found in Table 1.

4   Method

4.1   Patch Generation

Firstly, different types of image frames, that are provided on pre-detected RoIs, are generated resulting in raw and stain colour normalised image frames as:

Raw image frames: Raw data contains the image databases introduced in Sect. 3. These two sets of data are named AM-raw (Fig. 1b) and IC-raw (Fig. 1d) which represent images from the AMIDA13 and ICPRT14 databases, respectively.

Stain normalised image frames: As explained, histological images are prone to variations caused by different scanners used for digitisation, different staining appearance characteristics and magnification factors. Proposed methods are often hampered by these tissue variabilities. We used the RGB Histogram Specification [8] as a normalisation approach which is simple and fast. In this method, colour values of a target image are readjusted so that its histogram matches the source image histogram distribution. So, two new stain normalised image frames are generated using a source image that is selected experimentally from the AMIDA13 dataset considering how much it contains various cellular compartments:

Table 1. Two publicly available databases for mitosis detection and implementation assignment information in our experiments.

Database	AMIDA13 [2]	ICPR14 [7]
Number of cases used	21	11
Image format	.TIF	.TIFF
Magnification	x40	x10, x20, x40
Slide scanner	Aperio ScanScope XT	Hamamatsu 2.0HT
Spatial resolution	0.25 μm/pixel	0.227 μm/pixel
Image size	2000 × 2000	1663 × 1485
Origin of data	The Netherlands	France
Year announced	2013	2014
Implementation assignments	Train: 14 cases (65%), with 448 image frames	Train: 5 cases (46%), with 624 image frames
	Validation: 4 cases (20%), with 135 image frames	Validation: 3 cases (27%), with 288 image frames
	Test: 3 cases (15%), with 80 image frames	Test: 3 cases (27%), with 288 image frames

1. AM-AM-Norm: AMIDA13 images are histogram matched to a AMIDA13 source image (see Fig. 1c).
2. IC-AM-Norm: ICPR14 images are histogram matched to a AMIDA13 source image (see Fig. 1e).

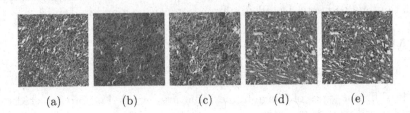

(a)   (b)   (c)   (d)   (e)

Fig. 1. Image frames at x40 magnification from the AMIDA13 and ICPR14 databases with the annotated mitosis figures shown as green circles. (a) the source image for the RGB Histogram Specification concept; (c) AM-AM-Norm and (e) IC-AM-Norm are stain normalised images of (b) AM-raw and (d) IC-raw. (Color figure online)

Using modified image frames, patches are extracted to construct training and validation sets. To ensure the creation of sufficiently rich representation sets, the patch extraction approach proposed in [9] is employed. The combination of blue ratio segmentation [4] and annotation mask is used to create patch candidates for training. To capture sufficient context and surrounding area of a mitotic figure, we extract 64×64 patches from x40 frames. Data augmentation is performed including random cropping and rotations of (0:45:315 degrees) for positive and (0:90:270 degrees) for negative exemplars to provide a balanced dataset.

4.2 CNN Architecture

Inspired by the work in [9], we use the same architecture but modify the last layers by removing the last convolutional layer and replacing it with a fully connected layer. The architecture consists of 6 convolutional layers followed by batch normalisation and ReLu layers and a combination of fully connected layer with dropout and Relu layers followed by the softmax loss function for creating smoother gradient. This way, the high-level invariant features captured in the convolutional layers are flattened and fed to the fully connected layer to learn non-linear combinations of these features and prevent inter-dependencies between nodes. In this architecture, the input size of each layer is decreased by convolutional layers instead of pooling layers. These characteristics allow the network to learn a more robust relationship. We use the same weight initialisation scheme as in [9] and a learning rate multiplier of 1.0 while training from scratch and 0.1 while using pre-trained weights. The fully connected layer is randomly Gaussian initialised in all implementations. The network is trained with an Adaptive Gradient solver, base learning rate of 0.001 with step down policy and dropout ratio of 50%. Other parameters are set as: weight decay 0.005, momentum 0.9, training batch size 128 and validation batch size 64 for a maximum of 10 epochs. This network results in about 103 K parameters to be learnt.

4.3 Training and Testing Workflow

Once the training and validation patches for each of the four prepared image sets from Subsect. 4.1 are created, the modified CNN architecture with the defined hyperparameters is used separately on each set to train different networks for classifying patches into mitosis and non-mitosis classes. These networks are trained in two ways: (i) from scratch and (ii) using transfer learning. In order to investigate the effect of performing stain colour normalisation on the classification improvement, we train six networks as:

- **Network-1**: is trained on the AM-raw dataset from scratch.
- **Network-2**: is trained on the AM-AM-Norm dataset from scratch.
- **Network-3**: is trained on the IC-raw dataset from scratch.
- **Network-4**: is trained on the IC-AM-Norm dataset from scratch.
- **Network-5**: fine-tuned on the IC-raw dataset based on Network-1.
- **Network-6**: fine-tuned on the IC-AM-Norm dataset based on Network-2.

Accordingly, a series of mean subtracted image patches are fed to the network over a series of epochs and the error derivative is calculated. Performing backpropagation through the network and shrinking the learning rate over time, the local minimum is obtained.

We test the performances of Networks 1, 3, 5 on the test image set from IC-raw database. Similarly, the performances of Networks 2, 4, 6 are evaluated on the test image set from IC-AM-Norm database. For this, we only consider the test candidates resulted from the blue ratio segmentation mask. Using the centres of these candidates, several patches can be extracted near to each candidate. Feeding these test patches to a trained model, a label (0 corresponding to non-mitosis and 1 corresponding to mitosis) is assigned to each patch. The details of what is counted as detected mitosis is described in the competition rules [7]. A voting scheme is applied to the labels given to different patches of a candidate and the final class is decided. A patch is decided as mitosis if more than three positive votes (experimentally determined) are assigned to it. So, the networks are compared based on the common evaluation metrics for the task of classification in histology images and the network with the best average F-score is reported.

Using the best Network from the first part, in the second part of our implementations, we address the effect of concatenating both datasets and training based on a large but miscellaneous normalised samples. This network is trained on the combination of AM-AM-Norm and IC-AM-Norm datasets and fine tuned using the best network's weights. The final trained network (called Network-7) is tested on additional data. Finally, we compare and report our outcomes with the best results of mentioned competitions.

5   Results and Discussion

The first aim of this study is to investigate the effect of performing stain colour normalisation in order to bring the image histograms in a similar range. This effect is investigated by evaluating the classification performance of six networks trained on different datasets. Empirical evaluation of the network ability for detecting mitoses on image frames of x40 magnification from the ICPR14 dataset is presented in detail in Table 2 using evaluation metrics: D = number of detected mitoses/number of annotated mitoses; TP = number of truly detected mitoses; FN = number of falsely detected non-mitoses; FP = number of falsely detected mitoses; recall = $\frac{TP}{(TP+FN)}$; precision = $\frac{TP}{(TP+FP)}$ and F-score= $\frac{2\times(precision\times sensitivity)}{(precision+sensitivity)}$.

In our experiments, an image from the AMIDA13 dataset is selected as a source image and all the ICPR14 images are mapped to this source image. As expected, when the network is trained on AM-raw data and tested on the IC-raw test set, favourable classification results can not be obtained due to different inherent image characteristics with respect to various stain concentration values. In addition, if the network is trained on IC-raw and tested on the same dataset, again good results are not obtained because of the small amount of data used while training (see Table 2). However, the results on Network-2 and Network-4 confirm that if the images in a dataset are all stain normalised to the source image, then various stain concentrations, that exist in different images, can be mapped to a unique source and the classification performance improves. The proposed CNN classifier not only extracts low to high level shape and morphology features but it considers the colour distribution as well. An important outcome of performing stain colour normalisation, as a preprocessing step, is that a model can be trained and then used for performing classification on additional test data which is also mapped to the source image. Moreover, this strategy leads to demonstrating stable performance which is not sensitive to imaging conditions in general and scanner variations in particular. Concatenating the two concepts of stain colour normalisation and transfer learning is reflected in Network-6, which shows improvement in the classification and detection performance.

The average of the resulting F-score for our best network (Network-6) is calculated as 0.78 which is superior to the top results reported in the Mitosis-Atypia-14 grand challenge competition on mitosis detection using data from

Table 2. Performance results on the test cases selected randomly from ICPR14.

| Networks | ICPR14 test cases | | | | | | | | | | | | | | |
| | Case-1: H03 | | | | | Case-2: H04 | | | | | Case-3: H07 | | | | |
	D	TP	FN	FP	F-score	D	TP	FN	FP	F-score	D	TP	FN	FP	F-score
Network-1	144/136	60	76	68	0.42	94/267	33	233	61	0.17	33/39	23	16	10	0.63
Network-2	165/136	112	24	53	0.73	151/267	100	167	51	0.47	45/39	34	5	11	0.80
Network-3	104/136	88	48	16	0.72	180/267	139	128	41	0.62	2/39	2	37	0	0.095
Network-4	108/136	92	44	16	0.74	131/267	113	154	18	0.56	35/39	28	11	7	0.75
Network-5	94/136	69	67	26	0.59	195/267	163	104	35	0.70	34/39	25	14	9	0.68
Network-6	133/136	116	20	17	**0.87**	164/267	159	108	15	**0.73**	36/39	29	10	7	**0.76**

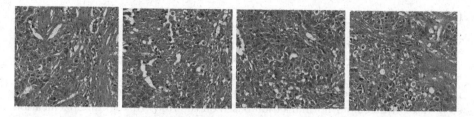

Fig. 2. Detection and classification results on ICPR14 test set fed to Network-6. The green, yellow and red circles represent TPs, FPs and FNs, respectively. (Color figure online)

Hamamatsu scanner [7]. In summary, the experimental results in the first part of our evaluations, implies that the concept of colour normalisation significantly improves the detection rate and this can be extended to other medical images. The reason is that medical images like H&E histology images share the same characteristics but in different ranges. So, being mapped to a source image, such differences can be removed successfully, which is specific to medical images. Figure 2 shows detection results on the ICPR14 test data for patch-wise mitosis classification using Network-6.

The second aim of our study is to investigate the performance of a network that uses a mixture of two training sets (AM-AM-Norm and IC-AM-Norm) and is fine-tuned using the weights of Network-6 (initialising convolution layers with pre-trained weights and retraining the fully connected layer from scratch). We test this network (Network-7) performance on several test data and present our results quantitatively in Table 3 which are stable, general and superior to the best reported F-score results in the AMIDA13 (61.1%) [2] and ICPR14 (71%) [7] challenges while the average F-score obtained in our experiments is 81.4%.

Table 3. Evaluation results of Network-7 performance, which is trained on the concatenated stain normalised training data (IC-AM-Norm and AM-AM-Norm), on various types of test data from different sources.

Source of test set	ID	D	Precision	Sensitivity	F-score
AMIDA13 test set	Case 01	59/74	0.79	1	0.88
	Case 02	35/37	0.94	1	0.96
	Case 11	9/15	0.60	1	0.75
ICPR14 test set	Case H03	168/136	0.85	0.69	0.76
	Case H04	186/267	0.61	0.88	0.72
	Case H07	32/39	0.82	0.80	0.80
Additional anonymous test set	-	14/21	0.66	1	0.79
	-	32/39	0.82	0.86	0.83
	-	21/25	0.84	0.84	0.84

6 Conclusion

Mitosis detection for assessing tumour proliferation is one of the three main components considered for grading breast cancer on histological images. In this work, a modified deep learning based network architecture is described to detect mitoses on RoI sections of breast histological images. We used a pre-processing step called stain colour normalisation based on the RGB Histogram Specification method to be employed on raw images that are captured from different sources or scanners. In this process, colour values of a target image are readjusted to match the colour distribution of a source image on a pixel-by-pixel basis. The goal of this strategy is to decrease stain concentration variation among different images and result in more stable and insensitive detection or classification performance to various imaging conditions. Eventually, concatenating different data sources using colour normalisation and transfer learning, a network is trained and tested on additional breast image data which shows significant robustness with respect to image variabilities.

References

1. Albarqouni, S., Baur, C., Achilles, F., Belagiannis, V., Demirci, S., Navab, N.: AggNet: deep learning from crowds for mitosis detection in breast cancer histology images. IEEE Trans. Med. Imaging **35**(5), 1313–1321 (2016)
2. Assessment of Mitosis Detection Algorithms (AMIDA13), MICCAI Grand Challenge (2013). http://amida13.isi.uu.nl
3. Bloom, H.J.G., Richardson, W.W.: Histological grading and prognosis in breast cancer: a study of 1409 cases of which 359 have been followed for 15 years. Br. J. Cancer. **11**(3), 359–377 (1957)
4. Chang, H., Loss, L.A., Parvin, B.: Nuclear segmentation in H&E sections via multi-reference graph cut (MRGC). In: International Symposium Biomedical Imaging (2012)
5. Chen, H., Dou, Q., Wang, X., Qin, J., Heng, P.-A.: Mitosis detection in breast cancer histology images via deep cascaded networks. In: Proceedings of the Thirtieth AAAI Conference on Artificial Intelligence, pp. 1160–1166 (2016)
6. Cireşan, D.C., Giusti, A., Gambardella, L.M., Schmidhuber, J.: Mitosis detection in breast cancer histology images with deep neural networks. In: Mori, K., Sakuma, I., Sato, Y., Barillot, C., Navab, N. (eds.) MICCAI 2013. LNCS, vol. 8150, pp. 411–418. Springer, Heidelberg (2013). doi:10.1007/978-3-642-40763-5_51
7. Detection of mitosis and evaluation of nuclear atypia score in Breast Cancer Histological Images. In: International Conference for Pattern Recognition (ICPR) (2014). https://mitos-atypia-14.grand-challenge.org/
8. Jain, A.K.: Fundamentals of digital image processing. Prentice-Hall Inc., Upper Saddle River (1989)
9. Janowczyk, A., Madabhushi, A.: Deep learning for digital pathology image analysis: a comprehensive tutorial with selected use cases. J. Pathol. Inf. **7** (2016)
10. Stewart, B.W.K.P., Christopher P.W.: International Agency for Research on Cancer, WHO: World Cancer Report (2014)
11. Tajbakhsh, N., Shin, J.Y., Gurudu, S.R., Hurst, R.T., Kendall, C.B., Gotway, M.B., Liang, J.: Convolutional neural networks for medical image analysis: full training or fine tuning? IEEE Trans. Med. Imaging **35**(5), 1299–1312 (2016)

12. Veta, M., Pluim, J.P., van Diest, P.J., Viergever, M.A.: Breast cancer histopathology image analysis: a review. IEEE Trans. Biomed. Eng. **61**(5), 1400–1411 (2014)
13. Wang, H., Cruz-Roa, A., Basavanhally, A., Gilmore, H., Shih, N., Feldman, M., Tomaszewski, J., Gonzalez, F., Madabhushi, A.: Cascaded ensemble of convolutional neural networks and handcrafted features for mitosis detection. In: SPIE Medical Imaging, p. 90410B (2014). doi:10.1117/12.2043902

3D FCN Feature Driven Regression Forest-Based Pancreas Localization and Segmentation

Masahiro Oda[1]([☒]), Natsuki Shimizu[2], Holger R. Roth[1], Ken'ichi Karasawa[2],
Takayuki Kitasaka[3], Kazunari Misawa[4], Michitaka Fujiwara[5],
Daniel Rueckert[6], and Kensaku Mori[1,7]

[1] Graduate School of Informatics, Nagoya University, Nagoya, Japan
`moda@mori.m.is.nagoya-u.ac.jp`
[2] Graduate School of Information Science, Nagoya University, Nagoya, Japan
[3] School of Information Science, Aichi Institute of Technology, Toyota, Japan
[4] Aichi Cancer Center, Nagoya, Japan
[5] Nagoya University Graduate School of Medicine, Nagoya, Japan
[6] Department of Computing, Imperial College London, London, UK
[7] Strategy Office, Information and Communications,
Nagoya University, Nagoya, Japan

Abstract. This paper presents a fully automated atlas-based pancreas segmentation method from CT volumes utilizing 3D fully convolutional network (FCN) feature-based pancreas localization. Segmentation of the pancreas is difficult because it has larger inter-patient spatial variations than other organs. Previous pancreas segmentation methods failed to deal with such variations. We propose a fully automated pancreas segmentation method that contains novel localization and segmentation. Since the pancreas neighbors many other organs, its position and size are strongly related to the positions of the surrounding organs. We estimate the position and the size of the pancreas (localization) from global features by regression forests. As global features, we use intensity differences and 3D FCN deep learned features, which include automatically extracted essential features for segmentation. We chose 3D FCN features from a trained 3D U-Net, which is trained to perform multi-organ segmentation. The global features include both the pancreas and surrounding organ information. After localization, a patient-specific probabilistic atlas-based pancreas segmentation is performed. In evaluation results with 146 CT volumes, we achieved 60.6% of the Jaccard index and 73.9% of the Dice overlap.

Keywords: Segmentation · Pancreas · Fully convolutional network · Regression forest

1 Introduction

Medical images contain much useful information for computer-aided diagnosis and interventions. Organ segmentation from medical images is necessary to

© Springer International Publishing AG 2017
M.J. Cardoso et al. (Eds.): DLMIA/ML-CDS 2017, LNCS 10553, pp. 222–230, 2017.
DOI: 10.1007/978-3-319-67558-9_26

utilize such information for many applications. Many abdominal organ segmentation methods have been reported [1–8]. We used the pancreas segmentation results obtained from them in diagnosis, visualization, surgical planning, and surgical simulation. Pancreas segmentation accuracies from CT volumes are commonly lower than those of other abdominal organs in previous methods. The main cause of the difficulty comes from the large spatial (position and size) variation of the pancreas. Because it is small and neighbors many organs, its actions and shape are influenced by the movements and the deformations of its neighbor organs. Therefore, prior to pancreas segmentation, pancreas localization using global information, which might include the intensity, texture, appearance, or structure of the surrounding organs, should be performed for higher segmentation accuracies. Pancreas localization denotes a rough estimation of the pancreas position and size.

Most previous pancreas segmentation methods employ probabilistic atlas-based segmentation approaches [1–5,8]. A patient-specific-atlas [1,4,5,8] and a hierarchical-atlas [3] have been used. However, such atlas-based approaches cannot deal with an organ's spatial variation. Some methods localized pancreas atlases using manually specified information [1,4] or the segmentation results of other organs [2,3]. These methods depend on other organ segmentation results or require manual interactions. Recently, pancreas segmentation methods have been proposed with automated localization techniques [7,8]. Roth et al. [7] used deep convolutional neural networks with holistically-nested networks for localization. Oda et al. [8] used the regression forest technique to localize the pancreas atlas based on global feature values. Other methods [7,8] showed higher pancreas segmentation accuracies than the previous methods. Their results provided that an accurate localization technique is crucial for pancreas segmentation.

Pancreas localization also resembles the regression of the pancreas position and size from the information of the surrounding tissues or organs. Roth et al. [7] performed localization using small 2D patch appearances obtained from around the pancreas. The 3D information of pancreas surroundings, which has not adequately considered, is important for more accurate localization. Oda et al. [8] used global feature values for localization. A global feature value is the first-order difference of the CT values sampled at many local regions in CT volumes. Global feature values, which only capture the edge information from CT values, do not utilize the more useful information contained in the CT volumes for such localizations as texture or appearance. In these methods [7,8], the tissue or organ information surrounding the pancreas is not well utilized for localization.

We propose a pancreas localization and segmentation method from a CT volume utilizing 3D fully convolutional network (FCN) features for localization. We perform localization using the regression forest technique followed by segmentation with a patient-specific probabilistic atlas. In the localization, we introduce 3D FCN-based features obtained from a 3D FCN trained to segment abdominal organs including the pancreas. We use the 3D U-Net [9] as the 3D FCN structure. Regression forests perform regression of a pancreas-bounding box from 3D FCN-based features and the first- and second-order intensity difference features calculated in many local regions obtained all over the CT volume. The 3D FCN features capture statistically useful information for organ localization from a training dataset. We also introduce a new organ texture feature calculation method from the 3D FCN features. Since direct use of them at each voxel is ineffective for localization, we introduce a local binary pattern (LBP)-like texture feature calculation method from the 3D FCN features to improve the localization accuracy. After estimating the pancreas-bounding box, we generate a patient-specific probabilistic atlas and perform pancreas segmentation with it.

The following are the contributions of this paper: (1) introduction of an accurate pancreas localization technique utilizing deeply learned features, which are statistically useful for solving organ localization problems, (2) introduction of a new deep learned texture feature measure based on LBP-like feature calculation from deeply learned features, and (3) full automation of pancreas segmentation.

2   Method

2.1   Overview

Our method, which segments the pancreas region from an input CT volume, consists of localization and segmentation processes. In the localization process, a bounding box of the pancreas is estimated using the regression forest technique. Regression forests estimate a bounding box from the 3D FCN and the intensity differences features that contain the automatically extracted global features. After performing the localization process, we generate a patient-specific probabilistic atlas of the pancreas in the bounding box. Pancreas segmentation is performed using the generated atlas.

2.2   Pancreas Localization

3D FCN Feature Extraction: We train a 3D U-Net [9] using a two-stage, coarse-to-fine approach that trains a FCN model to roughly delineate the organs of interest in the first stage: looking at ∼40% of voxels within a simple, automatically generated binary mask of the patient's body. We then use the predictions of the first-stage FCN to define a candidate region that will be used to train a second FCN. This step reduces the number of voxels the FCN has to classify to ∼10% while keeping the recall rate high at >99%. This second-stage FCN can now focus on more detailed organ segmentation. The last layer of the 3D U-Net contains a $1 \times 1 \times 1$ convolution that reduces the number of output channels to the number of class labels which is eight in our case: artery, vein, liver, spleen, stomach, gallbladder, pancreas, and background. We respectively utilize training and validation sets consisting of 281 and 50 clinical CT volumes. The validation dataset is used to determine good training of the model and avoid overfitting. This dataset is independent of the testing data in the remainder of this paper. For feature extraction, we deploy the trained model using a non-overlapping tiles approach on the target volume and extract a 64-dimensional feature vector from

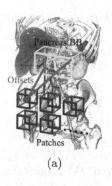

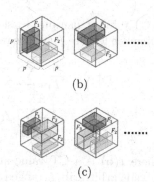

(b)

(c)

(a)

Fig. 1. (a) Schematic illustrations of positions of pancreas-bounding box, patch, and offset of patch in CT volume. (b), (c) Patch and cuboids (represented as cube frames and colored cuboids) used in difference feature value calculation from CT volume: (b) first-order and (c) second-order difference feature calculations. (Color figure online)

the pre-ultimate layer of the 3D U-Nnet for each voxel in the volume. For computational reasons and to increase the network's field of view when computing each tile, the resolution of all volumes is halved in each dimension. For implementation, we use the open-source distribution of 3D U-Net[1] based on the Caffe deep learning library [10].

Localization Using Regression Forests: The pancreas-bounding box is axis-aligned with a minimum size that includes a pancreas region, which can be represented as $\mathbf{b} = (b_{x1}, b_{x2}, b_{y1}, b_{y2}, b_{z1}, b_{z2})$. We define patches, whose size is $p \times p \times p$ voxels, and the center position is $\mathbf{v} = (v_x, v_y, v_z)$, allocated in the CT volume on a regular grid. The patch's offset with respect to $\mathbf{b}$ is represented as $\mathbf{d} = \mathbf{b} - \hat{\mathbf{v}}$, where $\hat{\mathbf{v}} = (v_x, v_x, v_y, v_y, v_z, v_z)$. They are also shown in Fig. 1(a).

We use a regression forest to estimate a patch's offset from the feature values calculated in the patches. A regression forest is constructed to estimate each face of a pancreas-bounding box. Six regression forests are constructed for a pancreas-bounding box. The feature values calculated in the patch include: (a) *first- and second-order differences of CT values*, (b) *3D FCN deep learned feature value-based LBP-like feature*, and (c) *organ likelihoods obtained from 3D FCN*.

(a) *Differences of CT values*: The first- and second-order differences of the CT values in a patch are used as feature values. The feature value from the first-order difference can be found in a previous work [8]. We newly introduce the second-order difference, which captures more useful information for localization, such

---

[1] http://lmb.informatik.uni-freiburg.de/resources/opensource/unet.en.html.

as CT value transitions. First-order difference feature $f_1(\mathbf{v})$ and second-order difference feature $f_2(\mathbf{v})$ are represented as

$$f_1(\mathbf{v}) = \frac{1}{|F_1|} \sum_{\mathbf{q} \in F_1} I(\mathbf{q}) - \frac{1}{|F_2|} \sum_{\mathbf{q} \in F_2} I(\mathbf{q}), \tag{1}$$

$$f_2(\mathbf{v}) = \frac{1}{|F_1|} \sum_{\mathbf{q} \in F_1} I(\mathbf{q}) + \frac{1}{|F_2|} \sum_{\mathbf{q} \in F_2} I(\mathbf{q}) - \frac{2}{|F_3|} \sum_{\mathbf{q} \in F_3} I(\mathbf{q}), \tag{2}$$

where $I(\mathbf{q})$ is a CT value at voxel $\mathbf{q}$ in a CT volume. F_1, F_2, and F_3 are cuboids arbitrarily arranged in a patch. $|F_1|, |F_2|$, and $|F_3|$ denote the number of voxels in the cuboids. The positions and the sizes of the cuboids are randomly selected. The arrangements of patches and cuboids are shown in Figs. 1(b) and (c).

(b) *3D FCN deep learned feature value-based LBP-like feature*: The deep learned feature values obtained from the last convolution layer of the trained 3D FCN contain useful information for segmentation. From the 3D U-Net output for each case, we get the feature volume, which contains 64-dimensional feature at each voxel. The local distribution of the intensity or texture in the feature volume represents crucial information for segmentation. We calculate the texture representing feature values based on the LBP feature from the feature volume and calculate the LBP-like features of a patch as follows. We obtain 2D images with the intensity values of the feature volume on three planes, which pass $\mathbf{v}$ and are parallel to the axial, coronal, and sagittal planes. One of the three planes is selected randomly. In the $p \times p$ voxels 2D image, we define the 3×3 regions of identical size and calculate the average intensity values in each region. The 3×3 average intensity values are used to calculate the LBP-like feature of the 2D image of the patch. The LBP-like feature is used as a feature value of the patch. This calculation process is shown in Fig. 2.

(c) *Organ likelihoods obtained from 3D FCN*: The trained 3D FCN outputs the likelihoods of eight organs (including pancreas) at each voxel. We use the likelihoods at $\mathbf{v}$ as a patch's feature values.

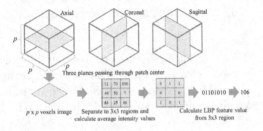

Fig. 2. Flow of 3D FCN deep learned feature value-based LBP-like feature calculation from patch in feature volume

The regression forest is trained using a training database that includes CT volumes and their corresponding manually segmented pancreas regions. A regression tree is constructed using patches obtained from the training database, as previously explained [8,11].

In the estimation process of a pancreas-bounding box from an unknown input CT volume, the patches obtained from the CT volume are input to the trained regression trees. Each regression tree outputs the offset from the patch position. We get an estimation result of the face of a bounding box as the sum of the offset and the patch position. The average estimation result of all the patches is the final estimation result of the face of a bounding box.

Table 1. Accuracies of proposed and previous pancreas segmentation methods.

Method	Data number	JI (%)	DICE (%)
Proposed	146	60.6 ± 16.5	73.9 ± 15.2
Okada et al. [1]	86	59.2	71.8
Chu et al. [2]	100	54.6 ± 15.9	69.1 ± 15.3
Wolz et al. [3]	150	55.0 ± 17.1	69.6 ± 16.7
Karasawa et al. [4]	150	61.6 ± 16.6	74.7 ± 15.1
Tong et al. [5]	150	56.9 ± 15.2	71.1 ± 14.7
Saito et al. [6]	140	62.3 ± 19.5	74.4 ± 20.2
Roth et al. [7]	82	–	78.0 ± 8.2
Oda et al. [8]	147	62.1 ± 16.6	75.1 ± 15.4

2.3 Patient-Specific Probabilistic Atlas Generation and Pancreas Segmentation

A blood vessel information-based patient-specific probabilistic atlas [8] is quite effective for pancreas segmentation. We generate this probabilistic atlas using a calculated bounding box.

For the input CT volume, we perform a rough segmentation using the patient-specific probabilistic atlas. Then precise segmentation is done using the graph-cut method [12]. We obtain the final pancreas segmentation results from this precise segmentation process.

3 Experiments and Discussion

Our proposed method was evaluated using 146 cases of abdominal CT volumes. The following are the acquisition parameters of the CT volumes: image size, 512×512 voxels, 263–1061 slices, 0.546–0.820 mm pixel spacing, and 0.40–0.80 mm slice spacing. The ground truth organ regions were semi-automatically made by three trained researchers. All of the ground truth regions were checked by a medical doctor. Our method's parameter was experimentally selected as $p = 25$.

We evaluated localization accuracy by measuring the average distance between six faces of estimated and ground truth bounding boxes. The average distance was 11.4 ± 4.5 mm when we use only the second-order difference of CT values as regression forest features. The average distance was improved to 11.0 ± 4.0 mm when we use all regression forest features proposed in this paper.

Pancreas segmentation accuracy was evaluated by five-fold cross validation by the following evaluation metrics: Jaccard Index (JI) and Dice Overlap (DICE). Table 1 compares accuracies of the proposed and previous methods. Parts of CT volumes used in [8] and our method are same. Figure 3 shows results.

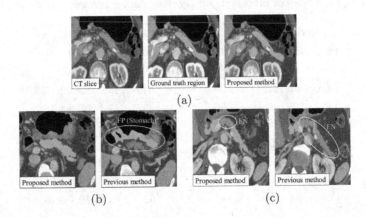

Fig. 3. (a) Pancreas segmentation results on axial slices in estimated bounding boxes. Ground truth and segmented regions are shown as blue regions. (b) and (c) compare segmentation results with previous method [8] using the same cases. DICE improved from 28.0% to 72.7% in (b) and 25.4% to 60.3% in (c). (Color figure online)

We newly introduced the 3D FCN-based features in the regression forest to improve pancreas localization accuracy. From the experimental result, both of the average and standard deviation of the localization accuracy was improved when we use the new features. The regression forest estimates a bounding box based on feature values calculated in patches located all over a CT volume. The difference of CT value feature captures only edge information. Our new 3D FCN-based features capture more useful information including texture, appearance, and relationships to other organs. Handcrafted features are useful if they are carefully designed for purpose of use. Our 3D FCN-based features are automatically designed for organ segmentation. They also contain useful information for organ localization. Combinational use of the handcrafted and deep learned features improved the pancreas localization accuracy.

We used the LBP-like feature calculation scheme to extract feature from the 3D FCN feature volume. Direct use of the voxel value in the feature volume as a feature value for the regression forest is one possible way to perform localization. Feature value used in the regression forest should contain position-related information because it can be considered as a clue to find a bounding box position. If we use the feature volume voxel value, many similar values are obtained at many positions in the volume. Such feature value is not effective for localization. Our LBP-like feature value captures local texture in the feature volume. Therefore, calculated feature value is position-related and effective for localization.

We also performed pancreas segmentation using the localization result. The segmentation accuracy was good among the recent segmentation methods. Especially, segmentation accuracies were improved among four "difficult to segment" cases whose DICE were lower than 30% in [8]. Among them, 28.0%, 25.4%, 29.1%, 18.9% of DICEs in [8] were improved to 72.7%, 60.3%, 35.6%, 20.1%, respectively. Figures 3(b) and (c) show examples of such cases. The pancreas touching the surrounding organs in these cases. False positives (FPs) and false negatives (FNs) were greatly reduced by the proposed method. The regression forest feature values in [8] were not effective for them. The proposed deep learned feature values effectively represented localization information even in such cases. The proposed method accurately localized pancreas and it resulted in reduction of FPs and FNs.

We proposed a pancreas localization method utilizing 3D FCN-based features and intensity difference features. The regression forest calculates a bounding box of the pancreas from the features. Also, our method segments pancreas region using a patient-specific probabilistic atlas. Experimental results using 146 CT volumes showed comparable results to the other state-of-the-art methods. Future work includes weighting of contribution rates of patches based on their position and application to multi-organ segmentation.

Acknowledgments. Parts of this research were supported by the MEXT/JSPS KAKENHI Grant Numbers 25242047, 26108006, 17H00867, the JSPS Bilateral International Collaboration Grants, and the JST ACT-I (JPMJPR16U9).

References

1. Okada, T., et al.: Automated segmentation of the liver from 3D CT images using probabilistic atlas and multi-level statistical shape model. In: Ayache, N., Ourselin, S., Maeder, A. (eds.) MICCAI 2007. LNCS, vol. 4791, pp. 86–93. Springer, Heidelberg (2007). doi:10.1007/978-3-540-75757-3_11
2. Chu, C., et al.: Multi-organ segmentation based on spatially-divided probabilistic atlas from 3D abdominal CT images. In: Mori, K., Sakuma, I., Sato, Y., Barillot, C., Navab, N. (eds.) MICCAI 2013. LNCS, vol. 8150, pp. 165–172. Springer, Heidelberg (2013). doi:10.1007/978-3-642-40763-5_21

3. Wolz, R., Chu, C., Misawa, K., et al.: Automated abdominal multi-organ segmentation with subject-specific atlas generation. IEEE TMI **32**(9), 1723–1730 (2013)

4. Karasawa, K., et al.: Structure specific atlas generation and its application to pancreas segmentation from contrasted abdominal CT volumes. In: Menze, B., Langs, G., Montillo, A., Kelm, M., Müller, H., Zhang, S., Cai, W., Metaxas, D. (eds.) MCV 2015. LNCS, vol. 9601, pp. 47–56. Springer, Cham (2016), doi:10. 1007/978-3-319-42016-5_5

5. Tong, T., Wolz, R., Wang, Z., et al.: Discriminative dictionary learning for abdominal multi-organ segmentation. Med. Image Anal. **23**(1), 92–104 (2015)

6. Saito, A., Nawano, S., Shimizu, A.: Joint optimization of segmentation and shape prior from level set-based statistical shape model, and its application to the automated segmentation of abdominal organs. Med. Image Anal. **28**, 46–65 (2016)

7. Roth, H.R., Lu, L., Farag, A., Sohn, A., Summers, R.M.: Spatial aggregation of holistically-nested networks for automated pancreas segmentation. In: Ourselin, S., Joskowicz, L., Sabuncu, M.R., Unal, G., Wells, W. (eds.) MICCAI 2016. LNCS, vol. 9901, pp. 451–459. Springer, Cham (2016). doi:10.1007/978-3-319-46723-8_52

8. Oda, M., et al.: Regression forest-based atlas localization and direction specific atlas generation for pancreas segmentation. In: Ourselin, S., Joskowicz, L., Sabuncu, M.R., Unal, G., Wells, W. (eds.) MICCAI 2016. LNCS, vol. 9901, pp. 556–563. Springer, Cham (2016). doi:10.1007/978-3-319-46723-8_64

9. Çiçek, Ö., Abdulkadir, A., Lienkamp, S.S., Brox, T., Ronneberger, O.: 3D U-Net: learning dense volumetric segmentation from sparse annotation. In: Ourselin, S., Joskowicz, L., Sabuncu, M.R., Unal, G., Wells, W. (eds.) MICCAI 2016. LNCS, vol. 9901, pp. 424–432. Springer, Cham (2016). doi:10.1007/978-3-319-46723-8_49

10. Jia, Y., Shelhamer, E., Donahue, J., et al.: Caffe: convolutional architecture for fast feature embedding. In: 22nd ACM International Conference On Multimedia, pp. 675–678. ACM (2014)

11. Criminisi, A., Robertson, D., Konukoglu, E., et al.: Regression forests for efficient anatomy detection and localization in computed tomography scans. Med. Image Anal. **17**, 1293–1303 (2013)

12. Boykov, Y., Veksler, O., Zabih, R.: Fast approximate energy minimization via graph cuts. IEEE PAMI **23**(11), 1222–1239 (2001)

A Unified Framework for Tumor Proliferation Score Prediction in Breast Histopathology

Kyunghyun Paeng(✉), Sangheum Hwang, Sunggyun Park, and Minsoo Kim

Lunit Inc., Seoul, Korea
{khpaeng,shwang,sgpark,mskim}@lunit.io

Abstract. We present a unified framework to predict tumor proliferation scores from breast histopathology whole slide images. Our system offers a fully automated solution to predicting both a molecular
data-based, and a mitosis counting-based tumor proliferation score. The
framework integrates three modules, each fine-tuned to maximize the
overall performance: An image processing component for handling whole
slide images, a deep learning based mitosis detection network, and a proliferation scores prediction module. We have achieved 0.567 quadratic
weighted Cohen's kappa in mitosis counting-based score prediction and
0.652 F1-score in mitosis detection. On Spearman's correlation coefficient, which evaluates predictive accuracy on the molecular data based
score, the system obtained 0.6171. Our approach won first place in all of
the three tasks in Tumor Proliferation Assessment Challenge 2016 which
is MICCAI grand challenge.

Keywords: Tumor proliferation · Mitosis detection · Convolutional
neural networks · Breast histopathology

1 Introduction

Tumor proliferation speed is an important biomarker for estimating the prognosis
of breast cancer patients [12]. The mitotic count is part of the Bloom & Richardson
grading system [5], and a well-recognized prognostic factor [2]. However, the procedure currently used by pathologists to count the number of mitoses is tedious and
subjective, and can suffer from reproducibility problems [13]. Automatic methods
have rapidly advanced the state-of-the-art in mitosis detection [1,4,14]. In particular, systems based on deep convolutional neural networks (DCNN) have been
successfully adopted to mitosis detection [4]. The latter family of algorithms have
advanced the performance of automatic mitosis detection to near-human levels,
offering promise in addressing the problem of subjectivity and reproducibility [13].

In a practical scenario, it is desirable to provide tumor proliferation assessment at the whole slide image (WSI) level. The size of the WSI is exceedingly
large, posing additional challenges for applying automatic techniques. Previous
studies carried out mitosis detection within ROI patches from the WSI which

© Springer International Publishing AG 2017
M.J. Cardoso et al. (Eds.): DLMIA/ML-CDS 2017, LNCS 10553, pp. 231–239, 2017.
DOI: 10.1007/978-3-319-67558-9_27

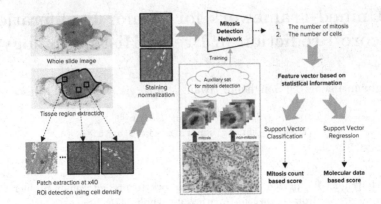

Fig. 1. Tumor proliferation score prediction framework. First, tissue region and patch extraction is performed and ROI patches are selected. After staining normalization, we count the number of mitoses using a DCNN based detection network. The mitosis detection network is trained by using a dataset containing annotated mitosis regions. The results of mitosis detection are converted to a feature vector in each WSI. Finally, the tumor proliferation scores are predicted by SVMs.

were pre-selected by a pathologist. On the contrary, automatic analysis of the WSI must resolve the issue of ROI selection without human guidance. An additional module is required to perform the final mapping between a mitosis counting result and the tumor proliferation score.

In this paper, we present our work in addressing each of the issues outlined above, in the form of a unified framework for predicting a tumor proliferation score directly from breast histopathology WSIs. Our pipeline system performs fully automatic prediction of tumor proliferation scores from WSIs. Our system participated in a well-recognized challenge and was able to outperform other systems in each subtask, as well as in the final proliferation score prediction task, validating the design choices of each module, as well as the capability of the integrated system.

2   Methodology

2.1   Whole Slide Image Handling

Tissue region and patch extraction: We first identify tissue blobs from WSI, through a combination of WSI resizing, tissue region extraction by Otsu's method [11], and binary dilation. Once the tissue blobs have been determined, we sample patches from them to be used in mitosis counting. There are many ways to perform the patch sampling, in terms of the choice of patch size and sampling step. Recalling the fact that Bloom & Richardson (B&R) grading depends on the number of mitoses within a 10 consecutive high power fields (HPFs) region, an area of approximately $2\,mm^2$, we extract patches corresponding to 10 HPFs. However, the definition of 10 consecutive HPFs region is highly variant, for instance, it may

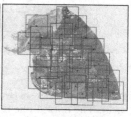

Fig. 2. The results of tissue region and patch extraction. The blue dots indicate the position of the sampling point (i.e., the centers of sampled patches). Each green square shows a 10 consecutive HPFs region. (Color figure online)

have only horizontal or vertical directions. For generality, we define 10 consecutive HPFs to be a square. The average number of patches extracted in each slide is roughly 300. The result of tissue region and patch extraction is shown in Fig. 2.

Region of interests detection: To select region of interests (ROIs) from the sampled patches, we utilize the fact that there are generally many mitoses in regions with high cell density. We use CellProfiler [3,7], to estimate the cell density of the patches. Finally, we obtain the number of cells in each patch, and after sorting patches from one WSI according to the number of cells, the top K patches are chosen as ROIs. We set K to 30 with consideration for computation cost.

Staining normalization: Because WSIs vary highly in appearance, it is beneficial to normalize the staining quality of ROI patches. In order to normalize the staining of patches, we apply the method described in [9]. We set α and β to 1.0 and 0.15 respectively.

2.2 Deep Convolutional Neural Networks Based Mitosis Detection

Training procedure: After selecting ROI patches, we detect mitoses using the trained model. The mitosis detector is trained on 128×128 patches extracted from the auxiliary dataset. We use a two step training procedure in order to reduce the false positive rate [4]. First, a first-pass training of the network is performed on an initial dataset. Then, a list of image regions that the network has identified as false positive mitoses cases is extracted by the trained network. After that, we build a new, second training dataset which consists of the same ground truth mitosis samples and normal samples from before, but with an additional 100,000 normal samples (i.e., false positives) generated with random translation augmentation from the initially trained network. In total, the new dataset consists of 70,000 mitosis patches and 280,000 normal patches. We retrain the network from scratch on this new dataset to obtain the final mitosis detection model.

Architecture: The architecture of the mitosis detection network is based on the Residual Network (ResNet) [6], and it consists of 6 or 9 residual blocks.

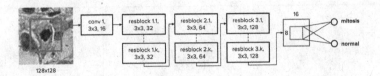

Fig. 3. The *L-view* network architecture for mitosis detection. k is a depth factor (2 or 3). The dotted line indicates that it is repeated k times. Two red rectangles indicate corresponding regions in the architecture. (Color figure online)

We train a network on 128×128 input patches, then convert the trained network to a fully convolutional network [8], which can then be applied directly to an entire ROI patch for inference. This method offers much greater computational efficiency over the alternative of subdividing the ROI patch and performing piecewise inference. However, although the fully convolutional approach is much faster, it fails to exactly match the performance of the standard subdivision approach. This problem arises from the large discrepancy between the input size of the network during training and inference when using the fully convolutional technique, exacerbated by the ResNet's large depth leading to a larger receptive field and corresponding larger zero-padding exposure. In order to alleviate this problem, we introduce a novel architecture named large-view (*L-view*) model. Figure 3 shows the *L-view* architecture. Although the *L-view* architecture has 128×128 input size, in the final global pooling layer, we only activate a smaller region corresponding to the central 64×64 region in the input patch. This allows the zero padding region to be ignored in the training phase and consequently solves the problem of the zero padding effect, leading to a sizable improvement in mitosis detection score. Figure 4 represents the examples of mitosis detection results.

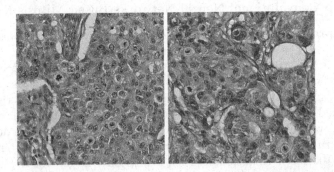

Fig. 4. Mitosis detection examples. The detected mitoses are highlighted with cyan circles. (Color figure online)

Table 1. Feature components for predicting tumor proliferation scores. # and mts represents the number and mitosis, respectively. B&R is Bloom & Richardson grading. Top 10% and 30–70% means rank 1–3 and rank 9–21 from sorted ROIs. min, avg, and max indicate the minimum, average, and maximum value in selected part from ROIs.

No.	Feat. description	No.	Feat. description	No.	Feat. description
0	avg. # mts	7	std. # cells	14	std. # mts in top 10%
1	max. # mts	8	ratio of avg. mts/cells	15	std. # cells in top 10%
2	std. # mts	9	ratio of max. mts/cells	16	min. # mts in top 10%
3	B&R of avg. # mts	10	avg. # mts in top 10%	17	min. # cells in top 10%
4	B&R of max. # mts	11	min. # mts	18	avg. # mts in 30–70%
5	avg. # cells	12	min. # cells	19	max. # mts in 30–70%
6	max. # cells	13	ratio of min. mts/cells	20	std. # mts in 30–70%

2.3 Tumor Proliferation Score Prediction

In the final module, we encode each slide into a feature vector and train an SVM to predict the tumor proliferation scores. The feature vector consists of the number of mitoses and cells in each patch through the steps previously described, in addition to 21 features based on statistical information, as shown in Table 1. Since we have no prior knowledge of which features are relevant for tumor proliferation score prediction, we find the best feature vector through cross validation from various feature combinations. We performed 10-fold cross validation using 500 training pairs with various C values of support vector machines (SVMs) with radial basis function (RBF) kernel in which the gamma of RBF kernel is fixed to 1/the dimension of feature. Finally, we selected combinations of features with the best performance. To reduce the search permutations, we reduced the dimension of the feature vector to 14 features.

3 Results

In this section, we introduce the dataset used for training and validation and show the performance of each individual component in the proposed system. In addition, all experimental results except for validation were evaluated in a public challenge.

3.1 Datasets

We used two datasets[1] for developing the fully automated tumor proliferation score prediction system. First, the auxiliary dataset containing annotated mitosis regions from two pathologists are used to build a mitosis detector. The training dataset includes 656 patches of 1 HPF or 10 HPFs from 73 patients from three

---

[1] http://tupac.tue-image.nl/node/3.

Table 2. F1 scores in mitosis detection. * indicates that additional data was used for training. Proposed method is based on L-view architecture with 128×128 input size.

Method	Validation	Test
Proposed	0.731	0.652
Team A*	-	0.648
Team B	-	0.616
Team C*	-	0.601

pathology centers, and 34 patches of 10 HPFs from 34 patients are used for testing. The training dataset is randomly split to validate the mitosis detector, and the validation dataset includes 142 mitoses from 6 patients. Secondly, to train the entire system, we used the main dataset which consists of 500 training WSIs and 321 testing WSIs. Each WSI has two corresponding scores, a mitosis counting based score indicating one of three classes and a molecular data based score [10] which is a continuous value. To validate the entire system, we used 10-fold cross validation.

3.2   Experiments

We evaluated on three tasks to validate the performance of the proposed system. Our results were compared with the other approaches in the challenge.[2] First, the mitosis detection performance is measured via F1-score. The performance of the entire system is based on two evaluation metrics: quadratic weighted Cohen's kappa for the mitosis counting based score and Spearman's correlation coefficient for the molecular based score.

Mitosis detection: We fixed the batch size to 128 and initial learning rate to 0.1, and applied the same two-step training procedure for all model configurations during training. The k of our *L-view* architecture is 3 in the first training phase, and color, brightness and contrast augmentation is used in various combinations. The learning rate is 0.1 for the first 8 epochs of the first training phase, and 0.1, 0.01 and 0.001 for 8, 12 and 14 epochs of the second training phase. The evaluation results are shown in Table 2. With the *L-view* architecture, we obtain 0.731 and 0.652 F1-score on the validation and test, respectively. In case of training without false positives, the performance degradation of 0.1 F1-score was observed in our validation set.

Mitosis counting based proliferation score prediction: After selecting the mitosis detector with the best performance, we trained the tumor proliferation score prediction module. We evaluated quadratic weighted Cohen's kappa score using 10-fold cross validation and found the best feature combination. The min related feature values were found to be unimportant, so features 11, 12, 13, 16, 17

---

[2] Compared methods are denoted by alphabet to anonymize the names of the participating teams.

Table 3. Mitosis counting based proliferation score prediction results in validation and test. The evaluation metric is Quadratic weighted Cohen's kappa score. * indicates that additional data was used for training. ** is the semi-automatic method where ROIs are selected by a pathologist.

Method	Validation	Test
Proposed	0.504	0.567
Team D**	-	0.543
Team B	-	0.534
Team E*	-	0.462

were removed from the combination list. Finally, we found the best performance to be shown by a 12 dimensional feature vector of the features 0, 1, 2, 3, 4, 5, 6, 7, 10, 15, 18, 20 from Table 1. The C value of SVMs was 0.03125. The evaluation results are shown in Table 3. The kappa score was 0.504 on the validation dataset and 0.567 on the test dataset, and the proposed system outperformed all other approaches. In addition, our system is even better than both the semi-automatic model and the model trained by using additional data.

Molecular data based proliferation score prediction: The same method is used to find the best feature combination for predicting the molecular data based score. The min related features were once again found to be unimportant and removed. The evaluation results are shown in Table 4. The 13 dimensional best feature vector includes features 0, 1, 2, 3, 4, 5, 6, 7, 8, 10, 14, 18, 20 from Table 1, and the C value of SVMs was 0.25. We obtained 0.642 Spearman score on the validation dataset and 0.617 on the test dataset. Our system significantly outperformed all other approaches. Through the best feature combination, we discovered that the values of features 8 and 14, rather than 15, are important in predicting the molecular data based score. In other words, the key factor for estimating this score is not cell information but the number of mitoses. This affirms the high correlation of mitotic count and tumor proliferation speed.

Table 4. Molecular data based proliferation score prediction results in validation and test. Evaluation metric is Spearman's correlation coefficient.

Method	Validation	Test
Proposed	0.642	0.617
Team F	-	0.516
Team B	-	0.503

4   Conclusion

We presented a fully automated unified system for predicting tumor proliferation scores directly from breast histopathology WSIs. The proposed system enables fully automated modular prediction of two tumor proliferation scores based on mitosis counting and molecular data. Our work confirms that a mitosis detection module could be integrated in prognostic grading system that is more practical in a clinical scenario. In addition, we demonstrated that our system achieved state-of-the-art performance in proliferation scores prediction.

References

1. Albarqouni, S., Baur, C., Achilles, F., Belagiannis, V., Demirci, S., Navab, N.: Aggnet: deep learning from crowds for mitosis detection in breast cancer histology images. IEEE Trans. Med. Imaging **35**(5), 1313–1321 (2016)
2. Baak, J.P., van Diest, P.J., Voorhorst, F.J., van der Wall, E., Beex, L.V., Vermorken, J.B., Janssen, E.A.: Prospective multicenter validation of the independent prognostic value of the mitotic activity index in lymph node-negative breast cancer patients younger than 55 years. J. Clin. Oncol. **23**(25), 5993–6001 (2005)
3. Carpenter, A.E., Jones, T.R., Lamprecht, M.R., Clarke, C., Kang, I.H., Friman, O., Guertin, D.A., Chang, J.H., Lindquist, R.A., Moffat, J., et al.: Cellprofiler: image analysis software for identifying and quantifying cell phenotypes. Genome Biol. **7**(10), R100 (2006)
4. Cireşan, D.C., Giusti, A., Gambardella, L.M., Schmidhuber, J.: Mitosis detection in breast cancer histology images with deep neural networks. In: Mori, K., Sakuma, I., Sato, Y., Barillot, C., Navab, N. (eds.) MICCAI 2013. LNCS, vol. 8150, pp. 411–418. Springer, Heidelberg (2013). doi:10.1007/978-3-642-40763-5_51
5. Elston, C.W., Ellis, I.O.: Pathological prognostic factors in breast cancer. I. The value of histological grade in breast cancer: experience from a large study with long-term follow-up. Histopathology **19**(5), 403–410 (1991)
6. He, K., Zhang, X., Ren, S., Sun, J.: Identity mappings in deep residual networks. arXiv preprint arXiv:1603.05027 (2016)
7. Kamentsky, L., Jones, T.R., Fraser, A., Bray, M.A., Logan, D.J., Madden, K.L., Ljosa, V., Rueden, C., Eliceiri, K.W., Carpenter, A.E.: Improved structure, function and compatibility for cellprofiler: modular high-throughput image analysis software. Bioinformatics **27**(8), 1179–1180 (2011)
8. Long, J., Shelhamer, E., Darrell, T.: Fully convolutional networks for semantic segmentation. In: CVPR, pp. 3431–3440 (2015)
9. Macenko, M., Niethammer, M., Marron, J.S., Borland, D., Woosley, J.T., Guan, X., Schmitt, C., Thomas, N.E.: A method for normalizing histology slides for quantitative analysis. In: ISBI, vol. 9, pp. 1107–1110 (2009)
10. Nielsen, T.O., Parker, J.S., Leung, S., Voduc, D., Ebbert, M., Vickery, T., Davies, S.R., Snider, J., Stijleman, I.J., Reed, J., et al.: A comparison of PAM50 intrinsic subtyping with immunohistochemistry and clinical prognostic factors in tamoxifen-treated estrogen receptor-positive breast cancer. Clin. Cancer Res. **16**(21), 5222–5232 (2010)
11. Otsu, N.: A threshold selection method from gray-level histograms. Automatica **11**(285–296), 23–27 (1975)

12. Van Diest, P., Van Der Wall, E., Baak, J.: Prognostic value of proliferation in invasive breast cancer: a review. J. Clin. Pathol. **57**(7), 675–681 (2004)
13. Veta, M., van Diest, P.J., Jiwa, M., Al-Janabi, S., Pluim, J.P.: Mitosis counting in breast cancer: object-level interobserver agreement and comparison to an automatic method. PLoS ONE **11**(8), e0161286 (2016)
14. Veta, M., Van Diest, P.J., Willems, S.M., Wang, H., Madabhushi, A., Cruz-Roa, A., Gonzalez, F., Larsen, A.B., Vestergaard, J.S., Dahl, A.B., et al.: Assessment of algorithms for mitosis detection in breast cancer histopathology images. Med. Image Anal. **20**(1), 237–248 (2015)

Generalised Dice Overlap as a Deep Learning Loss Function for Highly Unbalanced Segmentations

Carole H. Sudre[1,2(✉)], Wenqi Li[1], Tom Vercauteren[1], Sebastien Ourselin[1,2], and M. Jorge Cardoso[1,2]

[1] Translational Imaging Group, CMIC, University College London, London NW1 2HE, UK
[2] Dementia Research Centre, UCL Institute of Neurology, London WC1N 3BG, UK
carole.sudre.12@ucl.ac.uk

Abstract. Deep-learning has proved in recent years to be a powerful tool for image analysis and is now widely used to segment both 2D and 3D medical images. Deep-learning segmentation frameworks rely not only on the choice of network architecture but also on the choice of loss function. When the segmentation process targets rare observations, a severe class imbalance is likely to occur between candidate labels, thus resulting in sub-optimal performance. In order to mitigate this issue, strategies such as the weighted cross-entropy function, the sensitivity function or the Dice loss function, have been proposed. In this work, we investigate the behavior of these loss functions and their sensitivity to learning rate tuning in the presence of different rates of label imbalance across 2D and 3D segmentation tasks. We also propose to use the class re-balancing properties of the Generalized Dice overlap, a known metric for segmentation assessment, as a robust and accurate deep-learning loss function for unbalanced tasks.

1 Introduction

A common task in the analysis of medical images is the ability to detect, segment and characterize pathological regions that represent a very small fraction of the full image. This is the case for instance with brain tumors or white matter lesions in multiple sclerosis or aging populations. Such unbalanced problems are known to cause instability in well established, generative and discriminative, segmentation frameworks. Deep learning frameworks have been successfully applied to the segmentation of 2D biological data and more recently been extended to 3D problems [10]. Recent years have seen the design of multiple strategies to deal with class imbalance (e.g. specific organ, pathology...). Among these strategies, some focus their efforts in reducing the imbalance by the selection of the training samples being analyzed at the risk of reducing the variability in training [3,5], while others have derived more appropriate and robust loss functions [1,8,9]. In this work, we investigate the training behavior of three previously published loss functions in different multi-class segmentation problems in 2D and 3D while

© Springer International Publishing AG 2017
M.J. Cardoso et al. (Eds.): DLMIA/ML-CDS 2017, LNCS 10553, pp. 240–248, 2017.
DOI: 10.1007/978-3-319-67558-9_28

assessing their robustness to learning rate and sample rates. We also propose to use the class re-balancing properties of the Generalized Dice overlap as a novel loss function for both balanced and unbalanced data.

2 Methods

2.1 Loss Functions for Unbalanced Data

The loss functions compared in this work have been selected due to their potential to tackle class imbalance. All loss functions have been analyzed under a binary classification (foreground vs. background) formulation as it represents the simplest setup that allows for the quantification of class imbalance. Note that formulating some of these loss functions as a 1-class problem would mitigate to some extent the imbalance problem, but the results would not generalize easily to more than one class. Let R be the reference foreground segmentation (gold standard) with voxel values r_n, and P the predicted probabilistic map for the foreground label over N image elements p_n, with the background class probability being $1 - P$.

Weighted Cross-Entropy (WCE): The weighted cross-entropy has been notably used in [9]. The two-class form of WCE can be expressed as

$$\text{WCE} = -\frac{1}{N} \sum_{n=1}^{N} w r_n \log(p_n) + (1 - r_n) \log(1 - p_n),$$

where w is the weight attributed to the foreground class, here defined as $w = \frac{N - \sum_n p_n}{\sum_n p_n}$. The weighted cross-entropy can be trivially extended to more than two classes.

Dice Loss (DL): The Dice score coefficient (DSC) is a measure of overlap widely used to assess segmentation performance when a gold standard or ground truth is available. Proposed in Milletari et al. [8] as a loss function, the 2-class variant of the Dice loss, denoted DL_2, can be expressed as

$$\text{DL}_2 = 1 - \frac{\sum_{n=1}^{N} p_n r_n + \epsilon}{\sum_{n=1}^{N} p_n + r_n + \epsilon} - \frac{\sum_{n=1}^{N} (1 - p_n)(1 - r_n) + \epsilon}{\sum_{n=1}^{N} 2 - p_n - r_n + \epsilon}$$

The ϵ term is used here to ensure the loss function stability by avoiding the numerical issue of dividing by 0, i.e. R and P empty.

Sensitivity - Specificity (SS): Sensitivity and specificity are two highly regarded characteristics when assessing segmentation results. The transformation of these assessments into a loss function has been described by Brosch et al. [1] as

$$\text{SS} = \lambda \frac{\sum_{n=1}^{N} (r_n - p_n)^2 r_n}{\sum_{n=1}^{N} r_n + \epsilon} + (1 - \lambda) \frac{\sum_{n=1}^{N} (r_n - p_n)^2 (1 - r_n)}{\sum_{n=1}^{N} (1 - r_n) + \epsilon}.$$

The parameter λ, that weights the balance between sensitivity and specificity, was set to 0.05 as suggested in [1]. The ϵ term is again needed to deal with cases of division by 0 when one of the sets is empty.

Generalized Dice Loss (GDL): Crum et al. [2] proposed the Generalized Dice Score (GDS) as a way of evaluating multiple class segmentation with a single score but has not yet been used in the context of discriminative model training. We propose to use the GDL as a loss function for training deep convolutional neural networks. It takes the form:

$$\text{GDL} = 1 - 2\frac{\sum_{l=1}^{2} w_l \sum_n r_{ln}p_{ln}}{\sum_{l=1}^{2} w_l \sum_n r_{ln} + p_{ln}},$$

where w_l is used to provide invariance to different label set properties. In the following, we adopt the notation GDL_v when $w_l = 1/(\sum_{n=1}^{N} r_{ln})^2$. As stated in [2], when choosing the GDL_v weighting, the contribution of each label is corrected by the inverse of its volume, thus reducing the well known correlation between region size and Dice score. In terms of training with stochastic gradient descent, in the two-class problem, the gradient with respect to p_i is:

$$\frac{\partial \text{GDL}}{\partial p_i} = -2\frac{(w_1^2 - w_2^2)\left[\sum_{n=1}^{N} p_n r_n - r_i \sum_{n=1}^{N}(p_n + r_n)\right] + Nw_2(w_1 + w_2)(1 - 2r_i)}{\left[(w_1 - w_2)\sum_{n=1}^{N}(p_n + r_n) + 2Nw_2\right]^2}$$

Note that this gradient can be trivially extended to more than two classes.

2.2   Deep Learning Framework

To extensively investigate the loss functions in different network architectures, four previously published networks were chosen as representative networks for segmentation due to their state-of-the art performance and were reimplemented using Tensorflow.

2D Networks: Two networks designed for 2D images were used to assess the behaviour of the loss functions: UNet [9], and the TwoPathCNN [3]. The UNet architecture presents a U-shaped pattern where a step down is a series of two convolutions followed by a downsampling layer and a step up consists in a series of two convolution followed by upsampling. Connections are made between the downsample and upsample path at each scale. TwoPathCNN [3], designed for tumor segmentation, is used here in a fully convolutional 2D setup under the common assumption that a 3D segmentation problem can be approximated by a 2D network in situations where the slice thickness is large. This network involves the parallel training of two networks - a local and a global subnetwork. The former consists of two convolutional layers with kernel of size 7^2 and 5^2 with max-out regularization interleaved with max-pooling layers of size 4^2 and 2^2 respectively; while the latter network consists of a convolution layer of kernel size 13^2 followed by a max-pooling of size 2^2. The features of the local and global networks are then concatenated before a final fully connected layer resulting in the classification of the central location of the input image.

3D Networks: The DeepMedic architecture [4] and the HighResNet network [6] were used in the 3D context. DeepMedic consists in the parallel training of one network considering the image at full resolution and another on the down-sampled version of the image. The resulting features are concatenated before the application of two fully connected layers resulting in the final segmentation. HighResNet is a compact end-to-end network mapping an image volume to a voxel-wise segmentation with a successive set of convolutional blocks and residual connections. To incorporate image features at multiple scales, the convolutional kernels are dilated with a factor of two or four. The spatial resolution of the input volume is maintained throughout the network.

3   Experiments and Results

3.1   Experiments

The two segmentation tasks we choose to highlight the impact of the loss function target brain pathology: the first task tackles tumor segmentation, a task where tumor location is often unknown and size varies widely, and the second comprises the segmentation of age-related white matter hyperintensities, a task where the lesions can present a variety of shapes, location and size.

In order to assess each loss function training behavior, different sample and learning rates were tested for the two networks. The learning rates (LR) were chosen to be log-spaced and set to 10^{-3}, 10^{-4} and 10^{-5}. For each of the networks, three patch sizes (small: S, moderate: M, large: L), resulting in different effective field of views according to the design of the networks were used to train the models. A different batch size was used according to the patch size. Initial and effective patch sizes, batch size and resulting imbalance for each network are gathered in Table 1. In order to ensure a reasonable behavior of all loss functions, training patches were selected if they contained at least one foreground element. Larger patch sizes represent generally more unbalanced training sets. The networks were trained without training data augmentation to ensure more comparability between training behaviors. The imbalance in patches varied

Table 1. Comparison of patch sizes and sample rate for the four networks.

	UNet			TwoPathCNN			DeepMedic			HighResNet		
Batch size	5	3	1	5	3	1	5	3	1	5	3	1
Initial patch size	56	64	88	51	63	85	51	63	87	51	63	85
Effective patch size	16	24	48	19	31	53	3	15	39	15	27	49
Imbalance ratio	0.52	0.33	0.15	0.29	0.25	0.16	0.20	0.01	0.002	0.02	0.01	0.003

Table 2. Comparison of DSC over 200 last iterations in the 2D context for UNet and TwoPathCNN. Results are under the format median (interquartile range).

Patch	LR	UNet				TwoPathCNN			
		WCE	DL_2	SS	GDL_v	WCE	DL_2	SS	GDL_v
S	−5	0.71 (0.17)	0.73 (0.13)	0.37 (0.17)	0.75 (0.14)	0.56 (0.48)	0 (0)	0.53 (0.41)	0.49 (0.44)
	−4	0.77 (0.18)	0.76 (0.13)	0.74 (0.16)	0.80 (0.12)	0.80 (0.12)	0.79 (0.11)	0.81 (0.12)	0.80 (0.12)
	−3	0.70 (0.17)	0.72 (0.15)	0.39 (0.16)	0.72 (0.15)	0 (0)	0 (0)	0.77 (0.11)	0.72 (0.15)
M	−5	0.71 (0.23)	0.70 (0.22)	0.65 (0.25)	0.74 (0.19)	0 (0)	0.73 (0.18)	0.69 (0.21)	0.73 (0.19)
	−4	0.73 (0.18)	0.70 (0.22)	0.61 (0.25)	0.72 (0.19)	0.77 (0.16)	0.76 (0.17)	0.71 (0.18)	0.76 (0.17)
	−3	0.68 (0.23)	0.67 (0.21)	0.70 (0.26)	0.69 (0.22)	0 (0)	0.71 (0.22)	0.67 (0.21)	0.72 (0.19)
L	−5	0.63 (0.46)	0.62 (0.40)	0.49 (0.42)	0.56 (0.44)	0.62 (0.50)	0.50 (0.41)	0.50 (0.38)	0.56 (0.35)
	−4	0.68 (0.34)	0.64 (0.44)	0.18 (0.24)	0.66 (0.39)	0.64 (0.42)	0.59 (0.43)	0.52 (0.38)	0.64 (0.35)
	−3	0.59 (0.39)	0.57 (0.53)	0.16 (0.22)	0.59 (0.45)	0.77 (0.12)	0.77 (0.14)	0.79 (0.12)	0.79 (0.11)

greatly according to networks and contexts reaching at worst a median of 0.2% of a 3D patch.

The 2D networks were applied to BRATS [7], a neuro-oncological dataset where the segmentation task was here to localize the background (healthy tissue) and the foreground (pathological tissue, here the tumor) in the image. The 3D networks were applied to an in house dataset of 524 subjects presenting age-related white matter hyperintensities. In both cases, the T1-weighted, T2-weighted and FLAIR data was intensity normalized by z-scoring the data according to the WM intensity distribution. The training was arbitrarily stopped after 1000 (resp. 3000) iterations for the 2D (resp. 3D) experiments, as it was found sufficient to allow for convergence for all metrics.

3.2   2D Results

Table 2 presents the statistics for the last 200 steps of training in term of DSC for the four loss functions at the different learning rates, and different networks while

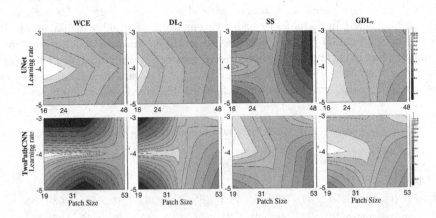

Fig. 1. Loss function behavior in terms of DSC (median over the last 200 iterations) under different conditions of effective patch size and learning rate in a 2D context. Isolines were linearly interpolated for visualization purposes.

Table 3. Comparison of DSC over 200 last iterations in the 3D context for DeepMedic and HighResNet. Results are under the format median (interquartile range).

Patch	LR	DeepMedic				HighResNet			
		WCE	DL_2	SS	GDL_v	WCE	DL_2	SS	GDL_v
S	-5	0.49 (0.17)	0.44 (0.19)	0.42 (0.14)	0.46 (0.17)	0 (0)	0 (0)	0.06 (0.15)	0.47 (0.32)
	-4	0.58 (0.20)	0.60 (0.15)	0.61 (0.22)	0.61 (0.18)	0 (0)	0.71 (0.18)	0.34 (0.20)	0.74 (0.15)
	-3	0.61 (0.12)	0.59 (0.14)	0.63 (0.15)	0.60 (0.15)	0 (0)	0 (0)	0 (0)	0 (0)
M	-5	0.05 (0.07)	0.05 (0.07)	0.05 (0.06)	0.04 (0.06)	0 (0)	0.60 (0.27)	0.15 (0.13)	0.64 (0.19)
	-4	0.09 (0.11)	0.07 (0.09)	0.08 (0.09)	0.08 (0.10)	0 (0)	0.71 (0.20)	0.20 (0.20)	0.69 (0.20)
	-3	0.45 (0.31)	0.42 (0.31)	0.17 (0.24)	0.48 (0.32)	0 (0)	0 (0)	0 (0)	0.65 (0.23)
L	-5	0.01 (0.03)	0.01 (0.03)	0.01 (0.03)	0.01 (0.03)	0 (0)	0.54 (0.27)	0.03 (0.06)	0.50 (0.32)
	-4	0.01 (0.04)	0.02 (0.04)	0.02 (0.04)	0.01 (0.04)	0 (0)	0.57 (0.32)	0.08 (0.19)	0.60 (0.30)
	-3	0.21 (0.33)	0.18 (0.30)	0.05 (0.12)	0.20 (0.33)	0 (0)	0.62 (0.18)	0.22 (0.15)	0.49 (0.34)

Fig. 1 shows the corresponding isolines in the space of learning rate and effective patch size illustrating notably the robustness of the GDL to the hyper-parameter space. The main observed difference across the different loss functions was the robustness to the learning rate, with the WCE and DL_2 being less able to cope with a fast learning rate (10^{-3}) when using TwoPathCNN while the efficiency of SS was more network dependent. An intermediate learning rate (10^{-4}) seemed to lead to the best training across all cases. Across sampling strategies, the pattern of performance was similar across loss functions, with a stronger performance when using a smaller patch but larger batch size.

3.3   3D Results

Similarly to the previous section, Table 3 presents the statistics across loss functions, sample size and learning rates for the last 200 iterations in the 3D experiment, while Fig. 2 plots the representation of robustness of loss function to the parameter space using isolines. Its strong dependence on the hyperparameters

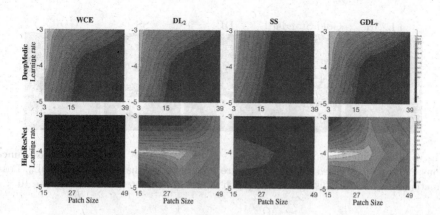

Fig. 2. Loss function behavior in terms of DSC (median over the last 200 iterations) under different conditions of effective patch size and learning rate in a 3D context. Isolines were linearly interpolated for visualization purposes.

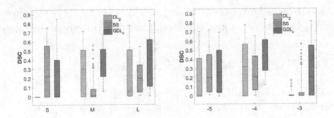

Fig. 3. Test set DSC for all loss functions across patch sizes (left) and across learning rates (right). WCE was omitted as it was unable to cope with the imbalance.

made DeepMedic agnostic to the choice of loss function. In the 3D context with higher data imbalance, WCE was unable to train and SS dropped significantly in performance when compared to GDL_v. DL_2 performed similarly to GLD_v for low learning rates, but failed to train for higher training rates. Similar patterns were observed across learning rates as for the 2D case, with the learning rate of 10^{-5} failing to provide a plateau in the loss function after 3000 iterations. We also observed that learning rates impacted network performance more for smaller patch sizes, but in adequate conditions (LR $= 10^{-4}$), smaller patches (and larger batch size) resulted in higher overall performance.

3D test set. For the 3D experiment, 10% of the available data was held out for testing purposes. The final HighResNet model was used to infer the test data segmentation. Figure 3 shows the comparison in DSC across loss functions for the different sampling strategies (right) and across learning rates (left). Overall, GDL_v was found to be more robust than the other loss functions across experiments, with small variations in relative performance for less unbalanced samples. Figure 4 presents an example of the segmentation obtained in the 3D experiment with HighResNet when using the largest patch size at a learning rate of 10^{-4}.

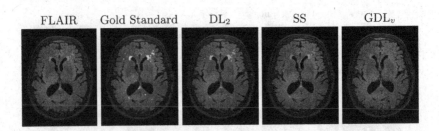

Fig. 4. The segmentation of a randomly selected 3D test set using different loss functions. Note the increased ability to capture punctuate lesions when using GDL_v. Loss functions were trained using a single patch of size 85^3 per step at learning rate 10^{-4}.

4   Discussion

From the observation of the training behavior of four loss functions across learning rates and sampling strategies in two different tasks/networks, it appears that a mild imbalance is well handled by most of the loss strategies designed for unbalanced datasets. However, when the level of imbalance increases, loss functions based on overlap measures appeared more robust. The strongest reliability across setups was observed when using GDL_v. Overall this work demonstrates how crucial the choice of loss function can be in a deep learning framework, especially when dealing with highly unbalanced problems. The foreground-background ratio in the most unbalanced case in this study was of 0.02% for the 3D experiment (white matter lesions). Future work will focus on more extreme imbalance situations, such as those observed in the case of the detection of lacunes and perivascular spaces (1/100000), where deep learning frameworks must find a balance between learning the intrinsic anatomical variability of all the classes and the tolerable level of class imbalance. The studied loss functions are implemented as part of the open source NiftyNet package (http://www.niftynet.io).

Acknowledgments. This work made use of Emerald, a GPU accelerated HPC, made available by the Science & Engineering South Consortium operated in partnership with the STFC Rutherford-Appleton Laboratory. This work was funded by the EPSRC (EP/H046410/1, EP/J020990/1, EP/K005278, EP/H046410/1), the MRC (MR/J01107X/1), the EU-FP7 project VPH-DARE@ IT (FP7-ICT-2011-9-601055), the Wellcome Trust (WT101957), the NIHR Biomedical Research Unit (Dementia) at UCL and the NIHR University College London Hospitals BRC (NIHR BRC UCLH/UCL High Impact Initiative- BW.mn.BRC10269).

References

1. Brosch, T., Yoo, Y., Tang, L.Y.W., Li, D.K.B., Traboulsee, A., Tam, R.: Deep convolutional encoder networks for multiple sclerosis lesion segmentation. In: Navab, N., Hornegger, J., Wells, W.M., Frangi, A.F. (eds.) MICCAI 2015. LNCS, vol. 9351, pp. 3–11. Springer, Cham (2015). doi:10.1007/978-3-319-24574-4_1
2. Crum, W., Camara, O., Hill, D.: Generalized overlap measures for evaluation and validation in medical image analysis. IEEE TMI **25**(11), 1451–1461 (2006)
3. Havaei, M., Davy, A., Warde-Farley, D., Biard, A., Courville, A., Bengio, Y., Pal, C., Jodoin, P.M., Larochelle, H.: Brain tumor segmentation with deep neural networks. MIA **35**, 18–31 (2017)
4. Kamnitsas, K., Ledig, C., Newcombe, V.F., Simpson, J.P., Kane, A.D., Menon, D.K., Rueckert, D., Glocker, B.: Efficient multi-scale 3D CNN with fully connected CRF for accurate brain lesion segmentation. MIA **36**, 61–78 (2017)
5. Lai, M.: Deep learning for medical image segmentation arXiv:1505.02000 (2015)
6. Li, W., Wang, G., Fidon, L., Ourselin, S., Cardoso, M.J., Vercauteren, T.: On the compactness, efficiency, and representation of 3D convolutional networks: brain parcellation as a pretext task. In: Niethammer, M., Styner, M., Aylward, S., Zhu, H., Oguz, I., Yap, P.-T., Shen, D. (eds.) IPMI 2017. LNCS, vol. 10265, pp. 348–360. Springer, Cham (2017). doi:10.1007/978-3-319-59050-9_28

7. Menze, B.H., et al.: The multimodal brain tumor image segmentation benchmark (BRATS). IEEE TMI **34**(10), 1993–2024 (2015)
8. Milletari, F., Navab, N., Ahmadi, S.A.: V-Net: fully convolutional neural networks for volumetric medical image segmentation. In: 2016 Fourth International Conference on 3D Vision (3DV), pp. 565–571. IEEE, October 2016
9. Ronneberger, O., Fischer, P., Brox, T.: U-Net: convolutional networks for biomedical image segmentation. In: Navab, N., Hornegger, J., Wells, W.M., Frangi, A.F. (eds.) MICCAI 2015. LNCS, vol. 9351, pp. 234–241. Springer, Cham (2015). doi:10.1007/978-3-319-24574-4_28
10. Zheng, Y., Liu, D., Georgescu, B., Nguyen, H., Comaniciu, D.: 3D deep learning for efficient and robust landmark detection in volumetric data. In: Navab, N., Hornegger, J., Wells, W.M., Frangi, A.F. (eds.) MICCAI 2015. LNCS, vol. 9349, pp. 565–572. Springer, Cham (2015). doi:10.1007/978-3-319-24553-9_69

ssEMnet: Serial-Section Electron Microscopy Image Registration Using a Spatial Transformer Network with Learned Features

Inwan Yoo[1], David G.C. Hildebrand[2], Willie F. Tobin[3], Wei-Chung Allen Lee[3], and Won-Ki Jeong[1]([⊠])

[1] Ulsan National Institute of Science and Technology, Ulsan, South Korea
{iwyoo,wkjeong}@unist.ac.kr
[2] The Rockefeller University, New York, USA
[3] Harvard Medical School, Boston, USA

Abstract. The alignment of serial-section electron microscopy (ssEM) images is critical for efforts in neuroscience that seek to reconstruct neuronal circuits. However, each ssEM plane contains densely packed structures that vary from one section to the next, which makes matching features across images a challenge. Advances in deep learning has resulted in unprecedented performance in similar computer vision problems, but to our knowledge, they have not been successfully applied to ssEM image co-registration. In this paper, we introduce a novel deep network model that combines a spatial transformer for image deformation and a convolutional autoencoder for unsupervised feature learning for robust ssEM image alignment. This results in improved accuracy and robustness while requiring substantially less user intervention than conventional methods. We evaluate our method by comparing registration quality across several datasets.

1 Introduction

Ambitious efforts in neuroscience—referred to as "connectomics"—seek to generate comprehensive brain connectivity maps. This field utilizes the high resolution of electron microscopy (EM) to resolve neuronal structures such as dendritic spine necks and synapses, which are only tens of nanometers in size [5]. A standard procedure for obtaining such datasets is cutting brain tissue into 30−50 nm-thick sections (e.g. ATUM [4]), acquiring images with 2−5 nm lateral resolution for each section, and aligning two-dimensional (2D) images into three-dimensional (3D) volumes. Though the tissue is chemically fixed and embedded in epoxy resin to preserve ultrastructure, several deformations occur in this serial-section EM (ssEM) process. These include tissue shrinkage, compression or expansion during sectioning, and warping from sample heating or charging due to the electron beam. Overcoming such non-linear distortions are necessary to reproduce a 3D image volume in a state as close as possible to the original biological specimen. Therefore, excellent image alignment is an important prerequisite for subsequent analysis.

© Springer International Publishing AG 2017
M.J. Cardoso et al. (Eds.): DLMIA/ML-CDS 2017, LNCS 10553, pp. 249–257, 2017.
DOI: 10.1007/978-3-319-67558-9_29

Significant research efforts in image registration have been made to address medical imaging needs. However, ssEM image registration remains challenging due to its image characteristics: large and irregular tissue deformations with artifacts such as dusts and folds, drifting for long image sequences alignment, and difficulty in finding the optimal alignment parameters. Several open-source ssEM image registration tools are available, such as bUnwarpJ [1] and Elastic alignment [9] (available via TrakEM2 [2]). They partially address the above issues, but some of them still remain, such as lack of global regularization and complicated parameter tuning.

Our work is motivated by recent advances in deep neural networks. Convolutional neural networks (CNNs) and their variants have shown unprecedented potential by largely outperforming conventional computer vision algorithms using hand-crafted feature descriptors, but their application to ssEM image registration has not been explored. Wu et al. [10] used a 3D autoencoder to extract features from MRI volumes, which are then combined with a conventional sparse, feature-driven registration method. Recent work by Jaderberg et al. [6] on the spatial transformer network (STN) uses a differentiable network module inside a CNN to overcome the drawbacks of CNNs (i.e., lack of scale- and rotation-invariance). Another interesting application of deep neural networks is energy optimization using backpropagation, as shown in the neural artistic style transfer proposed by Gatys et al. [3].

Inspired by these studies, we propose a novel deep network model that is specifically designed for ssEM image registration. The proposed model is a novel combination of an STN and a convolutional autoencoder that generates a deformation map (i.e., vector map) for the entire image alignment via backpropagation of the network. We propose a feature-based image similarity measure, which is learned from the training images in an unsupervised fashion by the autoencoder. Unlike other conventional hand-crafted features, such as SIFT and block-matching, the learned features used in our method significantly reduce the required user parameters and make the method easy to use and less error-prone. To the best of our knowledge, this is the first data-driven ssEM image registration method based on deep learning, which can easily extend to various applications by employing different feature encoding networks.

2   Method

2.1   Feature Generation Using a Convolutional Autoencoder

To compute similarities between adjacent EM sections, we generate data-driven features via a convolutional autoencoder, which consists of (1) a convolutional encoder comprised of convolutional layers with ReLU activations and (2) a deconvolutional decoder comprised of deconvolutional layers with ReLU activations that were symmetrical to the encoder without fully connected layers. Therefore, our method is applicable to any sized dataset (i.e., the network size is not constrained to the input data size). Our autoencoder can be formally defined as follows:

$$h = f_\theta(x) \tag{1}$$

$$y = g_\phi(h) \tag{2}$$

$$L_{\theta,\phi} = \sum_{i=1}^{N} ||x_i - y_i||_2^2 + \lambda \left(\sum_k ||\theta_k||_2^2 + \sum_k ||\phi_k||_2^2 \right) \tag{3}$$

where f_θ and g_ϕ are the encoder and the decoder and θ and ϕ are their parameters, respectively. The loss function (Eq. 3) consists of the reconstruction term minimizing the difference between the input and output images and the regularization terms minimizing the ℓ_2-norm of the weights of the network to avoid overfitting. Figure 1 shows that our autoencoder feature-based registration generates more accurate results compared to the conventional pixel intensity-based registration, i.e., (d) shows the smaller normalized cross correlation (NCC) error between aligned images than (c).

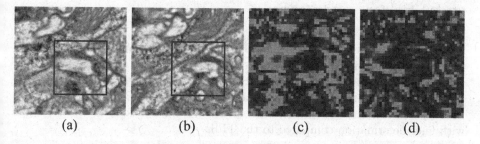

(a)   (b)   (c)   (d)

Fig. 1. Comparison between the pixel intensity-based and the autoencoder feature-based registration with backpropagation. (a) the fixed image, (b) the moving image, (c) the heat map of NCC of the pixel intensity-based registration result (NCC : 0.1670), and (d) the heat map of NCC of the autoencoder feature-based registration (NCC : 0.28) in red box region. (Color figure online)

2.2 Deformable Image Registration Using a Spatial Transformer Network

Upon completion of autoencoder training, a spatial transformer (ST) module T is attached to the front half (i.e., encoder) to form the proposed spatial transformer network (see Fig. 2, refer to [6] for the details of the ST module). This design is intended to find the proper deformation of the input image via an ST by minimizing the registration error measured by the pre-trained autoencoder. The objective function for registration errors between the reference and the moving images is formulated as Eq. 4. The reference image I_1 is fixed, and the moving image I_0 is deformed by the ST with the corresponding vector map v. Notably, the resolution of vector map v is usually coarser than that of the input image.

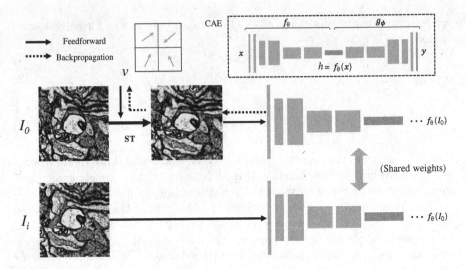

Fig. 2. The overview of our method. The upper right dashed box represents the pre-trained convolution autoencoder (CAE). The alignment is processed by backpropagation with loss of autoencoder features.

Therefore, we need smooth interpolation of a coarse vector map to obtain a per-pixel moving vector for actual deformation of the moving image. A thin plate spline (TPS) was used in the original STN for a smooth deformable transform, but other interpolation schemes, such as bilinear, bicubic, B-spline, etc., can be used as well. In our experiment, bilinear interpolation produced better results with finer deformation compared to the TPS.

$$L_v(I_0, I_1) = ||f_\theta(I_1) - f_\theta(T_v(I_0))||_2^2 + \alpha||v||_2^2 + \beta||\nabla v_x||_2^2 + \gamma||\nabla v_y||_2^2 \quad (4)$$

The first term of Eq. 4 measures how two images are contextually different via a trained autoencoder. We assumed that if the encoded features of two images are similar, then the images themselves are also similar and well-aligned. The rest of the terms in Eq. 4 reflect the regularization of vector map v, which penalizes large deformation while promoting smooth variation of the vector map, and α, β, γ are their corresponding weights. Because every layer is differentiable, including an ST, we directly optimize v by backpropagation with a chain rule, in which only v is updated and the weights in the autoencoder are fixed. We used the ADAM optimizer [7] for all our experiments.

The objective function using only adjacent image pairs could be vulnerable to imaging artifacts, which may result in drifting due to error accumulation when many sections are aligned. To increase the robustness of alignment, we extend the objective function (Eq. 4) to leverage multiple neighbor sections. Let the moving image be I_0, its neighbor reference n images be I_1 to I_n, and their corresponding weights be w_i. The proposed objective function (Eq. 5) combines the registration errors across neighbor images, which can lessen strong registration errors from images with artifacts and avoid large deformation.

To accumulate the registration error only within the image after deformation, we applied the *empty space mask* that represents the empty area outside the image. After image deformation, we collect the pixels outside the valid image region and make a binary mask image. We resize this mask image to match the size of the autoencoder feature map using a bilinear interpolation (shown as $M(T)$ in Eq. 5). Based on this objective function, the alignment of many EM sections is possible in an out-of-core fashion using a sliding-window method.

$$L_v(I_0, ..., I_n) = \sum_{i=1}^{n} w_i M(T_v) ||f_\theta(I_i) - f_\theta(T_v(I_0))||_2^2 + \alpha ||v||_2^2 + \beta ||\nabla v_x||_2^2 + \gamma ||\nabla v_y||_2^2$$

$$(5)$$

We also developed a technique for handling images with dusts and folds. Because the feature errors are high in the corrupted regions, we selectively ignore such regions during the optimization, which we call *loss drop*. This is similar to applying an empty space mask except that pixel selection is based on feature error. In our implementation, we first dropped the top 50% of high error features, and then reduced the dropping rate by half per every iteration. By doing this, we effectively prevented local minimums and obtained smoother registration results.

3   Results

We implemented our method using TensorFlow, and used a GPU workstation equipped with an NVIDIA Titan X GPU. We used three EM datasets: transmission EM (TEM) images of *Drosophila* brain, human-labeled TEM images of another *Drosophila* brain provided by CREMI challenge[1] (those two Drosophila images are collected independently on separate imaging systems), and mouse brain scanning EM (SEM) images with fold artifacts. We used two convolutional autoencoders: one is a deeper network (as shown in Fig. 2) with 3×3 filters used for the *Drosophila* TEM datasets, and the other is a shallower network with a larger filter size (i.e., 6 layers with 7×7 filters) used for the mouse SEM dataset. In bUnwarpJ and elastic alignment experiments, we performed various experiments to find the optimal parameters and selected the parameters that gave the best results.

Drosophila TEM data. The original volumetric dataset comprises the anterior portion of an adult female *Drosophila melanogaster* brain cut in the frontal plane. Each section was acquired at $4 \times 4 \times 40$ nm^{3}vx^{-1}, amounting to 4 million camera images and 50 TB of raw data. The original large-scale dataset was aligned with AlignTK (http://mmbios.org/aligntk-home) requiring extensive human effort and supercomputing resources. Although the alignment was sufficient for manual tracing of neurons, it must be improved for accurate and efficient automated segmentation approaches. Small volumes ($512 \times 512 \times 47$) were exported for re-alignment centered around synapses of identified connections between olfactory receptor neurons and second order projection neurons in

---

[1] https://cremi.org/.

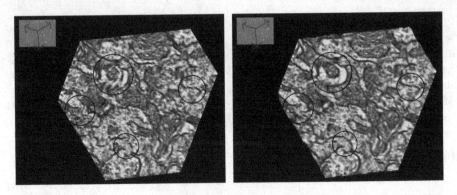

Fig. 3. Drosophila melanogaster TEM dataset. Left : Pre-aligned result. Right : After registration using our method. (Color figure online)

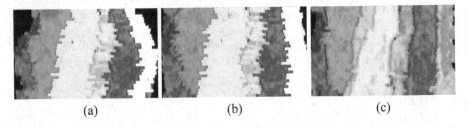

(a)   (b)   (c)

Fig. 4. Vertical view of the alignment result of the randomly deformed CREMI dataset. (a) bUnwarpJ, (b) elastic alignment, and (c) our method. Each neuron is assigned a unique color. (Color figure online)

the antennal lobe. Figure 3 shows the result of our registration method. Figure 3 left is the oblique (i.e., not axis-aligned) cross-sectional view of the original stack. Due to inaccurate pre-alignment, some discontinuous membranes are shown (see the red circle areas), which are corrected in the aligned result using our method (Fig. 3 right).

Labeled Drosophila TEM data from CREMI challenge. To quantitatively assess the registration quality, we used a small sub-volume ($512 \times 512 \times 31$) of registered and labeled TEM data from the CREMI challenge as a ground-truth. We first randomly deformed both the raw and the labeled images using a TPS defined by random vectors on random positions. The random positions were uniformly distributed in space, and the random vectors were sampled from the normal distribution with a zero mean value. Then we performed image registration using three methods (bUnwarpJ, elastic alignment and our method). Figure 4 shows the vertical cross section of each result. The bUnwarpJ result shows large deformation (i.e., drifting) across stacks (see the black regions on both sides). Although elastic alignment and our method show less deformation but our method clearly shows more accurate vertical membrane alignment. To quantitatively measure the registration accuracy, we selected the 50 largest neu-

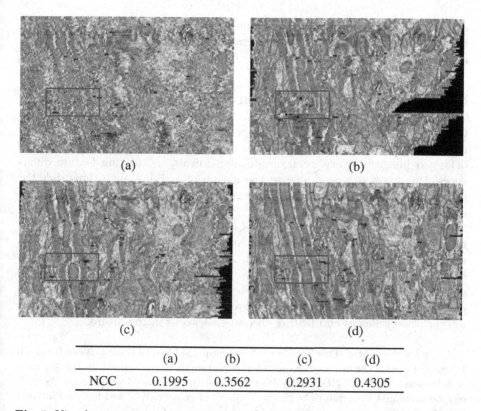

	(a)	(b)	(c)	(d)
NCC	0.1995	0.3562	0.2931	0.4305

Fig. 5. Visual comparison of mouse ssEM image registration results. (a) before alignment, (b) bUnwarpJ, (c) elastic alignment, and (d) our method. The red box is the region near the folds (shown as black spots). The below table shows NCC of the inner region in each aligned result (black backgroud regions are not counted for NCC computation). (Color figure online)

rons and calculated the average Dice coefficient for each result, which came to 0.60, 0.73, and 0.83 for bUnwarpJ, elastic, and our method, respectively. This result shows that our registration method is more robust and resilient to random deformation.

Mouse lateral geniculate nucleus SEM data with fold artifacts. We next sought to assess the applicability of our new alignment method to data acquired from different EM imaging methods, using different model organisms, and containing fold artifacts. A small volume ($1520 \times 2500 \times 100$) was selected from a mouse lateral geniculate nucleus dataset generously provided by the Lichtman laboratory [8]. This dataset was acquired using SEM with a resolution of $4 \times 4 \times 30$ nm^{3}vx^{-1}, and contains folds caused by cracks in the substrate onto which sections were collected. Figure 5 shows the vertical cross section of the registration results as compared to conventional registration methods. The overall registration quality of our method is higher than those of other methods, as indicated by clearer neuronal structures with low deformation. In particular, the red box shows a region containing warping due to folds, where our method is able to produce a smoother and more continuous result than others.

4   Discussion and Conclusion

One problem with the convolution operator is that it is neither scale- nor rotation-invariant. We addressed this problem by generating features on the deformed image in every iteration and dynamically calculating feature differences. Our method is slow due to the nature of learning algorithms, but the parameter tuning is much easier than existing methods, which makes it practically useful.

In this paper, we introduced a novel deep network for ssEM image registration that is easier to use and robust to imaging artifacts. The proposed method is a general learning-based registration model that can easily extend to various applications by modifying the network. Improving running time via parallel systems and deploying our method on tera-scale EM stacks would be an interesting and important future research direction. We also plan to employ various interpolation schemes and feature encoding networks in the future.

Acknowledgements. This work is partially supported by the Basic Science Research Program through the National Research Foundation of Korea funded by the Ministry of Education (NRF-2017R1D1A1A09000841) and the Software Convergence Technology Development Program through the Ministry of Science, ICT and Future Planning (S0503-17-1007).

References

1. Arganda-Carreras, I., Sorzano, C.O., Marabini, R., Carazo, J.M., Ortiz-de Solorzano, C., Kybic, J.: Consistent and elastic registration of histological sections using vector-spline regularization. In: International Workshop on Computer Vision Approaches to Medical Image Analysis, pp. 85–95. Springer, New York (2006)
2. Cardona, A., Saalfeld, S., Schindelin, J., Arganda-Carreras, I., Preibisch, S., Longair, M., Tomancak, P., Hartenstein, V., Douglas, R.J.: TrakEM2 software for neural circuit reconstruction. PloS One **7**(6), e38011 (2012)
3. Gatys, L.A., Ecker, A.S., Bethge, M.: Image style transfer using convolutional neural networks. In: Proceedings of the IEEE Conference on Computer Vision and Pattern Recognition, pp. 2414–2423 (2016)
4. Hayworth, K.J., Morgan, J.L., Schalek, R., Berger, D.R., Hildebrand, D.G.C., Lichtman, J.W.: Imaging ATUM ultrathin section libraries with WaferMapper: a multi-scale approach to EM reconstruction of neural circuits. Front. Neural Circuits **8**, 68 (2014)
5. Helmstaedter, M.: Cellular-resolution connectomics: challenges of dense neural circuit reconstruction. Nat. Methods **10**(6), 501–507 (2013)
6. Jaderberg, M., Simonyan, K., Zisserman, A., et al.: Spatial transformer networks. In: Advances in Neural Information Processing Systems, pp. 2017–2025 (2015)

7. Kingma, D., Ba, J.: Adam: a method for stochastic optimization arXiv preprint arXiv:1412.6980 (2014)
8. Morgan, J.L., Berger, D.R., Wetzel, A.W., Lichtman, J.W.: The fuzzy logic of network connectivity in mouse visual thalamus. Cell **1**, 192–206 (2017)
9. Saalfeld, S., Fetter, R., Cardona, A., Tomancak, P.: Elastic volume reconstruction from series of ultra-thin microscopy sections. Nat. Methods **9**(7), 717–720 (2012)
10. Wu, G., Kim, M.J., Wang, Q., Munsell, B., Shen, D.: Scalable high performance image registration framework by unsupervised deep feature representations learning (2015)

Fully Convolutional Regression Network for Accurate Detection of Measurement Points

Michal Sofka[(✉)], Fausto Milletari, Jimmy Jia, and Alex Rothberg

4Catalyzer, New York, USA
msofka@4catalyzer.com

Abstract. Accurate automatic detection of measurement points in ultrasound video sequences is challenging due to noise, shadows, anatomical differences, and scan plane variation. This paper proposes to address these challenges by a Fully Convolutional Neural Network (FCN) trained to regress the point locations. The series of convolutional and pooling layers is followed by a collection of upsampling and convolutional layers with feature forwarding from the earlier layers. The final location estimates are produced by computing the center of mass of the regression maps in the last layer. The temporal consistency of the estimates is achieved by a Long Short-Term memory cells which processes several previous frames in order to refine the estimate in the current frame. The results on automatic measurement of left ventricle in parasternal long axis view of the heart show detection errors below 5% of the measurement line which is within inter-observer variability.

1 Introduction

Regression modeling is an approach for describing a relationship between an independent variable and one or more dependent variables. In machine learning, this relationship is described by a function whose parameters are learned from training examples. In deep learning models, this function is a composition of logistic (sigmoid), hyperbolic tangent, or more recently rectified linear functions at each layer of the network. In many applications, the function learns a mapping between input image patches and a continuous prediction variable.

Regression has been used to detect organ [4] or landmark locations in images [2], visually track objects and features [8], and estimate body poses [14,15]. The deep learning approaches have outperformed previous techniques especially when a large annotated training data set is available. The proposed architectures used cascade of regressors [15], refinement localization stages [4,12], and combining cues from multiple landmarks [9] to localize landmarks. In medical images, the requirements on accurate localization are high since the landmarks or measurement points are used to help in diagnosis [11]. When tracking the measurements in video sequences, the points must be accurately detected in each frame while ensuring temporal consistency of the detections.

This paper proposes a Fully Convolutional Network for accurate localization of anatomical measurement points in video sequences. The advantage of the

M.J. Cardoso et al. (Eds.): DLMIA/ML-CDS 2017, LNCS 10553, pp. 258–266, 2017.
DOI: 10.1007/978-3-319-67558-9_30

Fully Convolutional Network is that the responses from multiple windows covering the input image can be computed in a single step. The network is trained end-to-end and outputs the locations of the points. The regressed locations are mapped at the last convolutional layer into a location using a new center-of-mass layer which computes mean position of the predictions. This approach has advantages to regressing heatmaps, since the predictions can have subpixel values and the regression objective can penalize measurement length differences from the ground truth. The temporal consistency of the measurements is improved by Convolutional Long Short-term Memory (CLSTM) cells which process the feature maps from several previous frames and produce updated features for the current frame in order to refine the estimate. The evaluation is fast to process each frame of a video sequence at near frame rate speeds.

2   Related Work

Regression forests were previously trained to predict locations and sizes of anatomical structures [2]. The initial estimates were refined via Hough regression forests [3] or local regressors guided by probabilistic atlas [4]. Automatic X-ray landmark detection in [1] estimated landmark positions via a data-driven non-convex optimization method while considering geometric constraints defined by relative positions.

Recently, deep learning approaches have been shown to effectively train representations that outperform traditional methods [7,10]. Multiple landmark localization in [9] was achieved by combining local appearance each landmark and spatial configuration of all other landmarks. The final combined heatmap of likely landmark location was obtained from appearance and spatial configuration heatmaps computed by convolutional layers. This approach requires to specify a hyperparameter of the heatmap Gaussian at the ground truth locations.

Long short-term memory (LSTM) architectures [5] were proposed to address the difficulties of training Recurrent Neural Networks (RNNs).

The regression capability of Long Short-Term Memory (LSTM) networks in the temporal domain can be used to concatenate high-level visual features produced by CNNs with region information [8]. The target coordinates are directly regressed taking advantage of the joint spatio-temporal model. Convolutional LSTMs [16] replace the matrix multiplication by the weight vector with a convolution. As a result, the model captures spatial context.

3   Regressing Point Locations

Denote an input image of width w and height h as $I \in \mathcal{R}^{w \times h}$ (independent variable) and the keypoint positions stacked columnwise into $\mathbf{p}$ (dependent variable). The goal of the regression is to learn a function $f(I; \theta) = \mathbf{p}$ parametrized by θ. We approximate f by a convolutional neural network and train the parameters θ using a database of images $\bar{\mathcal{I}} = \bar{I}_1, \ldots, \bar{I}_n$ and their corresponding annotations $\bar{\mathcal{P}} = \{\bar{\mathbf{p}}_1, \ldots, \bar{\mathbf{p}}_n\}$. Typically, a Euclidean loss $L(\mathcal{I}, \mathcal{P}; \theta) = \frac{1}{2N} \sum_{k=1}^{N} \|f(I_k; \theta) - \bar{\mathbf{p}}_k\|_2^2$ is employed to train f using each annotated image.

Previously, regression estimates were obtained directly from the last layer of the network, which was fully connected to previous layer. This is a highly non-linear mapping [14], where the estimate is computed from the fully connected layers after convolutional blocks.

3.1 Fully Convolutional Network with Center of Mass Layer

Instead of fully connected network, we propose to regress keypoint locations using a Fully Convolutional Network (FCN). FCNs have been previously used for image segmentation [6], for regressing heatmaps [9], and object localization [13]. Their advantage is that the estimates can be computed in a single evaluation step. In our architecture, we obtain point coordinate estimates at each image location.

The point coordinate predictions are computed in a new center of mass layer from input at each predicting location $\mathbf{l}_{ij}$ (see Fig. 1).

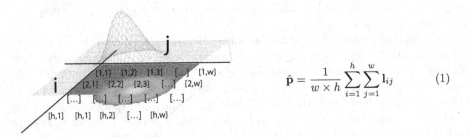

$$\hat{\mathbf{p}} = \frac{1}{w \times h} \sum_{i=1}^{h} \sum_{j=1}^{w} \mathbf{l}_{ij} \qquad (1)$$

Fig. 1. Center of Mass layer computes the estimate as a center of mass computed from the regressed location estimates at each location.

Center of mass layer makes it possible to design a loss function with a penalty on the error of the measurement line length. Our penalty is defined as an absolute value of the difference between estimated and ground truth lengths relative to the ground truth length. This penalty is combined with the Euclidean loss discussed above. The model is trained with an Adam optimizer with learning rate set as 0.0002 and converges within 100 epochs. The best model is selected based on the lowest error of the point location estimates.

3.2 Convolutional Long Short-Term Memory for Temporal Consistency

Recurrent neural networks (RNN) can learn sequential context dependencies by accepting input x_t and updating a hidden vector h_t at every time step t. The RNN network can be composed of Long-short Term Memory (LSTM) units, each controlled by a gating mechanism with three types of updates, $i_t, f_t, o_t \in R^n$ that range between 0 and 1. The value i_t controls the update of each memory cell, f_t controls the forgetting of each memory cell, and o_t controls the influence of the memory state on the hidden vector. In Convolutional LSTMs (CLSTMs), the input weights and hidden vector weights are convolved instead of multiplied to model spatial constraints. The function introduces a non-linearity which we chose as $tanh$. Denoting the convolutional operator as $*$, the values at the gates are computed as follows:

$$\text{forget gate:} \quad f_t = \text{sigm}(W_f * [h_{t-1}, x_t] + b_f) \tag{2}$$

$$\text{input gate:} \quad i_t = \text{sigm}(W_i * [h_{t-1}, x_t] + b_i) \tag{3}$$

$$\text{output gate:} \quad o_t = \text{sigm}(W_o * [h_{t-1}, x_t] + b_o) \tag{4}$$

The parameters of the weights W and biases b are learned from training sequences. In addition to the gate values, each CLSTM unit computes state candidate values

$$g_t = \tanh(W_g * [h_{t-1}, x_t] + b_g),$$

where $g_t \in R^n$ ranges between -1 and 1 and influences memory contents. The memory cell is updated by

$$c_t = f_t \odot c_{t-1} + i_t \odot g_t$$

which additively modifies each memory cell. The update process results in the gradients being distributed during backpropagation. The symbol $\odot$ denotes the Hadamard product. Finally, the hidden state is updated as:

$$h_t = o_t \odot \tanh(c_t).$$

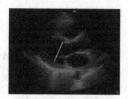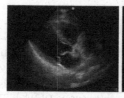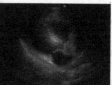

Fig. 2. (Left) Frame from an ultrasound sequence of the PLAx view of the left ventricle and overlaid measurement line. There is an ambiguity in the annotation points that can slide along the interface between myocardial wall and cavity and along the interface between wall and pericardium as reflected by aggregated prediction maps of the FCN regression model (Right).

In sequential processing of image sequences, the inputs into the LSTM consist of the feature maps computed from a convolutional neural network. In this work, we propose to use two architectures to compute the feature maps. The first architecture is a neural network with convolutional and pooling layers. After sequential processing the feature maps in CLSTM, the output is fed into fully connected layers to compute the point location estimate (Fig. 3). In the second architecture, the CLSTM inputs is the final layer of a convolutional path of the Fully Convolutional Network (FCN). The point location estimates are computed from the CLSTM output processed through the transposed convolutional part of the FCN network (Fig. 4). Similarly to [7,10], the feature maps are forwarded using connections from the previous layers.

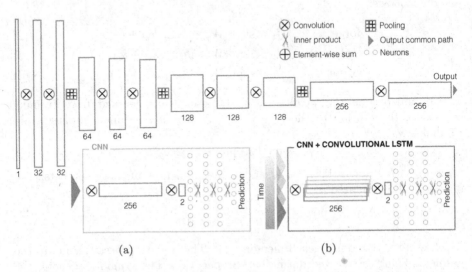

Fig. 3. (a) Convolutional Neural Network (CNN) architecture to regress the keypoint locations. (b) CNN with feature maps processed by a Convolutional LSTM to model temporal constraints. CLSTM processes 256 feature maps and its output is used to compute the point location estimate.

4   Results

We evaluated the proposed network architectures on a dataset of ultrasound videos showing parasternal long axis (PLAx) view of the heart (Fig. 2). The data was acquired in several clinics and hospitals with four ultrasound systems: Siemens Acuson Aspen 7.0 and X300, Phillips iE33, and Sonosite M-Turbo. A total of 4981 annotated video frames were used for training and 628 for validation (model selection). The testing data set had 90501 frames of which 2048 were annotated. Our datasets are substantially larger than data sets often used in the medical literature. Two experienced sonographers annotated the frames by manually placing two measurement line calipers (keypoints) perpendicular to the left ventricle (LV) long axis, and measured at the level of the mitral valve leaflet tips. Calipers were positioned on the interface between myocardial wall and cavity and the interface between wall and pericardium. Average locations across annotators were used for training.

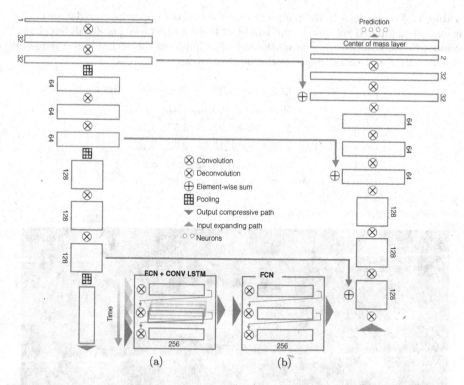

Fig. 4. (a) FCN with feature maps processed by a Convolutional LSTM to model temporal constraints. (b) Fully Convolutional Network (FCN) architecture to regress the keypoint locations. CLSTM processes 256 feature maps and its output is used to compute the point location estimate. In both cases, center of mass layer is used to compute the final estimate.

We computed the following error measures to compare the different architectures. Define the ground truth measurement line based on two keypoints as $\bar{l}_k = ||\bar{\mathbf{r}}_k - \bar{\mathbf{s}}_k||, \bar{\mathbf{p}}_k = (\bar{\mathbf{r}}_k, \bar{\mathbf{s}}_k)^\top$. Estimated measurement line $\hat{l}_k$ is defined similarly using points detected at $\hat{\mathbf{r}}_k$ and $\hat{\mathbf{s}}_k$. The length error is defined as $el_k = |\bar{l}_k - \hat{l}_k|/\bar{l}_k$. The temporal error is defined as $et_k = \big|||\hat{\mathbf{r}}_k - \hat{\mathbf{r}}_{k-1}|| + ||\hat{\mathbf{s}}_k - \hat{\mathbf{s}}_{k-1}||\big|/2\hat{l}_k$. We experimented with various frame sequence lengths and report results on sequences of 3 frames for the CNN + CLSTM model and 6 frames for the FCN + CSLSTM model. The results summarized in Table 1 show 50th, 75th, and 95th percentiles to present distribution of errors and evaluate difficult cases more directly.

Overall, the temporal modeling with CLSTM improves results to frame-wise processing. The final detection accuracy at the 50th percentile is 4.87% of the measurement length which is within average inter-observer error of 4.98%.

Table 1. Average length and temporal errors computed on the testing data set. The errors are computed relative to the length of the measurement line for different percentiles (50th, 75th, 95th). Convolutional LSTM improves the accuracy and temporal stability. FCN + CLSTM model performs best overall.

Error network	Length × 100%			Temporal × 100%	
	50th	75th	95th	75th	95th
CNN	5.67	10.02	21.10	3.52	8.73
CNN+CLSTM	4.89	8.68	17.51	2.92	7.13
FCN	5.00	9.36	19.32	3.36	8.28
FCN+CLSTM	4.87	8.86	18.27	2.67	6.70

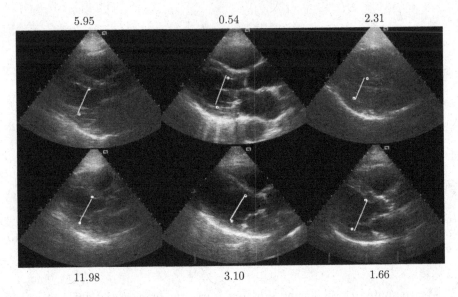

Fig. 5. Several examples of the detection results showing the measurement line (white) and ground truth annotations (red). Errors are shown as a percentage of the ground truth line length. Accurate measurements are obtained despite the shape and appearance variability and despite the ambiguity of the point annotations that can slide along the myocardial wall. (Color figure online)

5   Conclusion

This paper proposed to detect measurement keypoint locations by computing their regression estimates with a Fully Convolutional Network (FCN). The estimates at each pixel location are mapped into the predicted location with a new center-of-mass (CoM) layer. The CoM layer makes it possible to define penalty loss on the measurement line. Spatial context is modeled with Convolutional Long-Short Term Memory (CLSTM) cells.

The results showed errors below 5% of the left ventricle measurement which is within inter-observer variability. The automated measurement was computed in the Parasternal Long Axis (PLAx) view of the heart which has not been previously proposed in the literature. The measurement is an important indicator of the left ventricular function and can be used to compute ejection fraction. Our current work focuses on exploiting variance of the regressed predictions for regularization and on estimating additional measurements.

References

1. Chen, C., Xie, W., Franke, J., Grutzner, P., Nolte, L.P., Zheng, G.: Automatic X-ray landmark detection and shape segmentation via data-driven joint estimation of image displacements. Med. Image Anal. **18**(3), 487–499 (2014)
2. Criminisi, A., Robertson, D., Konukoglu, E., Shotton, J., Pathak, S., White, S., Siddiqui, K.: Regression forests for efficient anatomy detection and localization in computed tomography scans. Med. Image Anal. **17**(8), 1293–1303 (2013)
3. Donner, R., Menze, B.H., Bischof, H., Langs, G.: Global localization of 3D anatomical structures by pre-filtered hough forests and discrete optimization. Med. Image Anal. **17**(8), 1304–1314 (2013)
4. Gauriau, R., Cuingnet, R., Lesage, D., Bloch, I.: Multi-organ localization with cascaded global-to-local regression and shape prior. Med. Image Anal. **23**(1), 70–83 (2015)
5. Hochreiter, S., Schmidhuber, J.: Long short-term memory. Neural Comput. **9**(8), 1735–1780 (1997)
6. Long, J., Shelhamer, E., Darrell, T.: Fully convolutional networks for semantic segmentation. In: Proceedings of the IEEE Conference on Computer Vision and Pattern Recognition, pp. 3431–3440 (2015)
7. Milletari, F., Navab, N., Ahmadi, S.A.: V-net: Fully convolutional neural networks for volumetric medical image segmentation. In: 2016 Fourth International Conference on 3D Vision (3DV), pp. 565–571 (2016)
8. Ning, G., Zhang, Z., Huang, C., He, Z., Ren, X., Wang, H.: Spatially supervised recurrent convolutional neural networks for visual object tracking. arXiv preprint arXiv:1607.05781 (2016)
9. Payer, C., Štern, D., Bischof, H., Urschler, M.: Regressing heatmaps for multiple landmark localization using CNNs. In: International Conference on Medical Image Computing and Computer-Assisted Intervention, pp. 230–238. Springer, New York (2016)
10. Ronneberger, O., Fischer, P., Brox, T.: U-net: convolutional networks for biomedical image segmentation. In: International Conference on Medical Image Computing and Computer-Assisted Intervention, pp. 234–241 (2015)
11. Sofka, M., Zhang, J., Good, S., Zhou, S.K., Comaniciu, D.: Automatic detection and measurement of structures in fetal head ultrasound volumes using sequential estimation and integrated detection network (IDN). IEEE Trans. Med. Imaging **33**(5), 1054–1070 (2014)
12. Szegedy, C., Toshev, A., Erhan, D.: Deep neural networks for object detection. In: Advances in Neural Information Processing Systems, pp. 2553–2561 (2013)
13. Tompson, J., Goroshin, R., Jain, A., LeCun, Y., Bregler, C.: Efficient object localization using convolutional networks. In: Proceedings of the IEEE Conference on Computer Vision and Pattern Recognition, pp. 648–656 (2015)

14. Tompson, J.J., Jain, A., LeCun, Y., Bregler, C.: Joint training of a convolutional network and a graphical model for human pose estimation. In: Advances in Neural Information Processing systems, pp. 1799–1807 (2014)
15. Toshev, A., Szegedy, C.: Deeppose: human pose estimation via deep neural networks. In: Proceedings of the IEEE Conference on Computer Vision and Pattern Recognition, pp. 1653–1660 (2014)
16. Xingjian, S., Chen, Z., Wang, H., Yeung, D.Y., Wong, W.K., Woo, W.C.: Convolutional LSTM network: a machine learning approach for precipitation nowcasting. In: Advances in Neural Information Processing Systems, pp. 802–810 (2015)

Fast Predictive Simple Geodesic Regression

Zhipeng Ding[1]([✉]), Greg Fleishman[3,4], Xiao Yang[1], Paul Thompson[3],
Roland Kwitt[5], Marc Niethammer[1,2],
and The Alzheimer's Disease Neuroimaging Initiative

[1] Department of Computer Science, UNC Chapel Hill, Chapel Hill, USA
zp-ding@cs.unc.edu
[2] Biomedical Research Imaging Center, UNC Chapel Hill, Chapel Hill, USA
[3] Imaging Genetics Center, USC, Los Angeles, USA
[4] Department of Radiology, University of Pennsylvania, Philadelphia, USA
[5] Department of Computer Science, University of Salzburg, Salzburg, Austria

Abstract. Analyzing large-scale imaging studies with thousands of
images is computationally expensive. To assess localized morphological
differences, deformable image registration is a key tool. However, as registrations are costly to compute, large-scale studies frequently require
large compute clusters. This paper explores a fast predictive approximation to image registration. In particular, it uses these fast registrations to
approximate a simplified geodesic regression model to capture longitudinal brain changes. The resulting approach is orders of magnitude faster
than the optimization-based regression approach and hence facilitates
large-scale analysis on a single graphics processing unit. We show results
on 2D and 3D brain magnetic resonance images from OASIS and ADNI.

1 Introduction

Imaging studies on brain development, diseases, and aging will continue to dramatically increase in size: ADNI [10] and the Rotterdam study [9] contain thousands of subjects and the UK Biobank [1] targets an order of 100,000 images
once completed. Analyzing large-scale studies can quickly become computationally prohibitive; compute clusters are commonly used to parallelize analyses such
as deformable image registrations of 3D brain images. Looking ahead, very large-scale studies will require larger compute clusters or efficient algorithms to reduce
computational costs. Furthermore, an increasing number of longitudinal studies
require efficient algorithms for the analysis of longitudinal image data.

Geodesic regression (GR) [5,12,14] is an attractive approach to capture trends
in longitudinal imaging data. GR generalizes linear regression to Riemannian
manifolds. Applied to longitudinal image data it compactly expresses spatial image
transformations over time. Unfortunately, GR requires the solution of a computationally costly optimization problem. Hence, a simplified, approximate, GR approach (SGR) has been proposed [6] that allows for the computation of the regression

---

The original version of this chapter was revised: The affiliations of the authors were
corrected. An erratum to this chapter can be found at https://doi.org/10.1007/
978-3-319-67558-9_44

© Springer International Publishing AG 2017
M.J. Cardoso et al. (Eds.): DLMIA/ML-CDS 2017, LNCS 10553, pp. 267–275, 2017.
DOI: 10.1007/978-3-319-67558-9_31

geodesic via pairwise image registrations. But even SGR would require months of computation time on a single graphics processing unit (GPU) to process thousands of 3D image registrations for large-scale imaging studies.

However, efficient approaches for deformable image registration have recently been proposed. In particular, for the large displacement diffeomorphic metric mapping (LDDMM) model, which is the basis of GR approaches for images, registrations can be dramatically sped up by either working with finite-dimensional Lie algebras [18] or by fast predictive image registration (FPIR) [15,16]. FPIR learns a patch-based deep regression model to predict the initial momentum of LDDMM, which characterizes the spatial transformation. By replacing numerical optimization of standard LDDMM registration by a *single* prediction step followed by optional correction steps [16], FPIR can improve speed by orders of magnitude. Besides FPIR, other predictive image registration (i.e., optical flow) approaches exist [2,11,13]. However, FPIR is better suited for brain image registration than these optical flow approaches for the following reasons: (1) FPIR predicts the initial momentum of LDDMM and therefore inherits its theoretical properties. Consequentially, FPIR results in diffeomorphic transformations and geodesics even though predictions are patch-wise. This can not be guaranteed by optical flow prediction methods. (2) Patch-wise prediction allows training of the prediction models based on a very small number of images, utilizing a large number of patches. (3) By using patches instead of full images, predictions for large, high-resolution images are possible despite the memory constraints of current GPUs. (4) None of the existing prediction methods address longitudinal image data. Additionally, as both FPIR and SGR are based on LDDMM, they naturally integrate and hence result in our proposed *fast predictive simple geodesic regression* (FPSGR) approach.

Contributions. (1) *Predictive Geodesic Regression*: We introduce the first fast predictive geodesic regression approach for images. (2) *Large-scale dataset capability*: Our predictive approach facilitates large-scale image registration/regression within a day on a single GPU instead of months for optimization-based methods. (3) *Accuracy*: We show that FPSGR approximates the optimization-based simple GR result well. (4) *Validation*: We showcase the performance of our FPSGR approach by analyzing >6,000 images of the ADNI-1/ADNI-2 datasets.

Organization. Section 2 describes FPSGR. Section 3 discusses the experimental setup and the training of the prediction models. Section 4 presents experimental results for 2D and 3D magnetic resonance images (MRI) of the brain. The paper concludes with a summary and an outlook on future work.

2   Fast Predictive Simple Geodesic Regression (FPSGR)

Our fast predictive simple geodesic regression involves two main components: (1) the fast predictive image registration (FPIR) approach and (2) integration of FPIR into the simple geodesic regression (SGR) formulation. Both FPIR and SGR are based on the shooting formulation of LDDMM [14]. Figure 1 illustrates our overall approach. The individual components are described in the following.

LDDMM. Shooting-based LDDMM and geodesic regression minimize

$$E(I_0, m_0) = \frac{1}{2}\langle m_0, Km_0\rangle + \frac{1}{\sigma^2}\sum_i d^2(I(t_i), Y^i),$$

$$s.t. \quad m_t + \mathrm{ad}_v^* m = 0, I_t + \nabla I^T v = 0, m - Lv = 0, \tag{1}$$

where I_0 is the initial image (known for image-to-image registration and to be determined for geodesic regression), m_0 is the initial momentum, K is a smoothing operator that connects velocity v and momentum m as $v = Km$ and $m = Lv$ with $K = L^{-1}$, $\sigma > 0$ is a weight, Y^i is the measured image at time t_i (there will only be one such image for image-to-image registration at $t = 1$), and $d^2(I_1, I_2)$ denotes the image similarity measure between I_1 and I_2 (for example L_2 or geodesic distance); ad^* is the dual of the negative Jacobi-Lie bracket of vector fields: $\mathrm{ad}_v w = -[v, w] = Dvw - Dwv$ and D denotes the Jacobian. The deformation of the source image $I_0 \circ \Phi^{-1}$ can be computed by solving $\Phi_t^{-1} + D\Phi^{-1}v = 0$, $\Phi_0^{-1} = \mathrm{id}$, where id denotes the identity map.

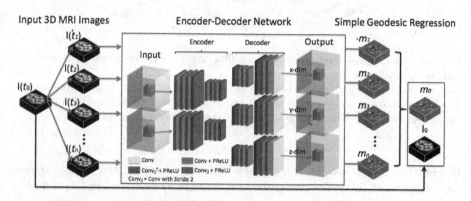

Fig. 1. Principle of Fast Predictive Simple Geodesic Regression (FPSGR). In the Encoder-Decoder network, the inputs are patches from source images and target images at the same position; the outputs are the predicted initial momenta of the corresponding patches. In simple geodesic regression, all the pairwise initial momenta are averaged according to Eq. 2 to produce the initial momentum of the regression geodesic.

FPIR. Fast predictive image registration [15,16] aims at predicting the initial momentum, m_0, between a source and a target image patch-by-patch. Specifically, we use a deep encoder-decoder network to predict the patch-wise momentum. As shown in Fig. 1, in 3D the inputs are two layers of $15 \times 15 \times 15$ image patches (15×15 in 2D), where the two layers are from the source and target images respectively. Two patches are taken at the same position by two parallel encoders, which learn features independently. The output is the predicted initial momentum in the x, y and z directions (obtained by numerical optimization for the training samples). The network can be divided into Encoder and Decoder.

An encoder consists of 2 blocks of 3 3 × 3 × 3 convolutional layers with PReLU activations, followed by another 2 × 2 × 2 convolution+PReLU with a stride of two, serving as "pooling" operation. The number of features in the first convolutional layer is 64 and increases to 128 in the second. In the Decoder, three parallel decoders share the same input generated from the encoder. Each decoder is the inverse of the encoder except for using 3D transposed convolution layers with a stride of two to perform "unpooling", and no non-linearity layer at the end. To speed up computations, we use patch pruning (i.e., patches outside the brain are not predicted as the momentum is zero there) and a large pixel stride (e.g. 14 for 15 × 15 × 15 patches) for the sliding window of the predicted patches.

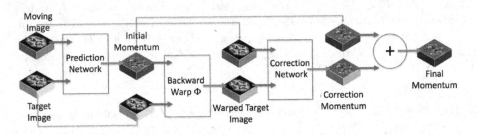

Fig. 2. Prediction + correction network architecture. (1) Predict initial momentum m_p and corresponding backward deformation, Φ. (2) Predict correction of initial momentum, m_c, based on the difference between moving image and warped-back target image. The final momentum is $m = m_p + m_c$. The correction network is trained with the moving images and warped-back target images of the training dataset as inputs.

Correction Network. We also use an additional correction step to improve prediction accuracy, by training a correction network as described in [16]. Figure 2 illustrates this approach graphically. The correction network has the same structure as the prediction network. Only the inputs and outputs change.

SGR. Determining the initial image, I_0, and the initial momentum, m_0, of Eq. 1 is computationally costly. In simple geodesic regression, the initial image is fixed to the first image of a subject's longitudinal image set. Furthermore, the similarity measure $d(\cdot, \cdot)$ is chosen as the geodesic distance between images and *approximated* so that the geodesic regression problem can by computed by computing pair-wise image registrations with respect to the first image. The approximated optimal m_0 of energy functional (1) for a fixed I_0 is then

$$m_0 \approx \frac{\sum (t_i - t_0)^2 m_0^{Y_i}}{\sum (t_i - t_0)^2} = \frac{\sum (t_i - t_0)\overline{m}_0^{Y_i}}{\sum (t_i - t_0)^2} \qquad (2)$$

where $\overline{m}_0^{Y_i}$ is obtained by registering I_0 to Y^i in unit time followed by a rescaling of the momentum to account for the original time duration: $m_0^{Y_i} = \frac{1}{t_i - t_0} \overline{m}_0^{Y_i}$. In our FPSGR approach we compute the momenta $\overline{m}_0^{Y_i}$ via FPIR.

3 Experimental Setup and Training of Prediction Models

We perform experiments on 2D axial MRI brain slices from the OASIS dataset and on 3D images from the ADNI dataset. **2D:** We verify our hypothesis that training FPIR on longitudinal data for longitudinal registrations is preferred over training using cross-sectional data; and that training FPIR on cross-sectional data for cross-sectional registrations is preferred over training using longitudinal data. Comparisons are with respect to registration results obtained by numerical optimization. **3D:** As the ADNI dataset is longitudinal and based on our findings for the 2D data, we train our models with longitudinal registrations only.

Training and Testing of the Prediction Models. We use a set of 150 patients' MRIs of the OASIS dataset for training the 2D model and testing the performance of FPIR. The resampled 2D images of size 128×128 are extracted from the same axial slice of the 3D OASIS images after affine registration. The ADNI data consists of 6471 3D MR brain images of size $220 \times 220 \times 220$ (3479 ADNI-1, 833 subjects; 2992 ADNI-2, 823 subjects). We use all of the ADNI data. **2D:** We used the first 100 patients for training and the last 50 for testing. For longitudinal training, we registered the first and the second time-point of a patient. For cross-sectional training, we registered a patient's first time point to another patient's second time point. To test longitudinal and cross-sectional registrations, we perform the same type of registrations on the 50 test sets. We compare the deformation error against the LDDMM solution obtained by numerical optimization. Table 1 shows the results which confirm our hypothesis that training the prediction model with longitudinal registration cases is preferred for longitudinal registration over training with cross-sectional data. The deformation error is very small for longitudinal training/testing and the predictive method has comparable performance to costly optimization-based LDDMM. These results indicate that it is beneficial to train a prediction model with deformations it is expected to encounter, i.e., relatively small deformations for longitudinal registrations and larger deformations for cross-sectional registration. Hence, for the ADNI data, we train the 3D models using longitudinal registrations only. **3D:** We randomly picked 1/6 patients from each diagnostic category to form a group of 139 patients for training in ADNI-1 and 150 in ADNI-2. The images of the first time-point were registered to all the later time-points within each patient.

Table 1. Deformation error of longitudinal and cross-sectional models tested on longitudinal and cross-sectional data. 2-norm deformation errors in pixels w.r.t. the ground truth deformation obtained by numerical optimization for LDDMM.

2D Longitudinal Test Case Deformation Error [pixel]						
Data Percentile	0.3%	5%	25%	50%	75%	95%	99.7%
Longitudinal Training	**0.0027**	**0.0112**	**0.0267**	**0.0425**	**0.0630**	**0.1222**	**0.2221**
Cross-sectional Training	0.0050	0.0201	0.0475	0.0744	0.1093	0.1862	0.2253
2D Cross-sectional Test Case Deformation Error [pixel]						
Data Percentile	0.3%	5%	25%	50%	75%	95%	99.7%
Longitudinal Training	0.0256	0.1068	0.2669	0.4552	0.7433	1.4966	1.9007
Cross-sectional Training	**0.0120**	**0.0508**	**0.1248**	**0.2047**	**0.3196**	**0.5781**	**0.6973**

Matching the distribution of diagnostic categories, we then randomly picked a subset of 165 of these registrations as the training set in ADNI-1 and 140 in ADNI-2. We trained 4 prediction models and their corresponding correction models totaling 8 prediction models, i.e. ADNI-1 Pred-1, ADNI-1 Pred+Corr-1, ADNI-1 Pred-2, ADNI-1 Pred+Corr-2 and analogously for ADNI-2. The training sets within ADNI-1/2 respectively were not overlapping, allowing us to compute predictions (not using the training data) for the *complete* ADNI-1/2 datasets.

Parameter Selection. We use the regularization kernel $K = L^{-1} = (-a\nabla^2 - b\nabla(\nabla\cdot) + c)^{-2}$ with $[a, b, c]$ set to $[0.05, 0.05, 0.005]$ for 2D and $[1, 0, 0.1]$ for 3D images; σ is 0.1 for 2D and 3D. We use *Adam* to optimize the network with 10 epochs and learning rates of 0.0005 and 0.0001 in 2D and 3D respectively.

Efficiency. Once trained, the prediction models allow fast computations of registrations. We use a TITAN X(Pascal) GPU and *PyTorch*. In 2D, FPIR took 4 seconds on average to predict 50 pairs of 128×128 MRIs (for the 4 longitudinal/cross-sectional experiments), while a GPU implementation of optimization-based LDDMM took 150 min. For the 3D ADNI-1 data ($220 \times 220 \times 220$ MRIs), FPSGR took about one day to predict 2646 pairwise registrations (25s/prediction) and to compute the regression result. Optimization-based LDDMM would require ≈40 days of runtime. Runtime for FPIR on ADNI-2 is identical to ADNI-1 as the images have the same dimension.

4   Discussion of Experimental Results for 3D ADNI Data

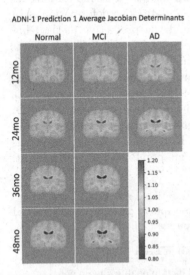

ADNI-1 Prediction 1 Average Jacobian Determinants

Normal   MCI   AD

12mo / 24mo / 36mo / 48mo

1.20 / 1.15 / 1.10 / 1.05 / 1.00 / 0.95 / 0.90 / 0.85 / 0.80

Fig. 3. Average Jacobian determinant over time and diagnostic category for one of the ADNI-1 groups.

We studied 10 experimental groups: all subjects from ADNI-1 using traditional LDDMM, two groups of ADNI-1 with different training data using FPSGR without correction network, and two groups of ADNI-1 using FPSGR with correction network. An analogous set of groups was studied for ADNI-2. We calculated the Jacobian determinant (JD) for every computed deformation. The JDs were then warped to a common coordinate system for the entire ADNI data set using existing deformations from [3,4]. Each such spatially normalized JD was then averaged within a region of interest (ROI). Specifically, we quantify atrophy as $(1 - \frac{1}{|\omega|}\int_\omega |D\phi(x)|dx) \times 100$, where $|\cdot|$ is the determinant, and ω is an area in the temporal lobes which was determined in prior studies [3,4] to be significantly associated with accelerated atrophy in Alzheimer's disease (AD). The resulting scalar value is an estimate of the relative volume change experienced by that region between the baseline and followup image acquisitions. Hence, its sign is positive when the region has lost volume over time and is negative if the region has gained volume over time. See Fig. 3.

Table 2. Slope and intercept values for simple linear regression of volume change over time [Lower end of 95% C.I., **point estimate**, Higher end of 95% C.I.]

ADNI-1		Slope	Intercept
Normal	LDDMM-1	[0.65, **0.73**, 0.81]	[−0.23, **−0.07**, 0.09]
	Pred-1	[0.39, **0.45**, 0.52]	[−0.20, **−0.08**, 0.05]
	Pred+Corr-1	[0.64, **0.71**, 0.77]	[−0.14, **−0.01**, 0.13]
	LDDMM-2	[0.63, **0.72**, 0.81]	[−0.21, **−0.03**, 0.15]
	Pred-2	[0.49, **0.56**, 0.63]	[−0.17, **−0.03**, 0.11]
	Pred+Corr-2	[0.56, **0.64**, 0.71]	[−0.14, **0.00**, 0.14]
MCI	LDDMM-1	[1.55, **1.68**, 1.82]	−[0.28, **−0.04**, 0.19]
	Pred-1	[1.03, **1.13**, 1.23]	[−0.24, **−0.07**, 0.10]
	Pred+Corr-1	[1.39, **1.50**, 1.61]	[−0.14, **0.05**, 0.25]
	LDDMM-2	[1.41, **1.54**, 1.66]	[−0.25, **−0.03**, 0.18]
	Pred-2	[1.15, **1.25**, 1.36]	[−0.20, **−0.02**, 0.16]
	Pred+Corr-2	[1.22, **1.33**, 1.43]	[−0.16, **0.02**, 0.20]
AD	LDDMM-1	[2.07, **2.42**, 2.77]	[−0.28, **0.14**, 0.57]
	Pred-1	[1.31, **1.57**, 1.84]	[−0.22, **0.11**, 0.43]
	Pred+Corr-1	[1.83, **2.13**, 2.42]	[−0.14, **0.22**, 0.58]
	LDDMM-2	[2.03, **2.39**, 2.75]	[−0.29, **0.14**, 0.57]
	Pred-2	[1.66, **1.94**, 2.23]	[−0.21, **0.13**, 0.47]
	Pred+Corr-2	[1.75, **2.03**, 2.32]	[−0.18, **0.16**, 0.51]

We limited our experiments to the applications in [7,8], wherein nonlinear registration/regression is used to quantify atrophy within regions known to be associated to varying degrees with AD (2), Mild Cognitive Impairment (MCI) (1) (including Late Mild Cognitive Impairment), and normal aging (0) in an elderly population.

Moreover, we are interested in the following two critical validations: (1) Are atrophy measurements derived from FPSGR biased to over- or underestimate change? (2) Are those atrophy measurements consistent with those derived from deformations given by the optimization procedure (LDDMM) which produced the training data set? If experiments show (1) and (2) resolve favorably, then the substantially improved computational efficiency of FPSGR may justify its use for some applications.

Bias. Estimates of atrophy are susceptible to bias [17]. We follow [8] by fitting a straight line (linear regression) through all such atrophy measurements over time in each diagnostic category. The intercept term is an estimate of the atrophy one would measure when registering two scans acquired on the same day; hence it should be near zero and its 95% confidence interval should contain zero. Table 2 shows the slopes, intercepts, and 95% confidence intervals for all five groups of ADNI-1. Results (not shown) are similar for ADNI-2. All of them have intercepts that are near zero relative to the range of changes observed and all prediction intercept confidence intervals contain zero. Further, all slopes are positive indicating average volume loss over time, consistent with expectations for an aging and neuro-degenerative population. The slopes capture increasing atrophy with disease severity. We conclude that neither LDDMM optimization nor FPSGR produced deformations with significant bias to over- or underestimate volume change. Note that our LDDMM optimization results and the prediction results show the same trends, and are directly comparable as the results are based on the same test images (same for atrophy measurement).

Atrophy. Atrophy estimates have also been shown to correlate with clinical variables. In accordance with validation (2) from above, we computed the correlation

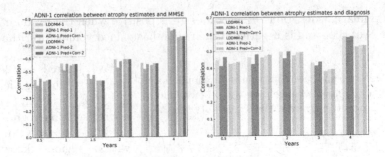

Fig. 4. FPSGR-derived correlations with clinical variables.

between our atrophy estimates and diagnosis, and also between our atrophy estimates and mini-mental state exam (MMSE) scores (Fig. 4). The magnitudes of correlations we observed for all eight prediction models (four with and four without correction networks) were in the range of -0.4 to -0.8 for MMSE and 0.2 to 0.6 for diagnosis. Previous studies have reported correlation between atrophy estimates and clinical variables as high as -0.7 for MMSE and 0.5 for diagnosis for 100 subjects [3,4]. All prediction results are comparable with baseline LDDMM results. The correction network generally improves prediction performance.

Jacobian. The average JD images qualitatively agree with prior results [7,8]: severity of volume change increases with severity of diagnosis and time. Change is most substantial in the temporal lobes near the hippocampus (see Fig. 3).

5   Conclusion and Future Work

We proposed a fast approach for geodesic regression (FPSGR) to study longitudinal image data. FPSGR is orders of magnitude faster than existing approaches and facilitates the analysis of large-scale imaging studies. Experiments on the ADNI-1/2 datasets demonstrate that FPSGR captures expected atrophy trends, exhibits negligible bias and shows high correlations with clinical variables.

Support. This work was supported by NSF grants EECS-1148870/1711776.

References

1. Biobank website. www.ukbiobank.ac.uk
2. Dosovitskiy, A., Fischer, P., Ilg, E., Hausser, P., Hazirbas, C., Golkov, V., van der Smagt, P., Cremers, D., Brox, T.: Flownet: learning optical flow with convolutional networks. In: ICCV, pp. 2758–2766 (2015)
3. Fleishman, G., Thompson, P.M.: Adaptive gradient descent optimization of initial momenta for geodesic shooting in diffeomorphisms. In: ISBI (2017)
4. Fleishman, G., Thompson, P.M.: The impact of matching functional on atrophy measurement from geodesic shooting in diffeomorphisms. In: ISBI (2017)

5. Fletcher, P.T.: Geodesic regression and the theory of least squares on Riemannian manifolds. IJCV **105**(2), 171–185 (2013)
6. Hong, Y., Shi, Y., Styner, M., Sanchez, M., Niethammer, M.: Simple geodesic regression for image time-series. In: Dawant, B.M., Christensen, G.E., Fitzpatrick, J.M., Rueckert, D. (eds.) WBIR 2012. LNCS, vol. 7359, pp. 11–20. Springer, Heidelberg (2012). doi:10.1007/978-3-642-31340-0_2
7. Hua, X., Ching, C.R.K., Mezher, A., Gutman, B., Hibar, D.P., Bhatt, P., Leow, A.D., Jack Jr., C.R., Bernstein, M.A., Weiner, M.W., Thompson, P.M., Alzheimer's Disease Neuroimaging Initiative: MRI-based brain atrophy rates in ADNI phase 2: acceleration and enrichment considerations for clinical trials. Neurobiol. Aging **37**, 26–37 (2016)
8. Hua, X., Hibar, D.P., Ching, C.R.K., Boyle, C.P., Rajagopalan, P., Gutman, B., Leow, A.D., Toga Jr., A.W., C.R.J., Harvey, D.J., Weiner, M.W., Thompson, P.M.: Unbiased tensor-based morphometry: improved robustness & sample size estimates for Alzheimer's disease clinical trials. NeuroImage **66**, 648–661 (2013)
9. Ikram, M.A., van der Lugt, A., Niessen, W.J., Koudstaal, P.J., Krestin, G.P., Hofman, A., Bos, D., Vernooij, M.W.: The Rotterdam scan study: design update 2016 and main findings. Eur. J. Epidemiol. **30**(12), 1299–1315 (2015)
10. Jack, C.R., Barnes, J., Bernstein, M.A., Borowski, B.J., Brewer, J., Clegg, S., Dale, A.M., Carmichael, O., Ching, C., DeCarli, C., et al.: Magnetic resonance imaging in ADNI 2. Alzheimer's Dement. **11**(7), 740–756 (2015)
11. Liu, Z., Yeh, R., Tang, X., Liu, Y., Agarwala, A.: Video frame synthesis using deep voxel flow. arXiv preprint arXiv:1702.02463 (2017)
12. Niethammer, M., Huang, Y., Vialard, F.-X.: Geodesic regression for image time-series. In: Fichtinger, G., Martel, A., Peters, T. (eds.) MICCAI 2011. LNCS, vol. 6892, pp. 655–662. Springer, Heidelberg (2011). doi:10.1007/978-3-642-23629-7_80
13. Schuster, T., Wolf, L., Gadot, D.: Optical flow requires multiple strategies (but only one network). arXiv preprint arXiv:1611.05607 (2016)
14. Singh, N., Hinkle, J., Joshi, S., Fletcher, P.T.: A vector momenta formulation of diffeomorphisms for improved geodesic regression and atlas construction. In: ISBI, pp. 1219–1222 (2013)
15. Yang, X., Kwitt, R., Niethammer, M.: Fast predictive image registration. In: Carneiro, G., et al. (eds.) LABELS/DLMIA -2016. LNCS, vol. 10008, pp. 48–57. Springer, Cham (2016). doi:10.1007/978-3-319-46976-8_6
16. Yang, X., Kwitt, R., Niethammer, M.: Quicksilver: fast predictive image registration-a deep learning approach. NeuroImage (2017, in press)
17. Yushkevich, P.A., Avants, B.B., Das, S.R., Pluta, J., Altinay, M., Craige, C.: Bias in estimation of hippocampal atrophy using deformation-based morphometry arises from asymmetric global normalization: an illustration in ADNI 3T MRI data. NeuroImage **50**(2), 434–445 (2010)
18. Zhang, M., Fletcher, P.T.: Finite-dimensional lie algebras for fast diffeomorphic image registration. In: Ourselin, S., Alexander, D.C., Westin, C.-F., Cardoso, M.J. (eds.) IPMI 2015. LNCS, vol. 9123, pp. 249–260. Springer, Cham (2015). doi:10.1007/978-3-319-19992-4_19

Learning Spatio-Temporal Aggregation for Fetal Heart Analysis in Ultrasound Video

Arijit Patra[✉], Weilin Huang, and J. Alison Noble

Institute of Biomedical Engineering, University of Oxford, Oxford, UK
arijit.patra@eng.ox.ac.uk

Abstract. We investigate recent deep convolutional architectures for automatically describing multiple clinically relevant properties of the fetal heart in Ultrasound (US) videos, with the goal of learning spatio-temporal aggregation of deep representations. We examine multiple temporal encoding models that combine both spatial and temporal features tailored for US video representation. We cast our task into a multi-task learning problem within a hierarchical convolutional model that jointly predicts the visibility, view plane and localization of the fetal heart at the frame level. We study deep convolutional networks developed for video classification, and analyse them for our task by looking at the architectures and the multi-task loss in the specific modality of US videos. We experimentally verify that the developed hierarchical convolutional model that progressively encodes temporal information throughout the network is powerful to retain both spatial details and rich temporal features, which leads to high performance on a real-world clinical dataset.

1 Introduction

Fetal ultrasound (US) is a standard part of pre-natal care primarily due to its non-invasive nature and the unsuitability of other imaging modalities through the length of pregnancy. Understanding fetal cardiac screening US videos is important to diagnose congenital heart disease. Fetal heart conditions are often missed because of factors like the shortage of trained clinicians and the requirement of expertise and equipment for fetal cardiac screening which causes it to be excluded from compulsory screening requirements in the 20-week abnormality scans. Precisely analysing fetal cardiac US videos is a challenging task even for human experts due to the complex appearance of different anatomical structures in a small region. Furthermore, there exist speckle, shadowing, enhancement and variations in contrast levels in US images. In addition, the clinician has to perform multiple activities during a typical US scan such as viewing plane identification, anomaly detection, gender identification and so on. In this work, we analyse the scope for using deep learning techniques with US videos for describing key parameters of the fetal heart, with a focus on aggregating spatio-temporal features within deep convolutional architectures.

Convolutional neural networks (CNNs) have been successfully applied to many computer vision tasks. Various CNN models have been developed recently

© Springer International Publishing AG 2017
M.J. Cardoso et al. (Eds.): DLMIA/ML-CDS 2017, LNCS 10553, pp. 276–284, 2017.
DOI: 10.1007/978-3-319-67558-9_32

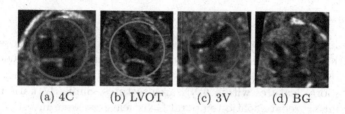

(a) 4C   (b) LVOT   (c) 3V   (d) BG

Fig. 1. Three different view planes of fetal heart: the four chamber (4C), the left ventricular outflow tract (LVOT), the three vessels (3V), and the background (BG).

for action recognition, with the goal of learning spatio-temporal features from videos [5,8,9]. These models were mainly designed for video-level classification with a single action category assigned to a whole video. High-level global context information is important for the task of video classification, while our task is more challenging by jointly considering both classification and localization at the frame level, where detailed local information is vital for precise prediction of multiple parameters of the heart. Furthermore, action recognition is mostly built on RGB domain where a single image can include strong visual information for identifying an action category. By contrast, object structure in an US image is often weakly defined, and local temporal dynamics are critically important for producing a reliable prediction.

We approach the problem of automated analysis of the fetal heart by formulating it as a multi-task prediction, where the visibility, view plane and localization of the heart are jointly predicted at the frame level. We study the problem of aggregating spatio-temporal information in a short US video clip by investigating various temporal connectivities within the convolutional architecture. We examine the impact of temporal information on the domain-specific task of fetal heart analysis in US videos. Our main contributions are summarized below:

- We cast the problem of fetal heart analysis as a multi-task prediction in a hierarchical convolutional architecture that progressively encodes temporal information throughout the network. This is vital to retaining both spatial details and meaningful temporal patterns, which are key to accurate and automatic heart description in fetal US videos.
- We investigate multiple temporal encoding architectures that learn a strong spatio-temporal representation tailored for US video representation. We study these approaches by analysing the architecture, the specific US image modality, and the loss functions designed for joint classification and localization.
- We conducted experiments on a real-world clinical dataset. The results suggest the ability to encode temporal information, which is of particular importance for our task, where image-level information is relatively weak and can be ambiguous.

1.1   Related Work

Deep learning approaches have been recently applied for action recognition in videos, with the focus on learning spatio-temporal information efficiently and accurately. Karpathy et al. [5] developed a number of deep fusion methods that encode temporal features within CNN architectures. Similarly, various pooling methods that aggregate spatio-temporal CNN features, were investigated in [8] for action recognition. Two-stream CNNs were introduced in [9], where short-term temporal information is captured by using a separate CNN processing on optical flow. These methods were developed for action recognition in video clips where the CNNs were designed to compute global high-level context from sequences of RGB images. Our work focuses on aggregating spatio-temporal information for our US video analysis, and our task requires local detail in both the spatial and the temporal domains for joint classification and localisation.

Recent work on automatic US video analysis mainly focuses on image-level classification of anatomical structures, e.g., [7]. CNNs have also been applied to this task with transfer learning [3] and recurrent models [2]. Fetal heart description is a more complicated application that jointly predicts multiple parameters of the fetal heart. Our work is closely related to recent work in [1,4], where multiple properties of the fetal heart were estimated. In [1], a CRF-filter along with hand-crafted features were developed for predicting the defined parameters of the fetal heart. We build on advanced deep learning methods, which allow our model to learn stronger deep representations in an end-to-end fashion. In [4], a Temporal HeartNet was proposed by using a recurrent model to encode temporal information. In the current paper, we focus on aggregating spatio-temporal features directly within a CNN architecture, setting it apart from [4].

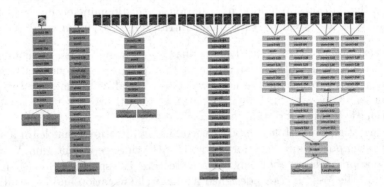

Fig. 2. CNN architectures for multi-task fetal heart prediction in US videos. Left: spatial baseline models based on AlexNet (SBM_{alex}) and VGGnet (SBM_{vgg}). Middle: Direct Temporal Encoding (DTE) models modified from AlexNet and VGGnet. Right: the developed Hierarchical Temporal Encoding (HTE) that progressively incorporates spatio-temporal information throughout the network.

2   Fetal Heart Description with Spatio-Temporal CNNs

Our task is to jointly predict the visibility, view plane, location of the fetal heart in US videos, which are similar to [1]. The visibility relates to the presence or absence of the heart in each frame (with partial visibility of less than 50% being deemed as not visible). When the heart is present, the view plane can be either the four chamber (4C) view, the left ventricular outflow tract (LVOT), or the three vessels (3V) view, as shown in Fig. 1. By jointly considering the visibility and the view-plane identification, we include a background class to define a 4-class classification problem. The location of the heart is defined by its center and radius, which can be cast as a 3-parameter regression problem.

In this work, both classification and localisation are formulated jointly as a multi-task learning problem within the convolutional architecture. The outputs of the two tasks are connected to the last fully-connected (FC) layer of a CNN, such as AlexNet [6] or VGGnet [10], as shown in Fig. 2. A cross-entropy loss with a softmax function is used for the classification task, while a l_2 loss is applied for regressing the values of heart centre and radius. These configurations customize a regular CNN to our task, and form our baseline spatial architecture where the input is a single US image. Our temporal models are extended from this baseline architecture by allowing input of a sequence of continuous frames. To investigate the model capability for learning spatio-temporal details, we extend recent temporal encoding methods, originally introduced to action recognition in [5], to multi-task prediction of the fetal heart in US videos.

2.1   Spatial Baseline Models

For the first step of creating spatial baseline models, we leverage the classical AlexNet [6] and the 16-layer VGGnet [10], both of which are modified to accept a single channel gray-scale image. The two spatial baseline models are referred to as SBM_{alex} and SBM_{vgg} respectively. The networks accept as inputs an US image with size of 224×224 pixels. Details of the architectures are presented in Fig. 2 (left). Notice that the number of neurons in the FC layers is changed from the original 4096 to 1024. This modification was made to avoid overfitting, as the number of predicted parameters in our task is significantly smaller than the 1000 categories of object classification in ImageNet. The last FC layer is connected to the output layer which computes the softmax probabilities of 4 classes (3 view planes and the background), with the predicted heart center and radius.

2.2   Direct Temporal Encoding

To compute temporal information, a straightforward approach is to modify the CNN architecture to allow for an input of multiple contiguous frames. Such a multi-frame input naturally includes temporal context which is important to produce a more reliable prediction. As the size of the CNN architecture is fixed by design, and the number of CNN parameters should be scalable, the input of the CNNs is configured and fixed to a short, fixed-length US video clip cropped from temporally contiguous frames. In experiments, we use an 8-frame video clip as input for all temporal models where the labels of the 4th frame are applied.

For model configurations, the main idea is to extend network connectivities in the time domain, allowing these connectivities to automatically learn temporal dynamics between contiguous US images. To this end, we extend our spatial baseline models to Direct Temporal Encoding (DTE) by introducing direct temporal filters with kernel size of $k \times k \times T$ in the first convolutional layer, where T is the temporal dimension which is set to $T = 8$ in our experiments. $k \times k$ is the spatial size of convolutional kernels which is identical to those of the spatial baseline models, e.g., 11×11 and 3×3 for the AlextNet and VGGnet respectively. These temporal filters are able to directly encode both spatial details and temporal connections of the continuous US frames. This enhances the low-level representation in the first convolutional layer, which in turn leads to stronger deep high-level representations by propagating them throughout the networks.

2.3 Hierarchical Temporal Encoding

The DTE approach directly computes local temporal features at the pixel level, and is powerful enough to capture detailed low-level motion characteristics. However, CNN predictions are built on high-level deep representation (e.g., the FC features). It is critically important to investigate: (i) how this low-level temporal information is propagated efficiently and accurately to the final deep representation; and (ii) which level of the temporal information is crucial to the final prediction. Since US images often contain relatively weak visual information at the pixel level, temporal representation immediately computed at the first layer by the DTE may not be robust, and multi-layer propagations throughout the whole network may lead to a certain degree of information loss. The need is to develop a new temporal encoding approach that allows the temporal details of US videos to be computed more accurately and robustly, and to be propagated more efficiently throughout the network without significant information loss.

With these considerations, we have developed a new Hierarchical Temporal Encoding (HTE) approach, which is extended from the slow fusion model developed for action recognition in [5]. We present a number of technical improvements that elegantly tailor it towards our task of fetal US description: (i) we incorporate multi-task prediction into this hierarchical architecture for jointly estimating the view plane and localization of the fetal heart. (ii) and, more convolutional layers using small kernel size of 3×3 are applied, resulting in a deeper model able to capture more spatio-temporal details which are particularly important to our task. These design features aim to capture key characteristics of US video interpretation to encode the temporal domain and the spatial domain, and a finer feature extraction from the previous layer is more desirable to propagate the features throughout the network.

Details of the HTE model are presented in Fig. 2 (right), where frames from the video clip are gathered in groups of 3, and then fed into the first layers of multiple sets of CNNs which share the same parameters. These serve as quasi-independent CNNs until the 3rd pooling layer, where the outputs of the 3rd

pooling layer are taken in groups of 2, and further fed into separate sets of 2 CNN structures also sharing the same parameters. This is maintained until the final encoding happens just prior to the first FC layer, and subsequent FC layers effectively process the overall information from the video. Therefore, a slow and progressive combination of the learned features enables the propagation of the diversity of information encoded in the temporal space.

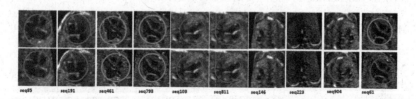

seq85   seq191   seq461   seq793   seq103   seq811   seq146   seq223   seq904   seq61

Fig. 3. The predicted results by the HTE (bottom) with GT (top). Color for different view planes. (Color figure online)

3 Implementation Details

Real-world clinical dataset. Our dataset consisted of 91 routine clinical cardiac screening videos from 12 subjects at gestational ages ranging from 20 and 35 weeks. Each video had a duration of between 2 and 10 s and a frame rate between 25 and 76 frames per second (39556 frames in total). It contained one or more of the three views of the fetal heart. Videos from 10 subjects were used for training, and the remaining 2 for test.

Data pre-processing. For the training step, we split available videos into frames and applied data augmentation by performing an up-down and a top-bottom flipping. Then, we chose 8 frames at a time in the sequence of their original occurrence in the videos and stitched them back into 8-frame sequences and repeated the same for the augmented data. This enabled us to obtain 9337 video clips with roughly equal samples from different view classes. The current frame is the 4th frame in a 8-frame video clip. All frames in the dataset of size 430×510 were cropped into 224×224 centered about the heart center.

Training details. All new layers across the networks used are initialized by using random weights. Our spatial baseline models are trained by using a 25-frame mini-batch with a learning rate of 0.01. The DTE models are trained with 8-frame chunks of videos using a batch size of 25 with a learning rate of 0.01. A batch size of 20 and a learning rate of 0.001 are used for the HTE.

4 Experimental Results and Comparisons

We performed multiple sets of experiments for both the classification and regression tasks. We start with the generation of a strong feature-based baseline using single frames of video sequences, followed by temporal encoding models, which are then augmented by a hierarchical transfer of spatio-temporal details through the network. We analyse our models for multi-task learning from shared features, achieved by focussing on requisite regions of interest for each task.

Table 1. Performances on real-world fetal cardiac screening videos, in accuracy rates (%). A correct localization is defined with an Intersection-over-Union ratio over 0.25.

Method	Classification					Localization
	4 chambers	LVOT	3 vessels	Background	Overall	
SBM_{alex}	89.22	67.60	59.11	90.36	78.95	69.71
SBM_{vgg}	90.84	69.83	62.07	91.37	80.73	72.29
DTE_{alex}	88.68	73.18	57.63	88.83	79.16	72.21
DTE_{vgg}	91.11	70.95	67.00	89.34	81.78	77.26
HTE	81.67	88.27	79.80	86.80	83.58	79.68

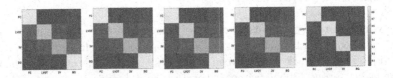

Fig. 4. Confusion matrices for classification. L to R: SBM_{alex}, SBM_{vgg}, DTE_{alex}, DTE_{vgg}, HTE.

Our models are evaluated on real-word fetal cardiac screening videos. Results on several exemplar frames are shown in Fig. 3, and full results are compared in Table 1, where temporal models obtain better performance, particularly on classification. Importantly, the improvements on the challenging classes are remarkable: 67.60% → 88.27% for the LVOT, and 59.11% → 79.80% for the 3 Vessels. The HTE outperforms DTE models by over 15% in both classes, which suggests that the HTE provides a more principled method for encoding detailed and meaningful spatio-temporal features for fetal heart description. This can be further verified by the confusion matrices shown in Fig. 4. The advantage of the HTE is also affirmed by gains seen in the localisation task. We compare our models with that in [1] which uses hand-crafted features and a CRF-filter. The IoU implementations of HTE achieve localisation errors of 20.32% compared to the best result of 34% in [1]. Our models explore stronger deep spatio-temporal representations learned end-to-end with features shared over multiple tasks. Furthermore, our model is able to predict both the centre and radius for heart localisation, while the radius is used as a strong prior information and only the centre is estimated in [1].

Finally, we present further analysis of spatio-temporal aggregation of US representations throughout the CNN architecture. The performance of the direct encoding architecture, DTE_{alex}, can approach that of the deeper baseline model SBM_{vgg} with VGGnet. This is indicative of the dynamical interplay between competing factors governing the ability of deep networks to optimise learning of spatial and temporal information. While an information fusion at the level of extracted features at different time-steps or frames could capture significantly more contextual detail than single frame approaches, the aggregation may cause an overlap between varying spatial features which tends to compromise the ability to learn distinct spatial features. This effect is pertinent to the ultrasound modality because of the non-trivial influence of imaging artefacts, shadows and speckle. Particularly, a pixel-level combination before the first convolution layer, as described in the DTE structures, tends to undermine the frame-wise variations in spatial information by directly fusing along the temporal direction. This trade-off is indirectly addressed in the HTE model, by feeding a relatively smaller temporal extent per convolutional network, and by aggregating multi-level spatio-temporal information progressively throughout the CNN architecture. This provides a more principled approach for spatio-temporal encoding that enables the final deep representation to essentially capture the entire temporal regime of the video segment. Thus, our HTE networks are able to achieve significant gains in accuracy for both the classification and localisation tasks. It is to be noted that the optimal number of frames to achieve the best gains in this trade-off would exhibit significant variations as a function of the modality, the size of datasets and the complexity of features in spatial and temporal domains.

5   Conclusions

We have presented and compared multiple CNN models for jointly predicting the visibility, view plane and localisation of the fetal heart in US videos. We developed a Hierarchical Temporal Encoding (HTE) with a multi-task CNN that allows spatio-temporal information to be aggregated progressively through the network. This approach is shown to efficiently retain meaningful spatial and temporal details which are critical to implementing multi-task prediction in US video. We also investigated different temporal encoding models on a real-world clinical dataset, where the proposed HTE achieved significant performance gains.

Acknowledgments. This work was supported by the EPSRC Programme Grant Seebibyte (EP/M013774/1). Arijit Patra is supported by the Rhodes Trust.

References

1. Bridge, C.P., Ioannou, C., Noble, J.A.: Automated annotation and quantitative description of ultrasound videos of the fetal heart. Med. Image Anal. **36**, 147–161 (2017)

2. Chen, H., Dou, Q., Ni, D., Cheng, J.-Z., Qin, J., Li, S., Heng, P.-A.: Automatic fetal ultrasound standard plane detection using knowledge transferred recurrent neural networks. In: Navab, N., Hornegger, J., Wells, W.M., Frangi, A.F. (eds.) MICCAI 2015. LNCS, vol. 9349, pp. 507–514. Springer, Cham (2015). doi:10.1007/978-3-319-24553-9_62
3. Gao, Y., Maraci, M.A., Noble, J.A.: Describing ultrasound video content using deep convolutional neural networks. In: ISBI (2016)
4. Huang, W., Bridge, C.P., Noble, J.A., Zisserman, A.: Temporal HeartNet: towards human-level automatic analysis of fetal cardiac screening video. In: MICCAI (2017)
5. Karpathy, A., Toderici, G., Shetty, S., Leung, T., Sukthankar, R., Fei-Fei, L.: Large-scale video classification with convolutional neural networks. In: CVPR (2014)
6. Krizhevsky, A., Sutskever, I., Hinton, G.E.: ImageNet classification with deep convolutional neural networks. In: NIPS (2012)
7. Maraci, M.A., Napolitano, R., Papageorghiou, A., Noble, J.A.: Searching for structures of interest in an ultrasound video sequence. Med. Image Anal. **37**, 22–36 (2017)
8. Ng, J.Y.H., Hausknecht, M., Vijayanarasimhan, S., Vinyals, O., Monga, R., Toderici, G.: Beyond short snippets: deep networks for video classification. In: CVPR (2015)
9. Simonyan, K., Zisserman, A.: Two-stream convolutional networks for action recognition in videos. In: NIPS (2014)
10. Simonyan, K., Zisserman, A.: Very deep convolutional networks for large-scale image recognition. In: ICLR (2015)

Fast, Simple Calcium Imaging Segmentation with Fully Convolutional Networks

Aleksander Klibisz[1(✉)], Derek Rose[1], Matthew Eicholtz[1], Jay Blundon[2], and Stanislav Zakharenko[2]

[1] Oak Ridge National Laboratory, Oak Ridge, TN 37831, USA
{klibisza,rosedc,eicholtzmr}@ornl.gov
[2] St. Jude Children's Research Hospital, Memphis, TN 38105, USA
{jay.blundon,stanislav.zakharenko}@stjude.org

Abstract. Calcium imaging is a technique for observing neuron activity as a series of images showing indicator fluorescence over time. Manually segmenting neurons is time-consuming, leading to research on automated calcium imaging segmentation (ACIS). We evaluated several deep learning models for ACIS on the Neurofinder competition datasets and report our best model: U-Net2DS, a fully convolutional network that operates on 2D mean summary images. U-Net2DS requires minimal domain-specific pre/post-processing and parameter adjustment, and predictions are made on full 512×512 images at $\approx$9K images per minute. It ranks third in the Neurofinder competition ($F_1 = 0.57$) and is the best model to exclusively use deep learning. We also demonstrate useful segmentations on data from outside the competition. The model's simplicity, speed, and quality results make it a practical choice for ACIS and a strong baseline for more complex models in the future.

Keywords: Calcium imaging · Fully convolutional networks · Deep learning · Microscopy segmentation

1 Introduction

Calcium imaging recordings show neurons from a lab specimen illuminating over time, ranging from minutes to hours in length with resolution on the order of 512×512 pixels. Identifying the neurons is one step in a workflow that typically

This manuscript has been authored by UT-Battelle, LLC under Contract No. DE-AC05-00OR22725 with the U.S. Department of Energy. The United States Government retains and the publisher, by accepting the article for publication, acknowledges that the United States Government retains a non-exclusive, paid-up, irrevocable, worldwide license to publish or reproduce the published form of this manuscript, or allow others to do so, for United States Government purposes. The Department of Energy will provide public access to these results of federally sponsored research in accordance with the DOE Public Access Plan (http://energy.gov/downloads/doe-public-access-plan).

© Springer International Publishing AG 2017
M.J. Cardoso et al. (Eds.): DLMIA/ML-CDS 2017, LNCS 10553, pp. 285–293, 2017.
DOI: 10.1007/978-3-319-67558-9_33

involves motion correction and peak identification with the goal of understanding the activity of large populations of neurons. Given a 3D spatiotemporal (height, width, time) series of images, an ACIS model produces a 2D binary mask identifying neuron pixels.

Until recently, popular ACIS models were mostly unsupervised and required considerable assumptions about the data when making predictions. For example, [8,16,17,20] each require the expected number of neurons and/or their dimensions to segment a new series. Moreover, many models were tested on different datasets, making objective comparison difficult.

The ongoing Neurofinder competition [4] has nineteen labeled training datasets and nine testing datasets. To our knowledge, this is the best benchmark for ACIS, and the quantity of data makes it appealing for deep learning. Our best model for the Neurofinder datasets is U-Net2DS, an adaptation of the fully convolutional U-Net architecture [21] that takes 2D summary images as input. Compared to other models that do not use deep learning, U-Net2DS requires minimal assumptions, parameter adjustment, and pre/post-processing when making predictions. The fully convolutional architecture enables training on small windows and making predictions on full summary images. It ranks third in the Neurofinder competition and also shows robustness to datasets for which it was not directly trained. The Keras [2] implementation and trained weights are available at https://github.com/alexklibisz/deep-calcium.

2   Related Work

Deep learning has been explored extensively for medical image analysis, covered thoroughly in [10]. Fully convolutional networks with skip connections were developed for semantic segmentation of natural images [11] and applied to 2D medical images [21] and 3D volumes [3,13]. An analysis of the role of skip connections in fully convolutional architectures for biomedical segmentation is offered by [5].

ACIS models can be grouped into several categories: matrix factorization, clustering, dictionary learning, and deep learning. In general, matrix factorization models [12,14,17–20] consider a calcium imaging series as a matrix of spatial and temporal components and propose methods for extracting the signals. Clustering models [8,22] define similarity measures for pairs of pixels and divide them using partitioning algorithms. Dictionary learning models [16,18] extract or generate neuron templates and use them to match new neurons. Finally, deep learning models learn hierarchies of relevant neuron features from labeled datasets. The deep networks proposed by [1] use 2D and 3D convolutions with some pre/post-processing, and [6] uses 2D convolutions. Both use fully-connected classification layers with fixed-size sliding window inputs and outputs.

3   Data and Metrics

3.1   Calcium Imaging Datasets

The Neurofinder [4] training datasets contain 60K images with 7K neurons labeled by researchers at four labs with various experimental settings. Testing labels are withheld and submissions are evaluated on the competition server. Dataset samples are shown in Figs. 1 and 3. To complement this data, we included our own calcium imaging datasets, referred to as the St. Jude datasets. They contain GCaMP6f-expressing neurons from the auditory cortex, a cortical region not represented in Neurofinder, and have bounding box ground truth masks instead of precise outlines. Neurofinder and St. Jude datasets are motion-corrected, 16-bit TIFFs with 512×512 resolution and length ranging from 1800 to 8000 images.

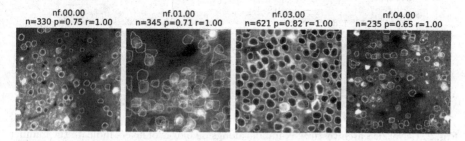

Fig. 1. 200×200 cropped mean summary images from four Neurofinder training datasets. Ground truth and predicted neurons are outlined in green and red, respectively. Titles denote the ground truth number of neurons and precision and recall metrics for predictions. (Color figure online)

We found several noteworthy challenges in the Neurofinder datasets. They are highly heterogeneous in appearance, making it difficult to define a characteristic neuron. Images have highly variable brightness with mean pixel values ranging from 57 to 2998 for individual datasets. The labels have a strong imbalance with an average of 12% of pixels labeled as neurons. Finally, labeling preference leaves the possibility for inconsistent ground truth. For example, the 04.00 labels were optimized for a particular type of experiment and might be considered inconsistent relative to other datasets.[1]

3.2   Summary Images

A common mode of visualization for calcium imaging is a 2D summary image, created by applying a function to each pixel across all frames to flatten the time dimension. Similar to [1,6], we found mean summary images produced a clear picture of most neurons with low computational cost. Examples are shown in Figs. 1 and 3. Other summary functions include the maximum, minimum, and standard deviation.

---

[1] See discussion: https://github.com/codeneuro/neurofinder/issues/25.

3.3   Evaluation Metrics

The Neurofinder competition measures F_1, precision, recall, inclusion, and exclusion, with submissions ranked by the mean F_1 score across all test datasets. Precision and recall are computed by matching each ground truth neuron to the spatially nearest predicted neuron without replacement. In contrast to the same pixelwise metrics, one predicted neuron may only correspond to a single ground truth neuron, even if the predicted neuron's region encompasses multiple ground truth neurons. Inclusion and exclusion quantify the structural similarity of matched ground truth and predicted neurons.[2]

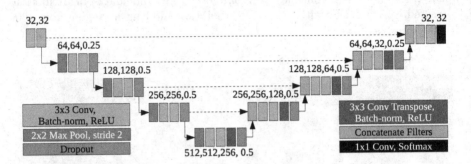

Fig. 2. U-Net2DS architecture. Numbers alongside blocks of layers indicate the number of convolutional filters, convolutional-transpose filters, and dropout probability.

4   Proposed Models

4.1   U-Net with 2D Summary Images

Architecture. The fully convolutional U-Net architecture [21] has four convolution/max-pooling blocks that reduce dimensionality followed by four convolution/transpose-convolution (deconvolution) blocks to restore input dimensionality. A ReLU activation follows each convolutional layer and skip-connections pass the output from each convolution/max-pooling block to corresponding convolution/transpose-convolution blocks via filter concatenation. U-Net2DS makes small modifications to the original U-Net to reduce overfitting. We added zero padding to each convolutional layer, reduced the number of filters at each convolutional layer by 50%, added batch normalization [7] after each convolutional layer, and increased dropout. The final layers are a 1×1 convolution with two filters followed by a softmax activation, producing a $h \times w \times 2$-channel mask, where the two channels are the probabilities of each pixel belonging to a neuron or background. U-Net2DS has 7.8M parameters and is illustrated in Fig. 2.

---

[2] Details and implementation: https://github.com/codeneuro/neurofinder-python.

Training. We trained a single U-Net2DS on all Neurofinder training datasets. Each summary image was normalized by subtracting its mean and dividing by its standard deviation. For each dataset, we combined all of the neuron masks into a single mask and removed overlapping and adjacent pixels belonging to different neurons to preserve the true number of independent neurons. We used the top 75% of each summary image for training and bottom 25% for validation. We sampled random batches of twenty 128×128 image and mask windows, ensuring each window contained a neuron, and applied random rotations and flips. We trained for ten epochs with 100 training steps per epoch using the Adam optimizer [9] with a 0.002 learning rate. After each epoch, we made full-image predictions on the validation data and computed the Neurofinder F_1 metric. We saved weights when the mean F_1 improved. Pixelwise F_1 plateaued around 0.7 and Neurofinder F_1 around 0.85. Training on an NVIDIA Titan-X GPU took under half an hour.

Loss Function. On average, the 128×128 training windows contained 9% neuron pixels. This imbalance led us to consider the modified Dice coefficient (MDC) loss function [13]. We compared MDC to standard logarithmic loss (LL) and found higher pixelwise recall with MDC, but the average test F_1 for LL was consistently better. LL computes each pixel's loss independently, which allows weighting for recall or precision. We experimented with this by multiplying false negative pixel losses between $2\times$ and $10\times$ and saw increased pixelwise recall but lower F_1. We ultimately found that standard LL resulted in the best F_1 scores.

Prediction. Full-image prediction was implemented by reshaping U-Net2DS to take 512×512 inputs. This change simplified the implementation and eliminated mask tiling artifacts that were present when using sliding windows. We found test-time augmentation consistently improved the mean F_1. This consists of averaging predictions for eight rotations and reflections of the summary image. Given raw TIFF files, the summaries and predictions are computed at $\approx$9K images per minute.

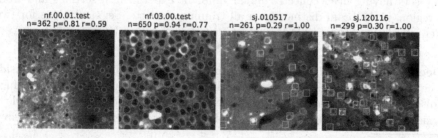

Fig. 3. 200×200 cropped mean summary images from Neurofinder test and St. Jude datasets with number of predicted neurons, precision, and recall in titles. Ground truth and predicted neurons are outlined in green and red, respectively. Ground truth for Neurofinder test data is not publicly available. (Color figure online)

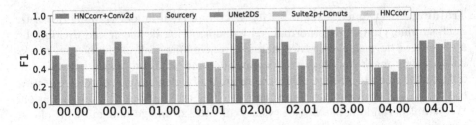

Fig. 4. F_1 for the top five models on each of the Neurofinder testing datasets.

Table 1. Mean and standard deviation of metrics for the top five Neurofinder models.

Neurofinder model	F_1	Precision	Recall
HNCcorr [22] + Conv2d [6]	0.617 ± 0.122	0.702 ± 0.170	0.602 ± 0.197
Sourcery [15]	0.583 ± 0.139	0.599 ± 0.197	0.629 ± 0.168
UNet2DS	0.569 ± 0.160	0.618 ± 0.235	0.609 ± 0.185
Suite2p [17] + Donuts [16]	0.550 ± 0.127	0.578 ± 0.156	0.568 ± 0.181
HNCcorr [22]	0.492 ± 0.180	0.618 ± 0.169	0.479 ± 0.268

Results on Neurofinder and St. Jude Datasets. As of July 2017, the top-scoring Neurofinder submission is an ensemble of HNCcorr [22], which clusters pixels based on correlation over time, and Conv2D [6], a convolutional neural network with 40 × 40 input and 20 × 20 output windows. Second and fourth place use matrix factorization methods from [16,17]. U-Net2DS ranks third and is the highest-scoring model to exclusively use deep learning. Table 1 and Fig. 4 compare the top models' F_1, precision, and recall metrics. Models that explicitly use temporal information considerably outperform those using summaries on the 02.00 and 02.01 datasets. This might correspond to differences in ground truth labeling techniques acknowledged by the Neurofinder competition.[3] To evaluate robustness beyond the Neurofinder datasets, we made predictions on the St. Jude datasets from the auditory cortex and had F_1 scores of 0.45 and 0.47 despite the datasets having been collected under different experimental settings and from a region of the brain distinct from all training datasets. Segmentations for two Neurofinder and two St. Jude datasets are shown in Fig. 3.

Deep Learning Insights for Calcium Imaging Data. We initially considered several pre-processing steps, including clipping maximum values to reduce noise and stretching each series' values to maximize contrast. These steps made it easier to recognize neurons visually but did not improve performance. The heavy

---

[3] From the website: For the 00 data, labels are derived from an anatomical marker that indicates the precise location of each neuron and includes neurons with no activity. For the 01, 02, 03, 04 data, labels were hand drawn or manually curated, using the raw data and various summary statistics, some of which are biased towards active neurons.

neuron-to-background class-imbalance required attention in training. Some combinations of learning rates and loss functions caused the network to predict 100% background early on and never recover. It helped to increase window size and monitor the proportion of neuron pixels predicted vs. the ground truth proportion. Finally, the training datasets have semantic differences (labeling technique, part of brain, etc.). In some cases we improved performance by training one model per dataset, but ultimately found it more practical to train a single model on all data.

4.2 Other Deep Learning Models

U-Net2DS with Temporal Summaries. By using mean summaries, U-Net2DS ignores temporal information and has difficulty separating overlapping neurons. One possible extension of U-Net2DS would add temporal inputs consisting of summary images along the time-axis (i.e., YZ, XZ summaries).

Segmentation with Better 2D Summary Images. A summary image that amplifies neurons and minimizes background should improve a model like U-Net2DS. Taking inspiration from [22], we experimented with summaries that compute the mean correlation and cosine distance of each pixel with its neighbors. Similar to [17], we found most neuron pixels had high similarity to nearby background pixels, making neighborhood similarity a poor summary function.

Frame-by-Frame Segmentation. Summary images can preserve unwanted background noise, so it is plausible that a model trained to segment the series frame-by-frame could recognize a neuron in frames where surrounding noise is minimized. Unfortunately, it is difficult to select frames for training as most neurons blend with background at arbitrary times and only 2D labels are given. We tried a contrast-based weighting scheme that decreased loss for neurons with low contrast to surrounding background and increased loss for high-contrast neurons. Still, we found most neurons were more distinguishable in a mean summary than in their highest-contrast frames. It seems the summary smooths areas that are otherwise noisy in single frames.

3D Convolutional Networks. Intuitively, directly including inputs from the time dimension should improve performance. We explored 3D convolutional networks, similar to [1,3,13]. Because series have variable length, we found it difficult to determine an input depth for all datasets. The 3D models were also prohibitively slow, taking at least an hour to segment a single series.

5 Conclusion

Large labeled datasets like Neurofinder enable the use of deep learning for automated calcium imaging segmentation. We demonstrated U-Net2DS, an adaptation of U-Net [21] that uses mean summaries of calcium imaging series to quickly produce full segmentation masks while minimizing parameter adjustment, pre/post-processing, and assumptions for new datasets. Despite its relative simplicity, it is the best-performing non-ensemble and deep learning model in the Neurofinder competition. We described several other deep learning formulations for this problem, and we believe a model that efficiently combines spatial and temporal information is a promising next step.

Acknowledgments. This work was supported in part by the Department of Developmental Neurobiology at St. Jude Children's Research Hospital and by the U.S. Department of Energy, Office of Science, Office of Workforce Development for Teachers and Scientists (WDTS) under the Science Undergraduate Laboratory Internship program.

References

1. Apthorpe, N., Riordan, A., Aguilar, R., Homann, J., Gu, Y., Tank, D., Seung, H.S.: Automatic neuron detection in calcium imaging data using convolutional networks. In: Advances in Neural Information Processing Systems, pp. 3270–3278 (2016)
2. Chollet, F., et al.: Keras (2015). https://github.com/fchollet/keras
3. Çiçek, Ö., Abdulkadir, A., Lienkamp, S.S., Brox, T., Ronneberger, O.: 3D U-net: learning dense volumetric segmentation from sparse annotation. In: Ourselin, S., Joskowicz, L., Sabuncu, M.R., Unal, G., Wells, W. (eds.) MICCAI 2016. LNCS, vol. 9901, pp. 424–432. Springer, Cham (2016). doi:10.1007/978-3-319-46723-8_49
4. CodeNeuro.org: The neurofinder challenge (2016). http://neurofinder.codeneuro.org/
5. Drozdzal, M., Vorontsov, E., Chartrand, G., Kadoury, S., Pal, C.: The importance of skip connections in biomedical image segmentation. In: Carneiro, G., et al. (eds.) LABELS/DLMIA -2016. LNCS, vol. 10008, pp. 179–187. Springer, Cham (2016). doi:10.1007/978-3-319-46976-8_19
6. Gao, S.: Conv2d (2016). https://github.com/iamshang1/Projects/tree/master/Advanced_ML/Neuron_Detection
7. Ioffe, S., Szegedy, C.: Batch normalization: accelerating deep network training by reducing internal covariate shift. arXiv preprint arXiv:1502.03167 (2015)
8. Kaifosh, P., Zaremba, J.D., Danielson, N.B., Losonczy, A.: Sima: Python software for analysis of dynamic fluorescence imaging data. Front. Neuroinform. **8**, 80 (2014)
9. Kingma, D., Ba, J.: Adam: a method for stochastic optimization. arXiv preprint arXiv:1412.6980 (2014)
10. Litjens, G., Kooi, T., Bejnordi, B.E., Setio, A.A.A., Ciompi, F., Ghafoorian, M., van der Laak, J.A., van Ginneken, B., Sánchez, C.I.: A survey on deep learning in medical image analysis. arXiv preprint arXiv:1702.05747 (2017)
11. Long, J., Shelhamer, E., Darrell, T.: Fully convolutional networks for semantic segmentation. In: Proceedings of the IEEE Conference on Computer Vision and Pattern Recognition, pp. 3431–3440 (2015)
12. Maruyama, R., Maeda, K., Moroda, H., Kato, I., Inoue, M., Miyakawa, H., Aonishi, T.: Detecting cells using non-negative matrix factorization on calcium imaging data. Neural Netw. **55**, 11–19 (2014)
13. Milletari, F., Navab, N., Ahmadi, S.A.: V-net: fully convolutional neural networks for volumetric medical image segmentation. In: 2016 Fourth International Conference on 3D Vision (3DV), pp. 565–571. IEEE (2016)

14. Mukamel, E.A., Nimmerjahn, A., Schnitzer, M.J.: Automated analysis of cellular signals from large-scale calcium imaging data. Neuron **63**(6), 747–760 (2009)
15. Pachitariu, M.: Sourcery (2016). https://github.com/marius10p/suite2p-for-neurofinder
16. Pachitariu, M., Packer, A.M., Pettit, N., Dalgleish, H., Hausser, M., Sahani, M.: Extracting regions of interest from biological images with convolutional sparse block coding. In: Advances in Neural Information Processing Systems, pp. 1745–1753 (2013)
17. Pachitariu, M., Stringer, C., Schröder, S., Dipoppa, M., Rossi, L.F., Carandini, M., Harris, K.D.: Suite2p: beyond 10,000 neurons with standard two-photon microscopy. bioRxiv, p. 061507 (2016)
18. Petersen, A., Simon, N., Witten, D.: Scalpel: extracting neurons from calcium imaging data. arXiv preprint arXiv:1703.06946 (2017)
19. Pnevmatikakis, E.A., Paninski, L.: Sparse nonnegative deconvolution for compressive calcium imaging: algorithms and phase transitions. In: Advances in Neural Information Processing Systems, pp. 1250–1258 (2013)
20. Pnevmatikakis, E.A., Soudry, D., Gao, Y., Machado, T.A., Merel, J., Pfau, D., Reardon, T., Mu, Y., Lacefield, C., Yang, W., et al.: Simultaneous denoising, deconvolution, and demixing of calcium imaging data. Neuron **89**(2), 285–299 (2016)
21. Ronneberger, O., Fischer, P., Brox, T.: U-net: convolutional networks for biomedical image segmentation. In: Navab, N., Hornegger, J., Wells, W.M., Frangi, A.F. (eds.) MICCAI 2015. LNCS, vol. 9351, pp. 234–241. Springer, Cham (2015). doi:10.1007/978-3-319-24574-4_28
22. Spaen, Q., Hochbaum, D.S., Asín-Achá, R.: HNCcorr: a novel combinatorial approach for cell identification in calcium-imaging movies. arXiv preprint arXiv:1703.01999 (2017)

Self-supervised Learning for Spinal MRIs

Amir Jamaludin[1]([⊠]), Timor Kadir[2], and Andrew Zisserman[1]

[1] VGG, Department of Engineering Science, University of Oxford, Oxford, UK
{amirj,az}@robots.ox.ac.uk
[2] Optellum, Oxford, UK
timor.kadir@optellum.com

Abstract. A significant proportion of patients scanned in a clinical setting have follow-up scans. We show in this work that such longitudinal scans alone can be used as a form of "free" self-supervision for training a deep network. We demonstrate this self-supervised learning for the case of T2-weighted sagittal lumbar Magnetic Resonance Images (MRIs). A Siamese convolutional neural network (CNN) is trained using two losses: (i) a contrastive loss on whether the scan is of the same person (i.e. longitudinal) or not, together with (ii) a classification loss on predicting the level of vertebral bodies. The performance of this pre-trained network is then assessed on a grading classification task. We experiment on a dataset of 1016 subjects, 423 possessing follow-up scans, with the end goal of learning the disc degeneration radiological gradings attached to the intervertebral discs. We show that the performance of the pre-trained CNN on the supervised classification task is (i) superior to that of a network trained from scratch; and (ii) requires far fewer annotated training samples to reach an equivalent performance to that of the network trained from scratch.

1 Introduction

A prerequisite for the utilization of machine learning methods in medical image understanding problems is the collection of suitably curated and annotated clinical datasets for training and testing. Due to the expense of collecting large medical datasets with the associated ground-truth, it is important to develop new techniques to maximise the use of available data and minimize the effort required to collect new cases.

In this paper, we propose a self-supervision approach that can be used to pre-train a CNN using the embedded information that is readily available with standard clinical image data. Many patients are scanned multiple times in a so-called longitudinal manner, for instance to assess changes in disease state or to monitor therapy. We define a pre-training scheme using only the information about which scans belong to the same patient. Note, we do not need to know the identity of the patient; only which images belong to the same patient. This information is readily available in formats such as DICOM (Digital Imaging and Communications in Medicine) that typically include a rich set of metadata such as patient identity, date of birth and imaging protocol (and DICOM

© Springer International Publishing AG 2017
M.J. Cardoso et al. (Eds.): DLMIA/ML-CDS 2017, LNCS 10553, pp. 294–302, 2017.
DOI: 10.1007/978-3-319-67558-9_34

anonymization software typically assigns the same 'fake-id' to images of the same patient).

Here, we implement this self-supervision pre-training for the case of T2-weighted sagittal lumbar MRIs. We train a Siamese CNN to distinguish between pairs of images that contain the same patient scanned at different points in time, and pairs of images of entirely different patients. We also illustrate that additional data-dependent self-supervision tasks can be included by specifying an auxiliary task of predicting vertebral body levels, and including both types of self-supervision in a multi-task training scheme. Following the pre-training, the learned weights are transferred to a classification CNN that, in our particular application of interest, is trained to predict a range of gradings related to vertebra and disc degeneration. The performance of the classification CNN is then used to evaluate the success of the pre-training. We also compare to pre-training a CNN on a large dataset of lumbar T2-weighted sagittal MRIs fully annotated with eight radiological gradings [5].

Related Work: Models trained using only information contained within an image as a supervisory signal have been proven to be effective feature descriptors e.g. [4] showed that a CNN trained to predict relative location of pairs of patches, essentially learning spatial context, is better at a classification task (after fine-tuning) then a CNN trained from scratch. The task of learning scans from the same unique identity is related to slow-feature learning in videos [7,10,11].

Transfer learning in CNNs has been found to be extremely effective especially when the model has been pre-trained on a large dataset like ImageNet [3], and there have been several successes on using models pre-trained on natural images on medical images e.g. lung disease classification [8]. However, since a substantial portion of medical images are volumetric in nature, it is also appropriate to experiment with transfer learning with scans of the same modality rather than using ImageNet-trained models.

2 Self-supervised Learning

This section first describes the details of the input volumes extraction: the vertebral bodies (**VBs**) that will be used for the self-supervised training; and the intervertebral discs (**IVDs**) that will be used for the supervised classification experiments. We then describe the loss functions and network architecture.

2.1 Extracting Vertebral Bodies and Intervertebral Discs

For each T2-weighted sagittal MRI we automatically detect bounding volumes of the (T12 to S1) VBs alongside the level labels using the pipeline outlined in [5,6]. As per [5], we also extract the corresponding IVD volumes (T12-L1 to L5-S1, where T, L, and S refer to the thoracic, lumbar, and sacral vertebrae) from the pairs of vertebrae e.g. a L5-S1 IVD is the disc between the L5 and S1 vertebrae. Figure 1 shows the input and outputs of the extraction pipeline. The slices of both the VB and IVD volumes are mid-sagittally aligned (to prevent

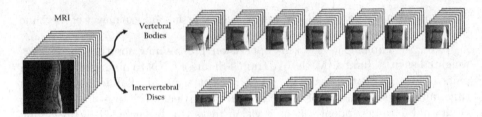

Fig. 1. Extracting VB and IVD Volumes: For each MRI, we extract 7 VB (T12 to S1) and 6 IVD (T12-L1 to L5-S1) volumes. The dimensions of the volumes are: (i) $224 \times 224 \times 9$ for the VBs, and (ii) $112 \times 224 \times 9$ for the IVDs. The whole volume is centred at the detected middle slice of the volume of the VB/IVD.

misalignment that can occur from scoliosis and other disorders) and zero-padded slice-wise if the number of slices is below the predefined 9 channels. The volumes are rescaled according to the width of the VB or IVD and normalized such that the median intensity of the VB above and below the current VB or IVD is 0.5.

2.2 Longitudinal Learning via Contrastive Loss

The longitudinal information of the scans is used to train a Siamese network such that the embeddings for scans of the same person are close, whereas scans of different people are not. The input is a pair of VBs of the same level; an S1 VB is only compared against an S1 VB and vice versa. We use the contrastive loss in [2], $\mathcal{L}_C = \sum_{n=1}^{N}(y)d^2 + (1 - y)\max(0, m - d)^2$, where $d = \|a_n - b_n\|_2$, and a_n and b_n are the 1024-dimensional **FC7** (embedding) vectors for the first and second VB in an input pair, and m is a predefined margin. Positive, $y = 1$, VB pairs are those that were obtained from a single unique subject (same VB scanned at different points in time) and negative, $y = 0$, pairs are VBs from different individuals (see Fig. 3).

We use the VBs instead of the IVDs due to the fact that vertebrae tend to be more constant in shape and appearance over time. Figure 2 shows examples of both VB and IVD at different points in time. In other medical tasks the pair

Fig. 2. VB and IVD Across Time: The VB appears unchanged but over time the IVD loses intensity of its nucleus and experiences a loss in height. Furthermore, in the IVD example, we can observe a vertebral slip, or spondylolisthesis, which does not change the appearance of the VBs themselves but significantly changes the IVD.

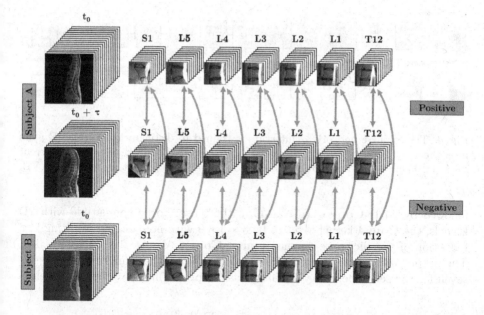

Fig. 3. Generating the positive/negative pairs: The arrows mark a pair of VBs, where blue arrows highlight positive pairs while negative are highlighted in red. Each pair is generated from two scans. t_0 refers to the time of the initial scan and τ is the time between the baseline and the follow-up scans, typically 10 years in our dataset. (Color figure online)

of VBs can easily be changed to other anatomies for example comparing lungs in chest X-rays.

2.3 Auxiliary Loss – Predicting VB Levels

In addition to the contrastive loss, we also employ an auxiliary loss to give complementary supervision. Since each VB pair is made up of VBs of the same level, we train a classifier on top of the **FC7** layer, i.e. the discriminative layer, to predict the seven levels of the VB (from T12 to S1). The overall loss can then be described as a combination of the contrastive and softmax log losses, $\mathcal{L} = \mathcal{L}_C - \sum_{n=1}^{N} \alpha_c \left(y_c(x_n) - \log \sum_{j=1}^{7} e^{y_j(x_n)} \right)$, where y_j is the jth component of the **FC8** output, c is the true class of x_n, and α_c is the class-balanced weight as in [5].

2.4 Architecture

The base architecture trained to distinguish VB pairs is based of the VGG-M network in [1] (see Fig. 4). The input to the Siamese CNN is the pairs of VBs, with dimension $224 \times 224 \times 9$. We use 3D kernels from **Conv1** to **Conv4** layers

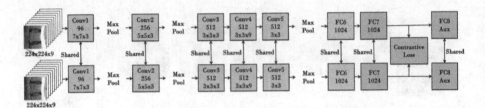

Fig. 4. The Siamese network architecture is trained to distinguish between VB pairs from the same subject (same VB scanned a decade apart) or VB pairs from different subjects. There is also an additional loss (shown as aux) to predict the level of the VB.

followed by a 2D **Conv5**. To transform the tensor to be compatible with 2D kernels, the **Conv4** kernel is set to be $3 \times 3 \times 9$ with no padding, resulting in a reduction of the slice-wise dimension after **Conv4**. We use 2×2 max pooling. The output dimension of the **FC8** layer depends on the number of classes i.e. seven for the self-supervisory auxiliary task of predicting the seven VB levels.

3   Dataset and Implementation Details

We experimented on a dataset of 1016 recruited subjects (predominantly female) from the TwinsUK registry (www.twinsuk.ac.uk) not using backpain as an exclusion or selection criterion. As well as a baseline scan for each subject, 423 subjects have follow-up scans taken 8–12 years after the original baseline. A majority of the subjects with the follow-up scans have only two scans (one baseline and one follow-up) while a minority have three (one baseline and two follow-up). The baseline scans were taken with a 1.0-Tesla scanner while the follow-up scans were taken with a 1.5-Tesla machine but both adhered to the same scanning protocol (slice thickness, times to recovery and echo, TR and TE). Only T2-weighted sagittal scans were collected for each subject.

3.1   Radiological Gradings

The subjects were graded with a measure of **Disc Degeneration**, not dissimilar to Pfirrmann Grading. The gradings were annotated by a clinician and were done on a per disc basis: from L1-L2 to L5-S1 discs (5 discs per subject). We use these gradings to assess the benefits of pre-training a classification network

Fig. 5. Disc Degeneration: a four-class grading system based on Pfirrmann grading that depends on the intensity and height of the disc.

on longitudinal information. Examples of the gradings can be seen in Fig. 5. 920 of the 1016 subjects are graded.

3.2   Training

Data augmentation: The augmentation strategies are identical to that described in [5] for the classification CNN while we use slightly different augmentations to train the Siamese network. The augmentations are: (i) rotation with $\theta = -15°$ to $15°$, (ii) translation of ± 48 pixels in the x-axis, ± 24 pixels in the y-axis, ± 2 slices in the z-axis, (iii) rescaling with a scaling factor between 90% to 110%, (iv) intensity variation ± 0.2, and (v) random slice-wise flip i.e. reflection of the slices across the mid-sagittal (done on a per VB pair basis).

Details: Our implementation is based on the MatConvNet toolbox [9] and the networks were trained using an NVIDIA Titan X GPU. The hyperparameters are: batch size 128 for classification and 32 for the Siamese network; momentum 0.9; weight decay 0.0005; learning rate $1e^{-3}$ (classification) and $1e^{-5}$ (self-supervision) and lowered by a factor of 10 as the validation error plateaus, which is also our stopping criterion, normally around 2000 epochs for the classification network and 500 epochs for the Siamese network.

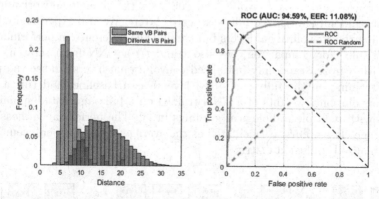

Fig. 6. Left: Histogram of VB pairs distances in the test set. Positive pairs in blue, negative in red. **Right:** The ROC of the classification of positive/negative VB pairs. (Color figure online)

4   Experiments and Results

4.1   Longitudinal Learning and Self-supervision

The dataset is split by subject 80:10:10 into train, validation, and test sets. For the 423 subjects with multiple scans this results in a 339:42:42 split. Note, a pair of twins will only be in one set i.e. one subject part of a twin pair can't be in training and the other in test. With the trained network, each input VB can be represented as a 1024-dimensional **FC7** vector. For each pair of VBs i.e. two **FC7** vectors, we can calculate the L_2 distance between two samples, which is the same distance function used during training. Figure 6 shows the histogram of the distances (both positive and negative) for all the VB pairs in the test set. In general, VB pairs that are from the same subject have lower distances compared to pairs from different subjects. Using the distances between pairs as classification scores of predicting whether the VB pairs are from the same or different subjects, we achieve an AUC of **94.6%**. We also obtain a very good performance of **97.8%** accuracy on the auxiliary task of predicting VB level.

4.2   Benefits of Pre-training

To measure the performance gained by pre-training using the longitudinal information we use the convolution weights learnt in the Siamese network, and train a classification CNN to predict the **Disc Degeneration** radiological grading (see Fig. 7). For this classification task we use the 920 subjects that possess gradings and split them into the following sets: 670 for training, 50 in validation, and 200 for testing. Subjects with follow-up scans (>1 scans) are only used in training and not for testing so, in essence, the Siamese network will never have seen the subjects in the classification test set.

We transfer and freeze convolutional weights of the Siamese network and only train the randomly-initialized fully-connected layers. We also experimented with fine-tuning the convolutional layers but we find the difference in performance to be negligible. For comparison, we also train: (i) a CNN from scratch, (ii) a CNN with a frozen randomly initialized convolutional layers (to see the power of just training on the fully-connected layers) as a baseline, and (iii) a CNN using convolutional weights of a CNN trained on a fully-annotated spinal MRI dataset with multiple radiological gradings in [5]. The performance measure is the average class accuracy, calculated as the average of the diagonal elements of the normalized confusion matrix.

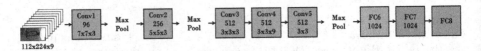

Fig. 7. The architecture of the classification CNN. The convolutional weights are obtained from the Siamese network.

Figure 8 shows the performance of all the models as the number of training samples is varied at [240, 361, 670] subjects or [444, 667, 976] scans. It can be seen that with longitudinal pre-training, less data is required to reach an equivalent point to that of training from scratch, e.g. the performance is **74.4%** using only 667 scans by pre-training, whereas training from scratch requires 976 scans to get to **74.7%**. This performance gain can also be seen with a lower amount of training data. As would be expected, transfer learning from a CNN trained with strong supervision using the data in [5] is better, with an accuracy at least **2.5%** above training from scratch. Unsurprisingly, a classifier trained on top of fully random convolutional weights performs the worst.

Since longitudinal information is essentially freely available when collecting data, the performance gain from longitudinal pre-training is also free. It is interesting to note that even though the Siamese network is trained on a slightly different input, VBs instead of IVDs, transferring the weights for classification task on IVDs still achieves better performance than starting from scratch. However, we suspect this difference in inputs is the primary reason why the performance gain is lower when the amount of training data is low as the network needs more data to adapt to the IVD input/task (in contrast, the strongly supervised CNN is trained on both VB and IVD classification).

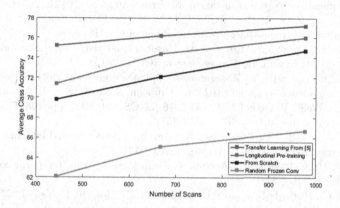

Fig. 8. Accuracy as the number of training samples for IVD classification is increased. Longitudinal pre-training improves over training from scratch, showing its benefit, and this performance boost persists even at 976 scans. Transfer learning from a CNN trained on a strongly-supervised dataset (including IVD classification) is better and provides an 'upper bound' on transfer performance. Note, even with 976 scans in the training set, the performance has not plateaued hinting at further improvements with the availability of more data.

5 Conclusion

We have shown that it is possible to use self-supervision to improve performance on a radiological grading classification task and we hope to explore the benefits of adding more auxiliary tasks in the near future e.g. predicting gender, age and weight. The performance improvement is nearing that of transfer learning from a model trained on a fully annotated dataset given that the target training set itself contains enough data. Furthermore, having a distance function between vertebral pairs opens up the possibility of identifying people using their MRIs.

Acknowledgments. We are grateful for discussions with Prof. Jeremy Fairbank, Dr. Jill Urban, and Dr. Frances Williams. This work was supported by the RCUK CDT in Healthcare Innovation (EP/G036861/1) and the EPSRC Programme Grant Seebibyte (EP/M013774/1). TwinsUK is funded by the Wellcome Trust; European Communitys Seventh Framework Programme (FP7/2007–2013). The study also receives support from the National Institute for Health Research (NIHR) Clinical Research Facility at Guys & St Thomas NHS Foundation Trust and NIHR Biomedical Research Centre based at Guy's and St Thomas' NHS Foundation Trust and King's College London.

References

1. Chatfield, K., Simonyan, K., Vedaldi, A., Zisserman, A.: Return of the devil in the details: delving deep into convolutional nets. In: Proceedings of BMVC (2014)
2. Chopra, S., Hadsell, R., LeCun, Y.: Learning a similarity metric discriminatively, with application to face verification. In: Proceedings of CVPR (2005)
3. Deng, J., Dong, W., Socher, R., Li, L.J., Li, K., Fei-Fei, L.: ImageNet: a large-scale hierarchical image database. In: Proceedings of CVPR (2009)
4. Doersch, C., Gupta, A., Efros, A.A.: Unsupervised visual representation learning by context prediction. In: Proceedings of ICCV (2015)
5. Jamaludin, A., Kadir, T., Zisserman, A.: SpineNet: automatically pinpointing classification evidence in spinal MRIs. In: Ourselin, S., Joskowicz, L., Sabuncu, M.R., Unal, G., Wells, W. (eds.) MICCAI 2016. LNCS, vol. 9901, pp. 166–175. Springer, Cham (2016). doi:10.1007/978-3-319-46723-8_20
6. Lootus, M., Kadir, T., Zisserman, A.: Vertebrae detection and labelling in lumbar MR images. In: MICCAI Workshop: CSI (2013)
7. Mobahi, H., Collobert, R., Weston, J.: Deep learning from temporal coherence in video. In: ICML (2009)
8. Shin, H.C., Roth, H.R., Gao, M., Lu, L., Xu, Z., Nogues, I., Yao, J., Mollura, D., Summers, R.M.: Deep convolutional neural networks for computer-aided detection: CNN architectures, dataset characteristics and transfer learning. IEEE Trans. Med. Imaging **35**(5), 1285–1298 (2016)
9. Vedaldi, A., Lenc, K.: MatConvNet - convolutional neural networks for MATLAB. CoRR abs/1412.4564 (2014)
10. Wang, X., Gupta, A.: Unsupervised learning of visual representations using videos. In: ICCV (2015)
11. Wiskott, L., Sejnowski, T.J.: Slow feature analysis: unsupervised learning of invariances. Neural Comput. **14**(4), 715–770 (2002)

Skin Lesion Segmentation via Deep RefineNet

Xinzi He, Zhen Yu, Tianfu Wang, and Baiying Lei[(✉)]

National-Regional Key Technology Engineering Laboratory for Medical
Ultrasound, Guangdong Key Laboratory for Biomedical Measurements and
Ultrasound Imaging, School of Biomedical Engineering, Health Science Center,
Shenzhen University, Shenzhen, China
leiby@szu.edu.cn

Abstract. Dermoscopy imaging has been a routine examination approach for
skin lesion diagnosis. Accurate segmentation is the first step for automatic
dermoscopy image assessment. The main challenges for skin lesion segmentation are numerous variations in viewpoint and scale of skin lesion region. To
handle these challenges, we propose a novel skin lesion segmentation framework via a very deep residual neural network based on dermoscopic images. The
deep residual neural network and generic multi-path Deep RefineNet are combined to improve the segmentation performance. The deep representation of all
available layers is aggregated to form the global feature maps using skip connection. Also, the chained residual pooling is leveraged to capture diverse
appearance features based on the context. Finally, we apply the conditional
random field (CRF) to smooth segmentation maps. Our proposed method shows
superiority over state-of-the-art approaches based on the public skin lesion
challenge dataset.

Keywords: Dermoscopy image · Skin lesion segmentation · Deep residual
network · Conditional random field · Deep RefineNet

1 Introduction

According to cancer statistics released by the American Cancer Society, melanoma
increasing at a growth rate of 14% and skin cancer has a death rate of 75% [1].
However, these diseases are curable if they are diagnosed timely and treated properly.
Clinically, dermoscopy is adopted to assist dermatologists in classifying melanoma
from nevi. The recent reports suggest that the human visual inspection only depends on
'Ugly Duckling' sign or experience [2]. So obtaining valid morphological features
conditioned on accurate segmentation of skin lesions. The manual segmentation process is often labor-intensive and subjective. However, clinicians who acquired adequate
levels of expertise are rare in unprivileged countries. Meanwhile, the blurry and
irregular boundary degrades the segmentation accuracy. Figure 1 illustrates the main
challenges for accurate diagnosis. Moreover, segmentation is critical in reducing
screening errors and to aid identification of benign and malignant melanoma. Automated dermoscopy image analysis based on numerous information is an effective way
to tackle these problems.

M.J. Cardoso et al. (Eds.): DLMIA/ML-CDS 2017, LNCS 10553, pp. 303–311, 2017.
DOI: 10.1007/978-3-319-67558-9_35

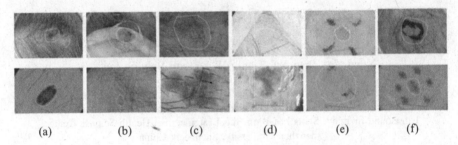

(a)   (b)   (c)   (d)   (e)   (f)

Fig. 1. Automatic segmentation of skin lesions in dermoscopy images is full of challenges. The main challenge includes distinguishable inter-class, indistinguishable intra-class variations, artifacts and inherent cutaneous features in natural images. (a–c) skin lesions are covered with hairs or exploded with blood vessels; (d) air bubbles and marks occlude the skin lesions; (e–f) dye concentration downgrades the segmentation accuracy. White contours indicate the skin lesions.

Deep learning has attracted significant attention and become a focus due to its ability to boost performance. Currently, the most popular deep learning method is convolutional network (CNN). However, the outputs of multi-scale CNN are too coarse to fulfill the requirement of segmentation. A new type of CNN, fully convolutional network (FCN) has witnessed a great success and achieved unprecedented progress in the development of segmentation [4] and removes the need of sophisticated pre-processing and post-processing algorithms, and allows researchers to concentrate on network architecture design. Despite the use of CNN, there are still significant differences between the result of automatic segmentation and the dermatologist's delineation.

Meanwhile, recent studies demonstrate increasing network depth can further boost performance due to the discriminative representations from deep layers of a network [6]. The latest generation CNN deep residual neural network (ResNet) proposed by He et al. [7] outperforms state-of-the-art techniques in classification task by solving vanishing/exploding gradient problem which inhibit convergence. However, ResNet has degraded segmentation accuracy due to the contradiction between classification and localization. When the network gets deeper, the training accuracy gets saturated and degrades rapidly. Hence, Deep RefineNet (DRN) is proposed. During the forward pass, the resolution of the feature maps (layer outputs) is decreased, the feature maps are deeper and the dimension is increased. To tackle this problem, we integrated a multi-path processing strategy in DRN to integrate the multi-level representation. Obtaining spatial transformation invariant features for a classifier needs to discard local information. To tackle this issue, Krahenbuhl and Koltun [5] proposed the conditional random field (CRF) algorithm, which integrates all pairs of pixels on an image to leverage spatial low-level information. Motivated by recent research of deep networks for obtaining high level features, we construct a network using multi-path features for segmentation with long-range skip connections. The combination of the RN network and CRF has been proposed for segmenting dermoscopy in this paper. Figure 2 illustrates the flowchart of our proposed method.

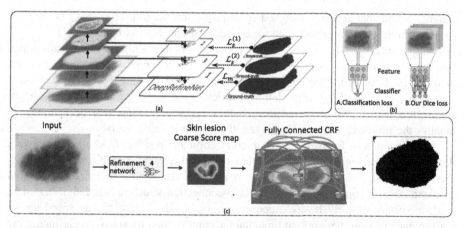

Fig. 2. Flowchart of proposed network architecture for melanoma segmentation; (a) multi-path processing to fuse various contrast information with deep supervision; (b) comparison of Dice loss and Classification loss; (c) a dense CRF is used to further refine the contour.

2 Methodology

2.1 Residual Learning

Residual Block: Our network starts with residual block inspired by ResNet. The first part of residual block contains an adaptive convolution layer via fine-tuning to maintain the weights of pretrained ResNet for our task. By directly applying skip-connection, our network suffers from the drawback of the small receptive field. To tackle this drawback, each input path is followed by two residual blocks, which are similar to that in the original ResNet with the removal of the batch-normalization layers. Since RN only accepts input from one path, the filter channel for each input path is set to 512, and the remaining ones are fixed to 256 for keeping dimension balanced in our experiments.

Multi-resolution Fusion: Since FCN and fully connected residual network (FCRN) still suffer from down-sampling and losing fine structure, a coarse-to-fine strategy is developed to recover high resolution. Instead of fusing all path inputs into the highest resolution feature map at once, we only integrate two features with adjacent size in each stage. This block first performs convolutions for input adaptation, which generates feature maps of the same number, and then interpolates smaller feature maps bilinearly with the largest resolution of the inputs. All feature maps are fused by summation.

Chained Residual Pooling: After the size observation and analysis of skin lesions, there is rich contextual information. Recently, pooling can utilize global image-level feature to support our research. Feature map passes sequentially to the chained residual pooling block, and is schematically depicted in Fig. 2. Fused feature map passes sequentially one max-pooling layer and one convolution layer. Nevertheless, pooling once needs large pooling window in network for the segmentation task. The proposed chained residual pooling method concatenates pooling layers as a chain with learnable weight. Noted that ReLu is adopted to improve the pooling efficiency. During the course of training, two pooling blocks are adopted and each is with stride 1.

The output of all pooling blocks is concatenated together with the input feature map through summation of skip-connection. We choose to employ skip-connections to facilitate gradient propagation during training. In one pooling block, each pooling operation is followed by convolutions, which serve as a weighting layer for the summation fusion. We expected that chained residual pooling will be learnable to identify the importance of pooling layers during the training course.

The main part of ResNet is the skip-connection and residual block. To train a very deep network for segmentation, we leverage a novel method named residual learning. The characteristics of residual learning is the skip connection to refine the flow of gradient. Also, a combination of multi skip-connection structure has demonstrated an evident activity of early layers. Accordingly, the residual learning can speed up the convergence of deep network and maintain accuracy gains by substantially increasing the network depth. By the dropout mechanism [8], batch normalization [9], and careful initialization, we address the problem of gradient vanishing. A residual block with identity mapping is formulated as

$$h_{l+1} = Relu((h_l) + \mathcal{F}(h_l, w_l)), \tag{1}$$

where h_{l+1} denotes the output of l-th residual block, h_l represents identical mapping and $\mathcal{F}(h_l, w_l)$ is the residual mapping.

2.2 Deep RefineNet

The first phase of segmentation is to obtain every pixel's prediction. Since the output of ResNet is a probability of each class being designed for classification, the single label prediction layer should be replaced by a multi-prediction layer. Hence, segmentation can be viewed as a dense prediction problem. The final prediction is based on feature maps from various receptive fields. The requirement of complex boundary delineation is met by local intensity information which only appears early in forward propagation. The main idea of the proposed Deep RefineNet is shown in Fig. 3.

Directly training such deep network may cause difficulty in optimization due to the issue of vanishing gradients. Motivated by previous studies on training neural networks with deep supervision, we utilize four side-output layers in our net to supervise early layers. Each side-output layer is in charge of one size of feature maps. However, explosion at high levels side-output layers does not work well. By probing the outputs from the first side-output layer associated with smallest feature maps, we cannot find any cue between the dense prediction and ground truth. The underlying reason for this is that the small skin lesions cannot be captured. Hence, we only exploit side-output layers on the last two layers. To address the above-mentioned problems, we combine all side-output layers and final-output layers, which is formulated as

$$\mathcal{L}(I, W) = \sum_s w_s \mathcal{L}_s(I, W) + \mathcal{L}_m w_m(I, W), \tag{2}$$

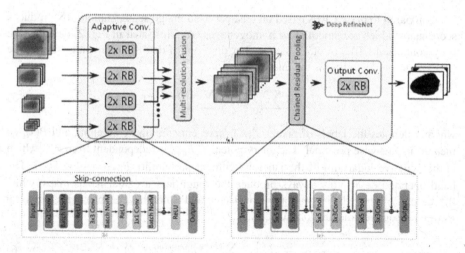

Fig. 3. Illustration of proposed method; (a) Deep RefineNet; (b) residual block; (c) chained residual pooling.

$$\mathcal{L}_s(I, W) = -log(p_0(x_{i,j}, t_{i,j})),\tag{3}$$

where the first part is side-output loss terms and another one is main function between the predicted results and ground truth, w_s and w_m are hyper-parameters for balancing the weight of loss layers, $p_0(x_{i,j}, t_{i,j})$ is predicted probability for true labels.

The skin lesions occupy only small regions, and the learning process is trapped into local minima. Therefore, we fit dermoscopy images' characteristics by taking different loss functions into account [10]. Therefore, our network is defined with per-pixel categorization. The main loss layer is formulated as

$$\mathcal{L}_m(I, W) = \frac{2\sum_i^N \sum_j^M x_{i,j} t_{i,j}}{\sum_i^N \sum_j^M x_{i,j}^2 + \sum_i^N \sum_j^M t_{i,j}^2},\tag{4}$$

where $x_{i,j}$ denote the predicted segmentation and $t_{i,j}$ denote the ground truth. $\mathcal{L}_m(I, W)$ is loss function based on dice coefficient between predicted results and ground-truth. Using dice loss can balance weights between different size of skin lesions.

2.3 Conditional Random Field for Boundary Refinement

Although the feature maps are integrated in our network from side-output to further refine the boundary of the outputs, score maps cannot fit heterogeneous ground truth. Obtaining too much low-level information is detrimental in generalizing the intricate validation set. Hence, we use CRF to refine our network's output rather than input high resolution images to overcome coarse outputs, as post-process outperforms high resolution inputs.

Compared to basic CRFs based on unsupervised image segmentation, RN produces score maps which are smooth and homogeneous. As illustrated in Fig. 2, we integrate fully-connected CRFs in our system rather than out-of-date fashion using short-range CRFs. The model employs the energy function

$$E(x) = \sum_i P(x_i) + \sum_{i,j} P(x_i, x_j), \tag{5}$$

where x denotes the labels of pixels, $P(x_i)$ represents the unary potential and $P(x_i, x_j)$ means the pair-wise potential. The unary potential $P_{i(x_i)}$ is independent from RN, which is the negative logarithm of the label assignment probability $P_{i(x_i)} = -\log S(x_i)$. The hand-operated descriptors always produce the rough and irregular labels maps, while the output of the RN is generally consistent and smooth. The pair-wise potential $P(x_i, x_j)$ in our regime are formulated as

$$P(x_i, x_j) = \mu(x_i, x_j) \sum_{m=1}^{K} \omega^{(m)} k^{(m)} (f_i, f_j), \tag{6}$$

where $k^{(m)}$ denotes the Gaussian kernel, $\mu(x_i, x_j)$ is a label compatibility function where $\mu(x_i, x_j) = 1$ if $x_i \neq x_j$ and 0 otherwise. The pair-wise potential connects each pair of pixel i and pixel j with the fully-connected graph, which are denoted as

$$\omega_1 \exp\left(-\frac{\|p_i - p_j\|^2}{2\sigma_a^2} - \frac{\|I_i - I_j\|^2}{2\sigma_b^2}\right) + \omega_2 \exp\left(-\frac{\|p_i - p_j\|^2}{2\sigma_{ac}^2}\right), \tag{7}$$

where p_i denotes the position of the pixel, I_i denotes the color intensities, hyper-parameters σ_a, σ_b, σ_c represent the "scale" of the Gaussian kernels.

3 Experiments

3.1 Dataset and Implementation

In this study, we perform the experiment to evaluate performance of our purposed method using the public challenge datasets - ISBI skin lesion segmentation dataset released in 2016 and 2017. The skin images are based on the International Skin Imaging Collaboration (ISIC) Archive, acquired from various international clinical centers' devices. The dataset released in 2016 consists of 900 images for training and 350 images for validation, which are extended to 2000 and 600, respectively, in 2017. Due to independent evaluation, the organizer excludes the ground truth of the validation part.

Our algorithm is implemented in MATLAB R2014b based on MatConvNet with a NVDIA TITAN X GPU. We utilize stochastic gradient descent (SGD) to optimize our objective function. The mini-batch involves 20 images. We adopt a weight decay of 0.0001 and a momentum of 0.9. The learning rate is 0.00001 for first 300 epochs and 0.000001 for the next 300 epochs. Apart from the hyper-parameters set above, we incorporate the dropout layers (dropout rate is set as 0.5) in our network to prevent co-adaptation of feature detectors. For computational efficiency, we decouple our algorithm to two stages. We train our network without CRF. For the CRF parameters, we assume that the unary terms are fixed.

3.2 Results

Due to rich parameters to learn, our network cannot train efficiently unless enough number of training images is provided. Despite the skin lesion images provided by the ISBI 2016 challenge, the segmentation performance is still greatly affected by image processing due to the huge intra-class variations. Labeling extra medical images are extremely tedious. Accordingly, data enlarging or augmentation are of great importance to boost the segmentation performance. Performing spatial transformation is one of the most effective ways to solve this problem. Motivated by this, lesion images are rotated to four degrees with (0°, 90°, 180°, 270°) to enlarge the dataset. In the challenge dataset, the segmentation errors are mainly caused by illumination variances and dye concentration. Meanwhile, the huge interclass variation further aggravates this issue. We utilize per-image-mean method instead of all-image-mean, thus each image is normalized to zero before processing our network, which alleviates bad influence caused by unobvious inter-class variations.

We apply our proposed method and evaluate our models in ISBI skin lesions datasets released in both 2016 and 2017. We use the models pretrained in ImageNet and all settings are consistent to assess our method fairly. Models are trained with and without deep supervised method, dice loss and CRF for performance comparison. The segmentation performance is evaluated based on Dice coefficient (DC), jaccard index (JA), accuracy (AC). Table 1 shows the segmentation results of various methods, where ML is multi-path loss. We can see that our proposed RN-ML-CRF method achieves the best result due to the multi-path information exploration and CRF. From the comparison, we can see that our proposed method overcomes the shortcoming of CN-CRF method due to multi-path information. In addition, CRF reuses the pixel-pair information to make one more step in skin-lesions' recovery, which boosts the segmentation performance. In Table 2, we provide the segmentation algorithm comparison based on ISBI 2016 and 2017 dataset. Compared with the listed methods, our proposed method gets the best segmentation performance. The main explanation is that we integrate local information, global information and up-to-down information. Our proposed method achieved considerable improvement in terms of good DC and JA values.

Table 1. Segmentation results of ISBI 2016 and 2017 dataset.

Network	Parameter	DC	JA	AC	Dataset	DC	JA	AC	Dataset
FCRN	91M	0.889	0.816	0.905	2016	0.814	0.721	0.928	2017
RN	410M	0.890	0.824	0.941	2016	0.828	0.735	0.931	2017
RN-CRF	410M	0.908	0.841	0.952	2016	0.830	0.741	0.934	2017
RN-ML-CRF	410M	**0.924**	**0.860**	**0.956**	2016	**0.843**	**0.758**	**0.938**	2017

Table 2. Segmentation algorithm comparison based on ISBI 2016 and 2017 dataset.

Method	DC	JA	AC	Dataset	Method	DC	JA	AC	Dataset
EXB	0.910	0.843	0.953	2016	USYD-BMIT	0.842	0.758	0.934	2017
CUMED	0.897	0.829	0.949	2016	RECOD	0.839	0.754	0.931	2017
Mahmudur	0.895	0.822	0.952	2016	FCN	0.837	0.752	0.930	2017
SFU-mial	0.885	0.811	0.944	2016	SMCP	0.839	0.749	0.930	2017
UiT-Seg	0.881	0.806	0.939	2016	INESC TECNALIA	0.810	0.718	0.922	2017
Our proposed	**0.924**	**0.860**	**0.956**	2016	Our proposed	**0.843**	**0.758**	**0.938**	2017

Figure 4 shows segmentation results of various methods. From the comparison of the proposed automatic method and the provided doctor's ground truth, we can see that the proposed segmentation method is consistent with ground truth. Also, our proposed method outperforms the traditional FCRN method.

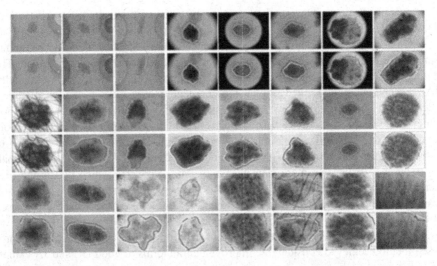

Fig. 4. Segmentation results of various methods. The pink and white contours denote the segmentation results of our method and ground truth, respectively. The upper row is FCRN method and the bottom row is the proposed method. (Color figure online)

4   Conclusion

In this paper, we proposed a deep learning framework for dermoscopy image segmentation based on RN with multi-path processing. The advantage of our proposed network based on multi-path has been fully demonstrated. Meanwhile, we performed extensive experiments on the publicly available ISBI 2016 and 2017 challenge skin lesion datasets. Experiments showed that our proposed method outperforms state-of-the-arts methods. Our future work will focus on how to integrate texture information in our network.

References

1. Siegel, R.L., Miller, K.D., Jemal, A.: Cancer Statistics, 2017. CA-Cancer J. Clin. **67**, 7–30 (2017)
2. Gaudy-Marqueste, C., Wazaefi, Y., Bruneu, Y., Triller, R., Thomas, L., Pellacani, G., Malvehy, J., Avril, M.F., Monestier, S., Richard, M.A., Fertil, B., Grob, J.J.: Ugly duckling sign as a major factor of efficiency in melanoma detection. JAMA Dermatol. **153**, 279–284 (2017)
3. Eigen, D., Puhrsch, C., Fergus, R.: Depth map prediction from a single image using a multi-scale deep network. In: NIPS, pp. 2366–2374 (2014)
4. Long, J., Shelhamer, E., Darrell, T.: Fully convolutional networks for semantic segmentation. In: CVPR, pp. 3431–3440 (2015)
5. Krähenbühl, P., Koltun, V.: Efficient inference in fully connected CRFs with Gaussian edge potentials. In: NIPS, pp. 109–117 (2011)
6. Szegedy, C., Liu, W., Jia, Y., Sermanet, P., Reed, S., Anguelov, D., Erhan, D., Vanhoucke, V., Rabinovich, A.: Going deeper with convolutions. In: CVPR, pp. 1–9 (2015)
7. He, K., Zhang, X., Ren, S., Sun, J.: Deep residual learning for image recognition. In: CVPR, pp. 770–778 (2016)
8. Hinton, G.E., Srivastava, N., Krizhevsky, A., Sutskever, I., Salakhutdinov, R.R.: Improving neural networks by preventing co-adaptation of feature detectors. CoRR 3, pp. 212–223 (2012)
9. Ioffe, S., Szegedy, C.: Batch normalization: accelerating deep network training by reducing internal covariate shift. arXiv preprint arXiv:1502.03167 (2015)
10. Milletari, F., Navab, N., Ahmadi, S.A.: V-net: fully convolutional neural networks for volumetric medical image segmentation. arXiv preprint arXiv:1606.04797 (2016)

Multi-scale Networks for Segmentation of Brain Magnetic Resonance Images

Jie Wei[1] and Yong Xia[2(✉)]

[1] Shaanxi Key Lab of Speech and Image Information Processing (SAIIP),
School of Computer Science and Engineering, Northwestern Polytechnical
University, Xi'an 710072, People's Republic of China
[2] School of Computer Science and Technology, Centre for Multidisciplinary
Convergence Computing (CMCC), Northwestern Polytechnical University,
Xi'an 710072, People's Republic of China
yxia@nwpu.edu.cn

Abstract. Measuring the distribution of major brain tissues using magnetic resonance (MR) images has attracted extensive research efforts. Due to its remarkable success, deep learning-based image segmentation has been applied to this problem, in which the size of patches usually represents a tradeoff between complexity and accuracy. In this paper, we propose the multi-size-and-position neural network (MSPNN) for brain MR image segmentation. Our contributions include (1) jointly using U-Nets trained on large patches and back propagation neural networks (BPNNs) trained on small patches for segmentation, and (2) adopting the convolutional auto-encoder (CAE) to restore MR images before applying them to BPNNs. We have evaluated this algorithm against five widely used brain MR image segmentation approaches on both synthetic and real MR studies. Our results indicate that the proposed algorithm can segment brain MR images effectively and provide precise distribution of major brain tissues.

Keywords: Brain MR image segmentation · Deep learning · U-Net · Convolutional Auto-Encoder (CAE) · Back Propagation Neural Network (BPNN)

1 Introduction

The segmentation of brain magnetic resonance (MR) images aims to classify brain voxels into the gray matter (GM), white matter (WM) and cerebrospinal fluid (CSF) and plays an essential role in clinical applications and research settings, such as abnormality detection, surgical planning, progress assessment and human brain mapping. Since manual segmentation is time-consuming, expensive and subject to operator variability, automated algorithms have been thoroughly studied and proposed in the literature [1–3]. Among them, the atlas-based registration and comparison, statistical model-based segmentation [1], fuzzy clustering and its variants [4, 5], combined statistical and fuzzy methods [6] and deformable model based methods [7] are the most commonly used ones.

© Springer International Publishing AG 2017
M.J. Cardoso et al. (Eds.): DLMIA/ML-CDS 2017, LNCS 10553, pp. 312–320, 2017.
DOI: 10.1007/978-3-319-67558-9_36

Recently, deep learning has become the most successful tool for solving many computer vision problems [8–10], due to its distinct advantages over traditional methods in learning image representation and classifying patterns simultaneously. When applying deep learning techniques to image segmentation, it is intuitive to view this problem as patch-based classification of voxels. For instance, Zhang et al. [8] extracted 2D patches on transverse slices of T1-weighted, T2-weighted and fractional anisotropy (FA) images to train a DCNN for infant brain MR image segmentation, and Moeskops et al. [10] used patches and kernel with multiple sizes to train a multi-scale deep convolutional neural network (DCNN) for adult brain MR image segmentation. However, most patches extracted around the boundary contain voxels from multiple classes, classifying such a patch is troublesome. This intrinsic drawback greatly limits the performance of the patch-based voxel classification strategy.

To overcome this drawback, "end-to-end" models, such as the fully convolutional network (FCN) [11], have been proposed. In these models, both the input and output are image patches. Nie et al. [12] jointly used T1-weighted, T2-weighted and FA images to train multi-fully convolutional networks (mFCNs) for pixel-level infant brain MR image segmentation. Chen et al. [13] proposed an auto-context version of deep voxelwise residual network (VoxResNet) by seamlessly integrating the low-level image appearance features, implicit shape information and high-level context together for volumetric brain segmentation in MR images. Despite their success, these methods have high computational complexity. Ronneberger et al. [14] proposed another form of end-to-end FCN called U-Net, which can be trained highly efficiently from very few images and performs well on neuronal structures segmentation in electron microscopic stacks and cell segmentation on transmitted light microscopy images.

In most deep learning-based segmentation approaches, the patch size has a major impact on the performance. Small patches may not be able to train deep enough models and are prone to be affected by artifacts; whereas large patches contain too complex image information and may lead to less accurate results. In this paper, we jointly use small patches to train a shallow model and large patches to train a deep FCN, and thus propose the multi-size-and-position neural network (MSPNN) for brain MR image segmentation. The uniqueness of this algorithm is to extract image patches with multiple sizes on transverse, sagittal and coronal slices, use large patches to train a U-Net, and use small patches to train back propagation neural networks (BPNNs) [15]. Moreover, since BPNN is relatively sensitive to noise and other artefacts in images, we adopt the convolutional auto-encoder (CAE) [16] to restore the quality of images before applying them to the BPNN. To further improve the performance, we also used the MRI slices extracted from multiple views and fused the segmentation results. We have evaluated the proposed MSPNN algorithm against four widely used brain MR image segmentation approaches on both synthetic and real MR images.

2 Algorithm

The proposed MSPNN algorithm extracts image patches on transverse (TRA), sagittal (SAG) and coronal (COR) MR slices to train a neural network module, respectively, which consists of three components: a multi-scale U-Net, CAE and multi-scale BPNNs. The diagram of this algorithm is illustrated in Fig. 1.

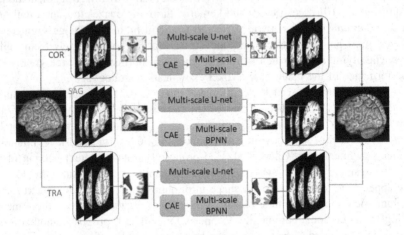

Fig. 1. Diagram of the proposed MSPNN algorithm

2.1 Multi-scale U-Nets for Voxel Classification

The U-Net used for this study consists of two parts: a typical convolutional network for downsampling and another convolutional network for unsampling. The architecture and major parameters of this network are shown in Fig. 2. To explore the information contained in patches of different size, we extract 48×48, 40×40 and 32×32 patches and use each type of patches to train a U-Net, respectively. Finally, we combine the outputs of three U-Nets to form the output of this multi-scale U-Nets component.

2.2 Multi-scale BPNNs for Voxel Classification

To complement the multi-scale U-Nets based segmentation, we extract small patches to train multiple BPNNs for voxel classification, too. On each brain voxel, let a 5×5, a 7×7 and a 9×9 window centered on it. The image patches within each type of windows, together with the corresponding center voxel's class label, are used to train a two-hidden layer BPNNs, respectively. We add a new output layer to combine three BPNNs. The weights between each BPNN and the new output layer are learned during the error back propagation process. The architecture of this multi-scale BPNNs component is shown in Fig. 4, where the figure on each layer is the number of neurons.

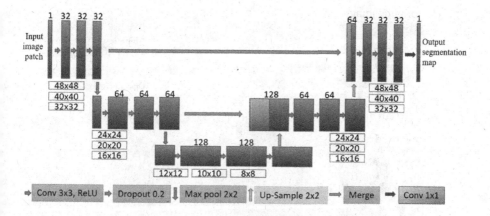

Fig. 2. Architecture of the U-Net used for this study

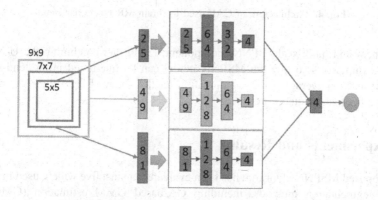

Fig. 3. Architecture of the multi-scale BPNN unit

2.3   CAE-Based Image Restoration

The BPNNs trained on small patches are sensitive to image quality. It is commonly acknowledged that MR images are usually corrupted by noise and bias field. Hence, we adopt CAE for image denoising and bias field correction.

The CAE used for this study consists of convolutional layers with dropout, 2×2 max pooling layers and 2×2 upsampling layers. The architecture and major parameters of this CAE are displayed in Fig. 3. We partition each MR slice into 20×20 patches and apply these patches to the CAE for image restoration.

2.4   Implementation

The proposed MSPNN algorithm can be implemented in four steps: (1) training the CAE, each U-Net and each BPNN, respectively, using image patches extracted on each cut-plan; (2) combining three U-Nets and three BPNNs to form the multi-scale U-Nets

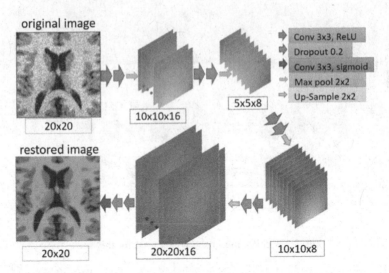

Fig. 4. Architecture of CAE used for brain MR image restoration

component and multi-scale BPNNs component; (3) further combining the U-Net and BPNN components so that the MSPNN model can be fine-tuned in an "end-to-end" manner; and (4) averaging the segmentation results obtained from three cut-plans to form the final segmentation result.

3   Experiments and Results

The proposed MSPNN algorithm has been evaluated against five widely used brain MR image segmentation methods, including GA-based GMM estimation (GMM-GA) algorithm [1], clonal selection algorithm based HMRF model estimation (HMRF-CSA) algorithm [2], variational expectation–maximization (VEM) for Bayesian inference algorithm [3] and the segmentation routines in the statistical parametric mapping (SPM) package [17] and FMRIB Software Library (FSL) package [18] on two datasets.

The BrainWeb dataset [19] consists of 18 simulated brain MRI images, each having a dimension of $181 \times 217 \times 181$ and a voxel size of $1 \times 1 \times 1$ mm^3. The noise level in these images ranges from 0% to 9% and the intensity inhomogeneity (INU) ranges from 0% to 40%. The internet brain segmentation repository (IBSR, V2.0) [20, 21] consists of 18 clinical brain MRI images, with a dimension of $256 \times 256 \times 128$ and a voxel size of $1.0 \times 1.0 \times 1.5$ mm^3. Skull stripping has been applied to these images for the removal of non-brain tissues by using the CMA "autoseg" routines [20, 21].

Since the proposed algorithm requires training cases, we adopted the leave-one-out validation strategy to test it on each dataset. Figure 5 shows the 88th transverse, 100th sagittal and 112th coronal slices of the simulated study with 7% noise and 40% INU and the corresponding segmentation results obtained by employing six segmentation approaches. The marked rectangles show the partial enlarged views of the corresponding segmentation regions. It reveals in the enlarged views of marked regions that

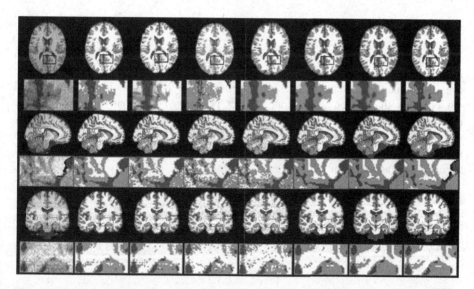

Fig. 5. Segmentation of the BrainWeb study with 7% noise and 40% INU: brain MR images (1st column), their segmentation results obtained by using the SPM (2nd column), FSL (3rd column), VEM (4th column), GMM-GA (5th column), HMRF-CSA (6th column), the proposed algorithm (7th column), and the ground truth (8th column).

our proposed algorithm is able to produce more accurate segmentation than other five methods.

The accuracy of delineating each type of brain tissue was assessed by using the Dice similarity coefficient (DSC) [22], and the overall accuracy of brain image segmentation was calculated as the percentage of correctly classified brain voxels. The performance of six algorithms was compared in Table 1, where the mean and standard deviation of DSC and overall segmentation accuracy were calculated across 18 BrainWeb studies and 18 IBSR studies, respectively. It shows that on average the proposed algorithm outperforms the other five methods on both simulated and real MR datasets, particularly on the real dataset.

Table 1. Performance of six algorithms (Mean ± std) on two datasets

Data	BrainWeb			IBSR V2.0		
Type	GM	WM	All	GM	WM	All
SPM	92.55 ± 2.37	95.82 ± 1.88	94.15 ± 1.88	84.42 ± 2.45	87.38 ± 1.30	81.26 ± 2.32
FSL	89.18 ± 2.95	95.82 ± 2.06	91.03 ± 1.89	77.35 ± 3.70	87.08 ± 3.09	75.06 ± 3.24
VEM	90.67 ± 2.52	90.62 ± 1.32	91.37 ± 1.66	86.67 ± 4.54	79.22 ± 5.22	82.55 ± 5.02
GMM-GA	88.59 ± 2.66	91.62 ± 1.59	92.31 ± 1.76	77.35 ± 6.01	87.23 ± 2.79	74.97 ± 6.09
HMRF-CSA	94.30 ± 1.58	**96.67 ± 0.94**	95.43 ± 1.20	84.81 ± 1.69	83.54 ± 2.26	82.79 ± 1.48
Proposed	**95.49 ± 1.04**	96.33 ± 0.83	**95.69 ± 1.03**	**96.10 ± 0.39**	**92.87 ± 1.04**	**94.92 ± 0.59**

4 Discussions

To demonstrate the contribution of U-Net and CAE, we chose five synthetic scans from BrainWeb as a case study and applied them to different segmentation approaches. The average segmentation accuracy was displayed in Table 2. It shows in the 2nd, 3rd and 4th columns that, when used alone, U-Net performs better than LeNet-5 and the one-hidden-layer BPNN. It shows in the 6th and 7th columns that, when used jointly with the CAE and BPNN, U-Net performs better than LeNet-5. Meanwhile, it shows in the 4th and 5th columns that using CAE to restore image quality can improve the segmentation accuracy of BPNN.

Table 2. Average accuracy of different segmentation approaches on five BrainWeb studies

Methods	U-Net	LeNet-5	BPNN	CAE + BPNN	LeNet-5 + BPNN + CAE	U-Net + BPNN + CAE
Average	94.09%	92.68%	90.24%	91.58%	93.78%	**95.28%**

The proposed MSPNN algorithm enables the automated segmentation of brain MR images. It, however, has extremely complex network architecture, and hence is very time consuming to train it, especially when applied to large 3D image volumes. Nevertheless, using the trained MSPNN model to segment a brain MR study is relatively efficient. The mean and standard deviation of the time cost of training and testing the proposed model (Intel E5-2600V4 CPU X2, NVIDIA TITAN X GPU, 512 GB memory and Python with Keras 1.1.0 and Theano) are listed in Table 3.

Table 3. Time cost of the proposed algorithm (Mean ± std) on two datasets

Dataset	Number of images	Image size	Training time (h)	Test time (s)
BrainWeb	18	$181 \times 217 \times 181$	37.6 ± 0.41	180 ± 1.22
IBSR V2.0	18	$256 \times 256 \times 128$	32.3 ± 0.35	146 ± 1.57

5 Conclusion

In this paper, we propose the MSPNN algorithm for brain MR image segmentation. We use multiple U-Nets to segment large patches and multiple BPNNs to segment small patches, which have been restored by CAE. With ensemble learning, the proposed model can be trained in an "end-to-end" manner. We have evaluated this algorithm against five widely used methods on synthetic and real MR image datasets. Our results suggest that the proposed algorithm outperforms other five methods on both datasets.

Acknowledgement. This work was supported by the National Natural Science Foundation of China under Grants 61471297.

References

1. Tohka, J., Krestyannikov, E., Dinov, I.D., Graham, A.M., Shattuck, D.W., Ruotsalainen, U., Toga, A.W.: Genetic algorithms for finite mixture model based voxel classification in neuroimaging. IEEE Trans. Med. Imaging **26**(5), 696–711 (2007)
2. Zhang, T., Xia, Y., Feng, D.D.: Hidden Markov random field model based brain MR image segmentation using clonal selection algorithm and Markov chain Monte Carlo method. Biomed. Signal Process. Control **12**(1), 10–18 (2014)
3. Tzikas, D.G., Likas, A.C., Galatsanos, N.P.: The variational approximation for Bayesian inference. IEEE Signal Process. Mag. **25**(6), 131–146 (2008)
4. Dubey, Y.K., Mushrif, M.M., Mitra, K.: Segmentation of brain MR images using rough set based intuitionistic fuzzy clustering. Biocybern. Biomed. Eng. **36**(2), 413–426 (2016)
5. Ouarda, A., Fadila, B.: Improvement of MR brain images segmentation based on interval type-2 fuzzy C-Means. In: Third World Conference on Complex Systems (2016)
6. Ji, Z., Xia, Y., Sun, Q., Chen, Q., Xia, D., Feng, D.D.: Fuzzy local Gaussian mixture model for brain MR image segmentation. IEEE Trans. Inf Technol. Biomed. **16**(3), 339–347 (2012). A Publication of the IEEE Engineering in Medicine & Biology Society
7. Su, C.M., Chang, H.H.: A level set based deformable model for segmentation of human brain MR images. In: IEEE International Conference on Biomedical Engineering and Informatics, pp. 105–109 (2014)
8. Zhang, W., Li, R., Deng, H., Wang, L., Lin, W., Ji, S., Shen, D.: Deep convolutional neural networks for multi-modality isointense infant brain image segmentation. Neuroimage **108**, 214–224 (2015)
9. Brébisson, A.D., Montana, G.: Deep neural networks for anatomical brain segmentation. In: Computer Vision and Pattern Recognition Workshops, pp. 20–28 (2015)
10. Moeskops, P., Viergever, M.A., Mendrik, A.M., Vries, L.S.D., Benders, M.J.N.L., Išgum, I.: Automatic segmentation of MR brain images with a convolutional neural network. IEEE Trans. Med. Imaging **35**(5), 1252–1261 (2016)
11. Long, J., Shelhamer, E., Darrell, T.: Fully convolutional networks for semantic segmentation. In: IEEE Conference on Computer Vision and Pattern Recognition, pp. 1337–1342 (2015)
12. Nie, D., Wang, L., Gao, Y., Sken, D.: Fully convolutional networks for multi-modality isointense infant brain image segmentation. In: IEEE International Symposium on Biomedical Imaging, pp. 1342–1345 (2016)
13. Chen, H., Dou, Q., Yu, L., Qin, J., Heng, P.A.: VoxResNet: deep voxelwise residual networks for brain segmentation from 3D MR images. Neuroimage (2017)
14. Ronneberger, O., Fischer, P., Brox, T.: U-Net: convolutional networks for biomedical image segmentation. In: International Conference on Medical Image Computing and Computer-Assisted Intervention, pp. 234–241 (2015)
15. Rumelhart, D., Mcclelland, J.: Parallel Distributed Processing: Explorations in the Microstructure of Cognition: Foundations. MIT Press, Cambridge (1986)
16. Masci, J., Meier, U., Dan, C., Schmidhuber, J.: Stacked convolutional auto-encoders for hierarchical feature extraction. In: International Conference on Artificial Neural Networks, pp. 52–59 (2011)
17. Dale, A.M., Liu, A.K., Fischl, B.R., Buckner, R.L., Belliveau, J.W., Lewine, J.D., Halgren, E.: Dynamic statistical parametric mapping: combining fMRI and MEG for high-resolution imaging of cortical activity. Neuron **26**(1), 55–67 (2000)

18. Smith, S.M., Jenkinson, M., Woolrich, M.W., Beckmann, C.F., Behrens, T.E.J., Johansen-Berg, H., Bannister, P.R., Luca, M.D., Drobnjak, I., Flitney, D.E.: Advances in functional and structural MR image analysis and implementation as FSL. Neuroimage **23** (Suppl. 1), S208–S219 (2004)
19. Collins, D.L., Zijdenbos, A.P., Kollokian, V., Sled, J.G., Kabani, N., Holmes, C.J., Evans, A.C.: Design and construction of a realistic digital brain phantom. IEEE Trans. Med. Imaging **17**(3), 463–468 (1998)
20. School, M.G.H.H.M.: The Internet Brain Segmentation Repository (IBSR). http://www.cma. mgh.harvard.edu/ibsr/index.html
21. Rohlfing, T.: Image similarity and tissue overlaps as surrogates for image registration accuracy: widely used but unreliable. IEEE Trans. Med. Imaging **31**(2), 153–163 (2012)
22. Bharatha, A., Hirose, M., Hata, N., Warfield, S.K., Ferrant, M., Zou, K.H., Suarez-Santana, E., Ruiz-Alzola, J., Amico, A.D., Cormack, R.A.: Evaluation of three-dimensional finite element-based deformable registration of pre- and intra-operative prostate imaging. Med. Phys. **28**(12), 2551–2560 (2001)

Deep Learning for Automatic Detection of Abnormal Findings in Breast Mammography

Ayelet Akselrod-Ballin$^{(\boxtimes)}$, Leonid. Karlinsky, Alon Hazan,
Ran Bakalo, Ami Ben Horesh, Yoel Shoshan, and Ella Barkan

IBM Research, Haifa, Israel
ayeletb@il.ibm.com

Abstract. Automatic identification of abnormalities is a key problem in medical imaging. While the majority of previous work in mammography has focused on classification of abnormalities rather than detection and localization, here we introduce a novel deep learning method for detection of masses and calcifications. The power of this approach comes from generating an ensemble of individual Faster-RCNN models each trained for a specific set of abnormal clinical categories, together with extending a modified two stage Faster-RCNN scheme to a three stage cascade. The third stage being an additional classifier working directly on the image pixels with the handful of sub-windows generated by the first two stages. The performance of the algorithm is evaluated on the INBreast benchmark and on a large internal multi-center dataset. Quantitative results compete well with state of the art in terms of accuracy. Computationally the methods runs significantly faster than current state-of-the art techniques.

1 Introduction

Numerous studies have shown that early detection of breast cancer can both increase treatment options and reduce mortality [1]. There are two main types of abnormal objects detected in mammograms (MG) – **Masses** are seen as compact areas that appear brighter than the embedding tissue and **Calcifications** [2]. **Macro calcifications (macro)** are usually benign bigger bits of calcium, appearing as coarse calcium deposits in the breast, such as Coarse or "popcorn-like", Dystrophic or Rim-like (eggshell). **Micro calcifications (micro)** are tiny specks of mineral deposits that can be distributed in various ways, clusters, specific patterns or scattered. Certain features and presentations of micro calcifications, specifically, amorphous, pleomorphic shapes and clustered distribution can be associated with malignant breast cancer. However, detection and identification are extremely difficult, due to the subtle fine-grained visual categories and large variability of appearance of the different categories. Therefore, automatic detection, localization and quantification of abnormal findings, has the potential of separating the normal-negative from positive exams in the screening

---

A. Akselrod-Ballin and L. Karlinsky contributed equally to this work.

© Springer International Publishing AG 2017
M.J. Cardoso et al. (Eds.): DLMIA/ML-CDS 2017, LNCS 10553, pp. 321–329, 2017.
DOI: 10.1007/978-3-319-67558-9_37

processes, allowing the radiologist to focus on the challenging cases and avoiding unnecessary breast biopsies.

Deep learning has shown remarkable performances in recognition tasks such as detection and classification [3–6]. Below we briefly review the dominant detection algorithms in the field, emphasizing the differences compared to our approach.

R-CNN and its variants use external region proposals instead of sliding windows to find objects in images. Commonly, classical methods are used to generate those region proposals. In R-CNN [5], Selective Search [7] is used to generate candidate object boxes, a CNN [3] is used to generate the feature vector for each box, an SVM is trained to classify the box feature vectors, and linear regression followed by non-maximal suppression is used to adjusts the bounding boxes and eliminate duplicate detections. Each stage of this complex pipeline must be precisely independently tuned and the resulting system is slow, taking more than 40 seconds per test on a GPU [8].

Other CNN based detectors, such as Deep MultiBox [9], Fast [8] and Faster [6] R-CNN, and YOLO [10], focus on speeding up the R-CNN framework by sharing computation and using Region Proposal Networks (RPN) – a CNN that effectively performs sliding window searches in a fully convolutional manner [11] over a grid of receptive fields, to generate object proposals instead of Selective Search [6–8]. YOLO [10] also puts spatial constraints on the grid cell proposals. While they offer speed and accuracy improvements over R-CNN, they still fall short of real-time performance, running in about 2–3 FPS similar to Faster-RCNN.

Advances in deep learning methods have recently been exploited successfully in the medical imaging domain [12, 13] and specifically in breast mammography (MG) for mass and calcification classification [14, 15]. Our study departs from the majority of previous work as it focuses on detection and localization of abnormalities rather than on binary classification of calcification or masses. An exception is the prominent work on mass detection of [16], which is based on a multiscale deep belief network classifier and is followed by a cascade of R-CNN and random forest classifiers. A recent study by the same authors [17] is currently considered the most effective study for detection of micro calcification, utilizing a cascade of boosting classifiers with shape and appearance features. Based on these approaches [18] recently presented a deep residual neural network aimed at classification of an MG as benign or malignant. The final segmentation maps of the masses and micro-calcifications are not evaluated, yet they demonstrate INBreast [19] separation into normal and benign with an ROC AUC of 0.8.

An important component of the Faster-RCNN framework differing it from similar detection frameworks like YOLO [10], is the classifier running after the RPN built on top of an ROI-pooling layer [6] that pools features from the top most layer of the base network shared between RPN and the classifier. This can be seen as a two stage cascade with RPN being the first stage and the classifier second. However, the representation generated by ROI-pooling is not specifically optimized for classification (being shared between the detector and the classifier). Therefore this work follows the Faster-RCNN architecture presented in [20] yet proposes to extend the scheme to a three stage cascade, the third stage being an additional classifier working directly on the image pixels with the handful of sub-windows generated by the first two stages.

Our contribution is three fold: First, to the best of our knowledge there are no studies utilizing a unified deep learning approach to combine both detection, localization and classification of multiple type of masses and calcifications. Second we utilize a three- stage cascade integrating a two-stage ensemble of RPN and Faster-RCNN models and a third CNN classifier and demonstrate its effectiveness for reducing the number of FP detections. Finally, our methodology competes well with state of the art in terms of inference time and accuracy even on a large dataset.

2 Methods

Problem Formulation: Given as input a set of training images, bounding boxes corresponding to abnormal findings $\{I_i, y_{1i}, x_{1i}, y_{2i}, x_{2i.}, c_i\}$ and a testing set of images, we seek to detect the abnormal findings in the test set and locate the bounding boxes with a corresponding confidence score.

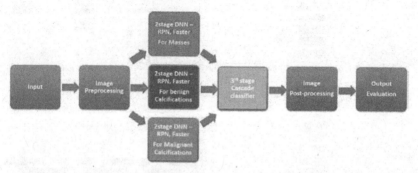

Fig. 1. The deep framework for detection and classification of abnormalities in Mammography.

The system preprocesses the input MG image by dividing the ~ 4 k $\times$ 3 k pixels onto an overlapping grid, utilized to train the three-stage cascade deep neural network (DNN) described below such that each subpart (grid cell) extracts candidate bounding boxes and predicts confidence scores for those boxes. Then, the image post-processing performs grid composition of all the DNN parts results. Figure 1 outlines the system architecture which is composed of three main components detailed below. Roughly speaking, the stages are 'initial detection', 'classification' and 'classification refinement'. In setting these stages we extend the Faster RCNN paradigm, by continuing the 're-classification' loop by adding one more classification iteration, and admitting the detected crops directly to the final classifier (rather than letting it look on them through the prism of the ROI pooling layer and a set of features optimized for detection).

(1) **A region proposal network (RPN):** a deep fully convolutional network that is trained to detect windows on the input image that are likely to contain objects of interest regardless of their class. The RPN simultaneously predicts objects bounds and objectness scores at each position on a wide range of scales and aspect ratios, following which top scoring 500 predictions are kept [11]. The way the network effectively operates on the image in a sliding window fashion, yet a lot of internal

computation are being inherently re-used due to local and hierarchical nature of the stacked network layers. The sliding window operates on a regular grid with the same step size in x and y equal to the final stride of the top most layer receptive field (32 pixels in our case). For each grid location 9 seed boxes (sampled with several different aspect ratios and sizes) are being classified. During training, these boxes, originally called anchors in [6], are being associated with ground truth object boxes according to IoU scores, where the most similar ones (according to IoU) are being labeled as object anchors as well as all those that pass a certain threshold (0.5) in IoU. For each anchor classified as being (part of, or containing) the object, a linear regression is employed to refine the anchor bounding box to the candidate object one.

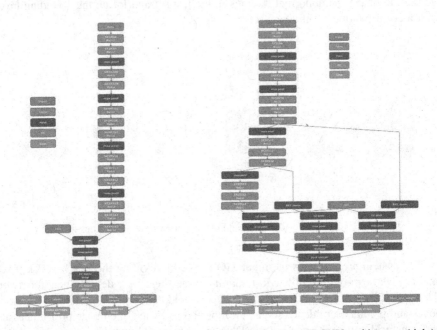

Fig. 2. Original Faster-RCNN architecture (left) Modified Faster-RCNN architecture (right)

(2) **A Fast R-CNN detection network:** that is trained to classify candidate object windows returned by the RPN, each window is classified into one of the classes of interest or rejected as a false alarm. Both RPN and Fast-RCNN share a large common set of bottom layers allowing computation re-use and hence a significant speed-up at test time. For both RPN and the classification network we employ a modified version of VGGNet by [4]. Originally trained on the ImageNet dataset [1], and fine-tuned by us on the task at hand. The system was trained on a single TitanX GPU with 12 GB on chip memory, and i7 Intel CPU with 64 GB RAM. Training times required ∼36 h, while testing takes 0.2 seconds per image. During training 2000 top-scoring boxes are sampled from the RPN, during testing top scoring 500 boxes are sampled using standard non-maximal suppression

(NMS) based on box overlap. We used the SGD solver with learning rate of 0.001, batch size 2, momentum 0.99, and 60 epochs. Figure 2 shows the modified Faster-RCNN architecture we followed (see details in [20]).

(3) **Third stage Cascade Classifier:** The true positive (TP) and false positive (FP) boxes candidates of the DNN are computed on the training set and selected to be the positive and negative samples for the next training phase respectively. These boxes are then cropped and used to train a VGG-16 classifier optimized to separate between the TP and the hard negatives of the last step (the classifier) of Faster-RCNN. During training the boxes are randomized and randomly re-cropped and augmented with random geometric and photometric transformations in order to enrich the training and improve generalization. The key idea behind adding this third stage to Faster-RCNN two stage cascade is to allow a separately optimized network to have an alternative look on the ROIs whose original Faster-RCNN internal representation is constrained to be shared between the first two stages (the detector and the classifier). The full training process details include: each batch contained four images chosen at random; out of each image up to three positive boxes (true detection of the previous stage) and one negative box (false detection of the previous stage) were chosen at random; each box was randomly rotated up to $30°$, randomly re-cropped by $\pm25\%$ of original size, randomly flipped, mean subtracted, and finally resized to 224×224 size as required by the standard VGG-16 classifier. During testing the original detected boxes from the previous stage were processed for speed (no augmentation, single crop).

3 Experiments and Results

We evaluated the algorithm for detection of masses and calcifications. The data set consists of (1) the publicly available INBreast dataset [19] (2) an Internal dataset consisting of approximately 3500 images, collected from several different medical centers with ground truth annotation by experts including 750, 360, 2400 images with masses, malignant calcifications, and benign calcifications respectively. The mass category includes Breast Imaging-Reporting and Data System (BI-RADS) subtypes of 2, 3, 4, 5 namely benign and malignant masses. The dataset was split into training 80% and testing 20% so that all test patients image were excluded from the training set.

Figure 3, reports the free response operating characteristic curve (FROC) calculating the number of true positive rate (TPR) as a function of false positives per image (FPI). The figure shows the results obtained by each class-model separately on one type of images, and also results of joining of all models on all images including normal images. We compare joining the ensemble by 'naïve' concatenation of results obtained by all models or by utilizing the 3rd cascade step on the ensemble of models. The results show clear reduction of FP's and performance improvement by the cascade approach.

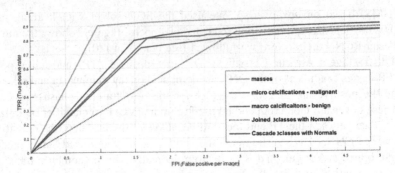

Fig. 3. Detection performance per class and entire set. FROC curves showing the result on various operating points with true positive rate (TPR) versus false positive per image (FPI) for mass images (green), malignant micro calcification images (magenta) and benign macro calcifications images (purple), for 'naïve' joining of all model results on all images (dotted blue), and for joining all models results on all images based on 3rd cascade (bold blue). (Color figure online)

Table 1 summarizes our results compared to the best state-of-the-art results (see details in [16, 17]). [17] Reported the leading results for detection of calcifications on the INBreast (INB) dataset, having a TPR for individual calcification of 40% at one FPI and a TPR of 80% at 10 FPI. The authors noted that these results are significantly better than the current state of the art, which has a TPR of less than 0.01@1 FPI and a TPR of 0.1@10 FPI. [16] Provided a detailed table for mass detection with the best TPR 0.96 ± 0.03@1.2 FPI and TPR = 0.87 ± 0.14@0.8 FPI for INBreast. The papers also obtained the best results in respect to running time of 20 s.

The INB results were obtained by testing separately on INB calcification and mass images with the models trained by our internal data. This is not optimal as there are significant differences in the images characteristics between the two sets. Accordingly, removal of small calcifications (radius < 10 pixels) yields a significant improvement in

Table 1. Presents our results compared to best state-of-art algorithms

Calcifications results	TPR@FPI	#images	Runtime
Ours on micro & macro calcifications	All: TPR 0.4@1 FPI No small: TPR 0.85@1.5 FPI	310/410 INB	5 s
Ours on macro benign calcifications	TPR 0.8@1.5 FPI TPR 0.52@1 FPI TPR 0.9@10 FPI	2400 internal	5 s
Ours on micro malignant calcifications	TPR 0.81@1.7 FPI TPR 0.48@1 FPI	360 internal	5 s
Micro & macro calcifications [17]	TPR 0.4@1 FPI TPR 0.8@10 FPI	410 INB	20 s
Mass results	TPR@FPI	#images	Runtime
Ours on masses	TPR 0.93@0.56 FPI	100/410 INB	5 s
Ours on masses	TPR 0.9@1 FPI	750 internal	5 s
Benign/Malignant [16]	TPR 0.96 ± 0.03@1.2 FPI TPR 0.87 ± 0.14@0.8 FPI	410 INB	20 s

performance. Comparison to published results in the field is difficult as most of the previous work focused on binary classification and also mainly report on DDSM. Yet, Table 1 and Fig. 3 demonstrate the high accuracy obtained by our approach, the advantage of adding the cascade to the Faster-RCNN model and the generalization ability of our model to different classes of abnormalities. The performance on calcifications, specifically the small ones remains to be further investigated and improved.

The visual results in Fig. 4, shows a zoom in on the malignant calcification detection results. A representative example of true-positive, false-negative and false positive detections is given in Fig. 5 for various abnormal subtypes including micro calcifications macro calcification and masses.

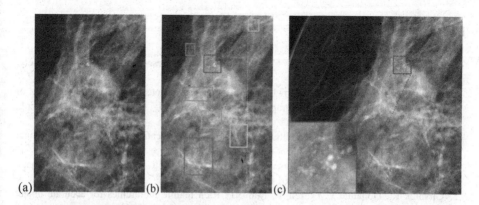

Fig. 4. Qualitative Detection Results displayed on a cropped image with clustered malignant micro calcifications (a) Original (b) With accurately detected boxes, where the lines in autumn colors correspond to score (red and yellow represents high and low score respectively) (c) zoom in on highest scoring detected box. (Color figure online)

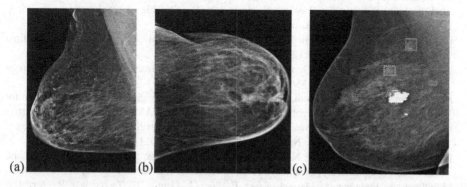

Fig. 5. Examples of detection results on full breast images. Ground truth annotation is highlighted in green for the class depicted while the candidates of automatic detection can be viewed in dashed line in autumn colors corresponding to score (red and yellow represents high and low score respectively). (a) Macro calcification (b) Micro calcification and (c) Masses. (Color figure online)

4 Summary

The paper presents an efficient DNN framework for detection of abnormalities in MG images focusing on the demanding task of detection localization and classification of masses and calcifications. The unified DNN built, is composed of a three-stage cascade, including RPN, a modified Faster-RCNN and a classifier CNN aimed at reducing the FP's. Our results are competitive with the state of the art in terms of efficiency and accuracy showing promising results for mass and calcification detection on large datasets. Future work will extend this approach to a multi view, bilateral approach to integrate information from the four mammography screening views of the right and left breast and generalize this approach to other abnormal classes.

References

1. American Cancer Society: 2015 Cancer Facts and Figures. American Cancer Society, Atlanta (2015)
2. Saranya, R., Bharathi, M., Showbana, R.: Automatic detection and classification of microcalcification on mammographic images. IOSR J. Electron. Commun. Eng. (IOSR-JECE) 9(3), 65–71 (2014)
3. Krizhevsky, A., Sutskever, I., Hinton, G.: ImageNet classification with deep convolutional neural networks. In: NIPS, vol. 25, pp. 1097–1105 (2012)
4. Simonyan K., Zisserman A.: Very Deep Convolutional Networks for Large-Scale Image Recognition. arXiv:1409.1556 (2014)
5. Girshick R., Donahue J., Darrell T., Malik J.: Rich feature hierarchies for accurate object detection and semantic segmentation. In: CVPR, pp. 580–587 (2014)
6. Ren, S., He, K., Girshick, R., Sun, J.: Faster R-CNN: towards real-time object detection with region proposal networks. In: NIPS. arXiv:1506.01497 (2015)
7. Uijlings, J.R., Van de Sande, K.E., Gevers, T., Smeulders, A.W.: Selective search for object recognition. Int. J. Comput. Vis. 104(2), 154–171 (2013)
8. Girshick R.: Fast R-CNN. In: ICCV, pp. 1440–1448 (2015)
9. Erhan, D., Szegedy, C., Toshev, A., Anguelov, D.: Scalable object detection using deep neural networks. In: CVPR, pp. 2155–2162 (2014)
10. Redmon, J., Divvala, S., Girshick, R., Farhadi, A.: You only look once: unified, real-time object detection. In: CVPR (2016)
11. Long, J., Shelhamer, E., Darrell, T.: Fully convolutional networks for semantic segmentation. In: CVPR, Boston, MA, pp. 3431–3440 (2015)
12. Gulshan, V., Peng, L., Coram, M., et al.: Development and validation of a DL algorithm for detection of diabetic retinopathy in retinal fundus photos. JAMA 316, 2402–2410 (2016)
13. Esteva, A., Kuprel, B., Novoa, R.A., et al.: Dermatologist-level classification of skin cancer with deep neural networks. Nature 2017(542), 115–118 (2017)
14. Giger, M.L., Karssemeijer, N., Schnabel, J.A.: Breast image analysis for risk assessment, detection, diagnosis, and treatment of cancer. Annu. Rev. Biomed. Eng. 15, 327–357 (2013)
15. Oliver, A., Freixenet, J., Marti, J., Perez, E., Pont, J., Denton, E., Zwiggelaar, R.: A review of automatic mass detection and segmentation in MG images. MIA 14, 87–110 (2010)
16. Dhungel, N., Carneiro, G., Bradley, A.P.: Automated mass detection in mammograms using cascaded deep learning and random forests. In: DICTA, pp. 1–8 (2015)

17. Lu, Z., Carneiro, G., Dhungel, N., Bradley, A.P.: Automated detection of individual microcalcifications from MG using a multistage cascade approach. arXiv:1610.02251v (2016)

18. Dhungelz, N., Carneiroy, G., Bradley, A.P.: Fully automated classification of mammograms using deep residual neural networks. In: ISBI 2017 (2017)

19. Moreira, I.C., Amaral, I., Domingues, I., Cardoso, A., Cardoso, M.J., Cardoso, J.S.: INBreast: toward a full-field digital MG database. Acad. Radiology. **19**(2), 236–248 (2012)

20. Akselrod-Ballin, A., Karlinsky, L., Alpert, S., Hasoul, S., Ben-Ari, R., Brakan, E.: A region based convolutional network for tumor detection and classification in breast mammography. In: DLMIA (2016)

Grey Matter Segmentation in Spinal Cord MRIs via 3D Convolutional Encoder Networks with Shortcut Connections

Adam Porisky[1]([✉]), Tom Brosch[1], Emil Ljungberg[1], Lisa Y.W. Tang[1],
Youngjin Yoo[1], Benjamin De Leener[2], Anthony Traboulsee[1],
Julien Cohen-Adad[2], and Roger Tam[1]

[1] MS/MRI Research Group, University of British Columbia, Vancouver, BC, Canada
adam.porisky@gmail.com
[2] NeuroPoly Lab, Institute of Biomedical Engineering, Ecole Polytechnique,
Montreal, QC, Canada

Abstract. Segmentation of grey matter in magnetic resonance images of
the spinal cord is an important step in assessing disease state in neurological disorders such as multiple sclerosis. However, manual delineation of
spinal cord tissue is time-consuming and susceptible to variability introduced by the rater. We present a novel segmentation method for spinal
cord tissue that uses fully convolutional encoder networks (CENs) for
direct end-to-end training and includes shortcut connections to combine
multi-scale features, similar to a u-net. While CENs with shortcuts have
been used successfully for brain tissue segmentation, spinal cord images
have very different features, and therefore deserve their own investigation. In particular, we develop the methodology by evaluating the impact
of the number of layers, filter sizes, and shortcuts on segmentation accuracy in standard-resolution cord MRIs. This deep learning-based method
is trained on data from a recent public challenge, consisting of 40 MRIs
from 4 unique scan sites, with each MRI having 4 manual segmentations
from 4 expert raters, resulting in a total of 160 image-label pairs. Performance of the method is evaluated using an independent test set of
40 scans and compared against the challenge results. Using a comprehensive suite of performance metrics, including the Dice similarity coefficient (DSC) and Jaccard index, we found shortcuts to have the strongest
impact (0.60 to 0.80 in DSC), while filter size (0.76 to 0.80) and the number of layers (0.77 to 0.80) are also important considerations. Overall,
the method is highly competitive with other state-of-the-art methods.

1 Introduction

A common metric used for assessing the disease state in conditions such as multiple
sclerosis (MS) is the cross sectional area (CSA) of the spinal cord [7]. Segmentation of the spinal cord in magnetic resonance images (MRIs) is an important step in
acquiring this measurement. However, CSA does not discriminate between white
matter (WM) and grey matter (GM). In the case of relapsing MS, significant GM

© Springer International Publishing AG 2017
M.J. Cardoso et al. (Eds.): DLMIA/ML-CDS 2017, LNCS 10553, pp. 330–337, 2017.
DOI: 10.1007/978-3-319-67558-9_38

atrophy can occur without a global reduction in CSA [1,10]. Thus, distinguishing between tissue types in MRIs allows a greater differentiation in disease conditions compared to a full cord area measurement. One method of acquiring this information is manual delineation of GM tissue on a slice-by-slice basis. However, this is time-consuming, and susceptible to inter- and intra-rater variability. As a result, there exists a need for an unbiased and automatic method to segment GM in MRIs of the spinal cord.

Here we present an automatic method for segmenting GM in axially oriented MRIs of the spinal cord using a fully convolutional encoder network (CEN) for direct end-to-end training [3]. Our network architecture is similar to a u-net [9] in that both consist of complementary contracting and expanding pathways, but our method uses deconvolution [13] instead of upsampling in the expanding pathway, which generates segmentations of the same resolution as the input images. Shortcuts have been explored for larger images [2,5], but not for small images like the cord. Thus, lessons learned by applying CENs with shortcuts to larger images like brain MRIs may not necessarily transfer to the cord. While a high-level description of a CEN model for cord segmentation was included in a publication of a public segmentation challenge [8], it did not address key design considerations such as the number of network layers, filter size, and inclusion of shortcuts.

Training of the network is done by unsupervised pre-training and supervised training with manual segmentations. Segmentation of new MRIs is accomplished through five steps: (1) Each slice of the original image is cropped to a size of 256×256 voxels. (2) The spinal cord is identified using a CEN with our proposed architecture, but trained on spinal cord segmentations instead of GM segmentations. (3) Each image slice is then centered and cropped to an in-plane size of 100×100 voxels. (4) A CEN trained on GM segmentations is applied to the cropped image and a probabilistic mask is generated. (5) A threshold is applied to the probabilistic mask to produce a binary mask.

The main contributions of our work are as follows. We (1) describe a deep learning-based pipeline for GM segmentation of the spinal cord; (2) investigate variations in the architecture of the network to optimize performance; and (3) compare the performance of our method against current methods.

2   Methods

2.1   Network

The overall architecture of the convolutional network is illustrated in Fig. 1. The network is separated into a contracting pathway that uses alternating convolution (w_c) and pooling layers (2D 2×2 average pooling) and an expanding pathway that applies deconvolution (w_d) and unpooling. Shortcut connections (w_s) exist between corresponding convolution layers in the contracting path and deconvolution layers in the expanding path. These connections allow for both low level and high level features to be considered at each layer. We evaluate the impact of the following design considerations for segmentation of spinal cord

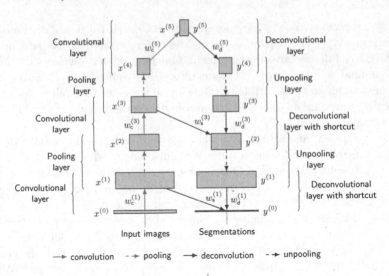

Fig. 1. Convolutional encoder network (11 layers with 9×9 convolutional filters) with shortcuts for spinal cord GM segmentation. Input images $x^{(0)}$ generate segmentations $y^{(0)}$ through a series of convolutions $w_c^{(L)}$, deconvolutions $w_d^{(L)}$, and shortcuts $w_s^{(L)}$.

tissue. Compared to brain images, spinal cord images have less complex features that are also less variable in scale, which brings into question whether shortcuts for multi-scale modeling make a significant contribution. Therefore, we test the network after removing the shortcuts. The smaller cord images allow a deeper model to be trained efficiently, compared to the 7-layer one used for brain by Brosch et al. [2], but it is unknown whether any additional layers would improve segmentation accuracy. We investigate the impact of network depth by adding two new layers to each of the pathways, making it an 11-layer network. Finally, the differences in brain and cord image features motivate the comparison of different filter sizes. We vary in the in-plane filter size between 5×5, 9×9, and 13×13. For all networks, an empirically determined kernel depth of 3 is used in the first layer, while a kernel depth of 1 is used in the subsequent layers.

2.2   Training

The weights for each layer of the CEN are initialized using the weights from a set of consecutive convolutional restricted Boltzmann machines (convRBMs) [6]. Recent work has shown that this pre-training procedure improves the performance of CENs [11]. The first convRBM was trained on the input images, while each convRBM after that receives the hidden activations of the previous layer as an input. Hyperparameter values were optimized via a trial run process. Several models were trained, each with different hyperparameters, for 100 epochs. Training was then restarted using the hyperparameters that yielded the lowest reconstruction error. Our best results were obtained using the hyperparameters listed in Table 1.

Table 1. Training methods and their hyperparameters for networks used in this paper.

Training method	Hyperparameter
Unit type	Rectified Linear Unit (ReLU)
Learning rate in pre-training (unsupervised)	AdaDelta: $\epsilon = 1e-10$
Learning rate in fine-tuning (supervised)	AdaDelta: $\epsilon = 1e-8$, $\rho = 0.95$
Error function (supervised)	Sensitivity Ratio [Refer to Eq. (1)]
Mini-batch size	20

Once weights for the network were initialized, the model was fine-tuned via supervised training. Rather than using the typical sum of squared differences as an error function, our method used a combination of sensitivity and specificity, shown in Eq. (1). Here $S(\boldsymbol{p})$ represents the binary value of the ground truth segmentation mask at point $\boldsymbol{p}$, $y^0(\boldsymbol{p})$ is the binary value of the proposed segmentation mask at point $\boldsymbol{p}$, and r is a tunable value representing the ratio between sensitivity and specificity.

$$E = r\frac{\sum_p (S(\boldsymbol{p}) - y^0(\boldsymbol{p}))^2 S(\boldsymbol{p})}{\sum_p S(\boldsymbol{p})} + (1-r)\frac{\sum_p (S(\boldsymbol{p}) - y^0(\boldsymbol{p}))^2 (1 - S(\boldsymbol{p}))}{\sum_p (1 - S(\boldsymbol{p}))} \quad (1)$$

This error function is more robust to unbalanced classes where the error measure is dominated by the majority class, and has been demonstrated to be effective in applications such as MS lesion segmentation in brain MRI [2]. GM voxels only make up about 2% of the total volume in cropped spinal cord MRIs. As such, it is also beneficial to use this more robust error function in our application. For all networks in our experiments, we set the value of r to 0.1. Our chosen gradient descent method is AdaDelta, which uses an adaptive learning rate [12]. We chose AdaDelta over comparable alternatives such as RMSprop, and Adam because recent work using a similar network configuration found it to be more robust to hyperparameter selection [2]. Different values for the hyperparameters, in this case the learning rate ϵ and decay rate ρ, were explored and optimized using a validation set to prevent overfitting. For supervised training, hyperparameters for each configuration were determined by using a hold-out validation set of 32 image-label pairs. Several models, each with different hyperparameters, were trained for 100 epochs using the remaining 128 image-label pairs. The hyperparameter value with the lowest training error was selected and training was restarted with the full training set of 160 image-label pairs. Hyperparameter values are summarized in Table 1. Training took approximately 3 h for 1000 epochs, using a single GeForce GTX 780 graphics card.

As part of our training procedure, a version of the model was saved periodically. This allowed us to test the performance after different numbers of epochs. We generated segmentations for both the training and test sets every 50 epochs, and calculated the DSC. The mean values for both the training and test data are shown in Fig. 2. It is clear that rapid improvement in DSC occurs in the first 200 epochs for both training and testing. Performance on test data appears

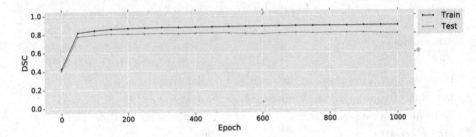

Fig. 2. Mean DSC computed for GM segmentations of axially oriented MRIs of the spinal cord, generated using a CEN. Performance on both the training and test set images was calculated as the number of training epochs increased.

to plateau at approximately 0.80 after 400 epochs with the exception of some minor fluctuations. Based on the similar performance for both training and test data, no overfitting is evident and suitable stopping criteria were used.

2.3 Segmentation

Once training was completed, our method was applied to new images to generate segmentation masks. GM segmentation was accomplished using two trained networks - one trained on spinal cord segmentations and the other trained on GM segmentations in images. First, the input images were cropped to an in-plane size of 256×256 pixels, from which the entire cord cross-section was identified using our CEN architecture. Alternative methods for cord segmentation, such as that available in the Spinal Cord Toolbox [4], could have been used, but using the same architecture was more convenient in our case. A centered version of the original scan was created by centering the mask and applying the same translation to the original image. Then each slice in the entire scan volume was cropped to a size of 100×100 voxels. This centered version of the cord ROI was used as an input for the second trained CEN, which generated a probabilistic mask of the GM. The probabilistic mask was converted to a binary mask using an optimized threshold value. This value was determined by calculating the mean error across all probabilistic training data masks at different threshold values. The threshold value that yielded the lowest mean error on the training set was

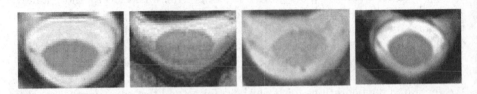

Fig. 3. Four sample MRIs of the spinal cord overlayed with segmentations generated using our method. One image from each scan site is included. GM segmentation is shown in red. (Colour figure online)

applied to the probabilistic masks for new scans. Finally, the binary mask was transformed back into the native space of the original image. A new MR image of the spinal cord can be segmented in less than one second with a trained model. Figure 3 shows four sample MRIs with GM segmentations overlaid.

3 Experiments and Results

For the purpose of our experiments, we used a publicly available dataset that was part of a GM segmentation challenge associated with the 3rd Annual Spinal Cord MRI workshop [8]. The dataset was divided into a training set and a test set. The training dataset consisted of 40 T2* MRIs acquired from 4 different scanners (10 scans from each). Each scan had 4 manual spinal cord and GM segmentations completed by 4 independent raters. Thus, a total of 160 image-label pairs were available to be used for training. The test dataset also contained 40 T2* scans from the same 4 scanners as the training set. These scans, however, did not have any publicly available manual segmentations. Resolution varied by site. A more detailed description of acquisition pararameters and demographic data can be found in [8]. Proposed segmentations were evaluated independently using an automatic tool available at http://cmictig.cs.ucl.ac.uk/niftyweb/.

Experiment I. Comparison of network architectures.
To optimize performance, several CENs were trained with variations in their network architecture. The configuration that achieved the best results in used 11 layers, 9×9 filters, and shortcut connections as described in our Methods section. We evaluated four architectural variations: (1) no shortcuts, (2) 7 network layers as previously used for brain lesion segmentation, (3) 5×5 filters, and (4) 13×13 filters. All configurations were evaluated using the online tool developed

Table 2. Mean (standard deviation) values from the online evaluation tool provided as part of a GM segmentation challenge for five variations of the CEN architecture. The best result for each metric is shown in bold. Performance metrics include Dice coefficient (DSC), mean surface distance (MSD), Hausdorff distance (HD), skeletonized Hausdorff distance (SHD), skeletonized median distance (SMD), true positive rate (TPR), true negative rate (TNR), positive predictive value (PPV), Jaccard index (JI), and conformity coefficient (CC). The symbol * denotes a statistically significant difference (paired t-test with $p < 0.05$) when compared with the highest value.

	DSC	MSD	HD	SHD	SMD	TPR	TNR	PPV	JI	CC
CEN with shortcuts 11-layer, 9×9 filter	**0.80** (0.06)	**0.53** (0.57)	**3.69** (3.93)	**1.22** (0.51)	0.44 (0.19)	**79.65** (9.56)	99.97 (0.04)	**81.29** (5.30)	**0.67** (0.07)	**48.79** (18.09)
No shortcuts	0.60* (0.10)	0.96* (1.03)	6.57* (5.42)	1.75* (0.90)	0.57* (0.37)	57.24* (12.05)	99.96* (0.05)	64.74* (8.90)	0.44* (0.10)	−43.52* (76.23)
7-layer	0.77* (0.07)	0.66 (0.80)	11.90* (8.76)	1.88* (0.93)	0.52* (0.25)	77.08* (10.38)	**99.97** (0.04)	76.89* (6.99)	0.63* (0.09)	35.69* (33.45)
5×5 filter	0.76* (0.10)	0.69* (0.74)	7.51* (7.08)	1.49* (0.80)	0.48 (0.26)	78.41 (14.96)	99.97 (0.04)	75.64* (6.82)	0.62* (0.11)	27.40* (67.82)
13×13 filter	0.76* (0.06)	0.60 (0.59)	3.86 (4.54)	1.25 (0.53)	**0.42** (0.19)	74.69* (8.47)	99.97 (0.04)	77.14* (6.08)	0.61* (0.08)	33.53* (22.61)

for a GM segmentation challenge [8]. Segmentations for the test set were uploaded to the tool, which returned the mean and standard deviation for 10 different performance metrics. A t-test was used to identify statistically significant differences, under the assumption that the data is normally distributed, because the online evaluation tool does not provide the detailed results needed for performing a non-parametric test. Table 2 shows an overview of these performance metrics for each configuration. The largest reduction in performance occurred when the shortcut connections were removed, resulting in the lowest Dice coefficient (DSC) out of all variants (0.80 to 0.60). Changes made to the convolutional filter size (increasing and decreasing) and decreasing the number of layers resulted in relatively small reductions in performance across all metrics with the exception of the skeletonized median distance (SMD), which had a small improvement (0.44 to 0.42) when using the 13×13 filter.

Experiment II. Comparison with current methods.

We compared the performance of our method against the publicly available results of the ISMRM Grey Matter segmentation challenge [8]. Table 3 shows an overview of these performance metrics. Other methods listed in the table and their associated results were all from the original segmentation challenge. While several methods performed well, two methods, our CEN and JCSCS, stood out by leading in several metrics. Our proposed method was competitive across all metrics, and even outperformed competing methods in key segmentation metrics like Dice coefficient (DSC), precision (PPV), and Jaccard index (JI).

Table 3. Mean (standard deviation) values from the online evaluation tool provided as part of a GM segmentation challenge for segmentations generated using our method, and the methods described in the challenge paper [8]. The best result for each metric is shown in bold. Titles shown in this table use the same abbreviations as Table 2. The full CEN with shortcuts approach is highly competitive with other state-of-the-art methods.

	DSC	MSD	HD	SHD	SMD	TPR	TNR	PPV	JI	CC
CEN with shortcuts 11-layer, 9×9 filter	**0.80** (**0.06**)	0.527 (0.572)	3.69 (3.93)	1.22 (0.51)	0.44 (0.19)	79.65 (9.56)	99.97 (0.04)	**81.29** (**5.30**)	**0.67** (**0.07**)	**48.79** (**18.09**)
JCSCS	0.79 (0.04)	**0.39** (**0.44**)	**2.65** (**3.40**)	**1.00** (**0.35**)	**0.37** (**0.18**)	77.89 (4.88)	**99.98** (**0.03**)	81.06 (5.97)	0.66 (0.05)	47.17 (11.87)
MGAC	0.75 (0.07)	0.70 (0.79)	3.56 (1.34)	1.07 (0.37)	0.39 (0.17)	**87.51** (**6.65**)	99.94 (0.08)	65.60 (9.01)	0.60 (0.08)	29.36 (29.53)
GSBME	0.76 (0.06)	0.62 (0.64)	4.92 (3.30)	1.86 (0.85)	0.61 (0.35)	75.69 (8.08)	99.97 (0.05)	76.26 (7.41)	0.61 (0.08)	33.69 (24.23)
SCT	0.69 (0.07)	0.69 (0.76)	3.26 (1.35)	1.12 (0.41)	0.39 (0.16)	70.29 (6.76)	99.95 (0.06)	67.87 (8.62)	0.53 (0.08)	6.46 (30.59)
VBEM	0.61 (0.13)	1.04 (1.14)	5.34 (15.35)	2.77 (8.10)	0.54 (0.25)	65.66 (14.39)	99.93 (0.09)	59.07 (13.69)	0.45 (0.13)	−44.25 (90.61)

4 Conclusions

We described in detail a novel, fully automatic, deep learning method for GM segmentation in MRIs of the spinal cord. Variations in network architecture were investigated, and we found that the best performance was achieved using an 11-Layer CEN with shortcut connections and 9×9 filters. Segmentations generated using our method were compared with publicly available results from the 2016 ISMRM Grey Matter Segmentation challenge. Our proposed network architecture outperformed other current methods in key segmentation metrics such as Dice coefficient (DSC) and positive predictive value (PPV). In the future, we aim to incorporate vertebral level information as well as investigate variations in the loss function. In addition, we plan to make our GM segmentation algorithm publicly available as part of the Spinal Cord Toolbox [4].

References

1. Bakshi, R., et al.: Measurement of brain and spinal cord atrophy by magnetic resonance imaging as a tool to monitor multiple sclerosis. J. Neuroimag. **15**(3) (2004). Offcial journal of the American Society of Neuroimaging
2. Brosch, T., et al.: Deep 3D convolutional encoder networks with shortcuts for multiscale feature integration applied to multiple sclerosis lesion segmentation. IEEE Trans. Med. Imaging **35**(5), 1229–1239 (2016)
3. Brosch, T., et al.: Convolutional encoder networks for multiple sclerosis lesion segmentation. In: Lecture Notes in Computer Science (including subseries Lecture Notes in Artificial Intelligence and Lecture Notes in Bioinformatics), vol. 9351, pp. 3–11 (2015). ISBN 9783319245737
4. De Leener, B., et al.: SCT: Spinal Cord Toolbox, an open-source software for processing spinal cord MRI data. NeuroImage **145** (2017)
5. Drozdzal, M., et al.: The Importance of skip connections in biomedical image segmentation. CoRR abs/1608.04117 (2016)
6. Lee, H., Ng, A.: Convolutional deep belief networks for scalable unsupervised learning of hierarchical representations (2009)
7. Lukas, C., et al.: Relevance of spinal cord abnormalities to clinical Disability in Multiple sclerosis: MR Imaging Findings in a Large Cohort of Patients. Radiology **269** (2013)
8. Prados, F., et al.: Spinal cord grey matter segmentation challenge. NeuroImage **152** (2017)
9. Ronneberger, O., Fischer, P., Brox, T.: U-Net: convolutional networks for biomedical image segmentation. In: Navab, N., Hornegger, J., Wells, W.M., Frangi, A.F. (eds.) MICCAI 2015. LNCS, vol. 9351, pp. 234–241. Springer, Cham (2015). doi:10.1007/978-3-319-24574-4_28
10. Schlaeger, R., et al.: Spinal cord gray matter atrophy correlates with multiple sclerosis disability. Ann. Neurol. **76**(4) (2014)
11. Tajbakhsh, N., et al.: Convolutional neural networks for medical image analysis: full training or fine tuning? IEEE Trans. Med. Imaging **35**(5), 1299–1312 (2016). ISSN 1558254X
12. Zeiler, M.: ADADELTA: an adaptive learning rate method. arXiv arXiv:1212.5701 (2012)
13. Zeiler, M., Taylor, G., Fergus, R.: Adaptive deconvolutional networks for mid and high level feature learning. In: Proceedings of the IEEE International Conference on Computer Vision (2011). ISSN 1550–5499

7th International Workshop on Multimodal Learning for Clinical Decision Support, ML-CDS 2017

Mapping Multi-Modal Routine Imaging Data to a Single Reference via Multiple Templates

Johannes Hofmanninger[1]([✉]), Bjoern Menze[2], Marc-André Weber[3,4], and Georg Langs[1]

[1] Department of Biomedical Imaging and Image-guided Therapy Computational Imaging Research Lab, Medical University of Vienna, Vienna, Austria
johannes.hofmanninger@meduniwien.ac.at

[2] Department of Computer Science & Institute for Advanced Study, Technical University of Munich, Munich, Germany

[3] Department of Diagnostic and Interventional Radiology, University of Heidelberg, Heidelberg, Germany

[4] Institute of Diagnostic and Interventional Radiology, Rostock University Medical Center, Rostock, Germany
https://www.cir.meduniwien.ac.at

Abstract. Population level analysis of medical imaging data relies on finding spatial correspondence across individuals as a basis for local comparison of visual characteristics. Here, we describe and evaluate a framework to normalize routine images covering different parts of the human body, in different modalities to a common reference space. The framework performs two basic steps towards normalization: (1) The identification of the location and coverage of the human body by an image and (2) a non-linear mapping to the common reference space. Based on these mappings, either coordinates, or label-masks can be transferred across a population of images. We evaluate the framework on a set of routine CT and MR scans exhibiting large variability on location and coverage. A set of manually annotated landmarks is used to assess the accuracy and stability of the approach. We report distinct improvement in stability and registration accuracy compared to a classical single-atlas approach.

1 Introduction

Analysis of medical routine imaging data is highly relevant, since they provide a realistic sample of the clinical population and are a key to modeling the natural variability of disease progression and treatment response. In order to compare local characteristics across the population, we need to establish spatial correspondence. Unlike in brain imaging studies, routine imaging is not harmonized

---

G. Langs—This research was supported by teamplay which is a Digital Health Service of Siemens Healthineers, by the Austrian Science Fund FWF (I2714-B31), by the WWTF (S14-069), and by the DFG (WE 2709/3-1, ME 3511/3-1).

M.J. Cardoso et al. (Eds.): DLMIA/ML-CDS 2017, LNCS 10553, pp. 341–348, 2017.
DOI: 10.1007/978-3-319-67558-9_39

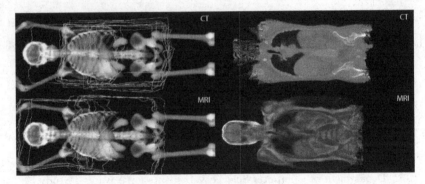

Fig. 1. Left: Visualization of the coverage of the images in the reference space and their center position. Right: Mean volume, generated by mapping intensities of 49 CT and 28 MR images to the reference space. High contrast of organ and bone borders indicate accurate registration.

by protocols, but applied and guided by indication and case specific needs. Here, we describe and evaluate a robust framework for multi-modal registration of routine imaging data. Many medical imaging applications and research studies rely on the alignment of volumetric images to a standard frame of reference. For example, structural and functional analysis of brain MR images [3] or atlas based labeling of anatomical structures [8]. However, medical routine images exhibit a wide range of inter-subject variability such as age, sex, disease and health status but also varying image characteristics such as quality and field of view not controlled by any study protocol. Due to these properties, spatial alignment of heterogeneous multi-subject datasets poses a challenging task.

Related Work. True correspondences between images are unknown and have to be inferred by matching boundaries of anatomical structures or visual landmarks [1]. Existing methods, producing high quality alignments, have been proposed for specific organs or organ parts. Especially the normalization of brain MR images is challenging, yet highly relevant. Advanced methods, that are capable to cope with structural variability in the brain have been proposed and vary from geodesic registration using anatomical manifold learning [6], feature based approaches [10] to multi-template alignments [13]. Studies of other organs typically aim for segmentation of distinct organ or organ-parts. Label fusion of *multi-atlas segmentations* is a popular approach that yields robust and accurate results using a set of multiple manually annotated atlases [8,9,12]. Multi-modal alignment of images is mostly relevant for intra-subject registration due to its role for image-guided interventions. Hence, work on inter-subject multi-modal registration is scarce. Mutual information (MI) [11] is often used as similarity function to align multi-modal images. Heinrich et al. propose point-wise matching of local descriptors [7] for matching of 3D ultrasound and MRI brain scans, but also for registration of inter-subject full body MR and CT volumes to generate pseudo-CT scans [2].

Contribution. In this work, we address the problem of aligning truncated inter-subject multi-modal images with widely different fields of view to a common whole body reference space (Fig. 1). The contribution of this study is threefold (1) we propose a novel framework, capable of reliably aligning routine images by (2) adapting the idea of multi-atlas segmentation towards multi-template localization and normalization and (3) evaluating the accuracy and effectiveness of the approach on a heterogeneous set of real world routine CT and MR images.

2   Method

The normalization framework is based on an offline and online phase. The offline phase is performed in advance and requires carefully supervised registration of multiple pre-defined template images to a common reference space (Sect. 2.2). During the online phase, no manual interaction with the data is required. Novel images, for which no landmarks are available, are registered to the templates (Sect. 2.1) and the best matching template is selected (Sect. 2.2). An overview of the multi-modal multi-template framework is given in Fig. 2. The framework allows for normalization of routine images $\mathcal{I} = \{\mathbf{I}_1, \ldots, \mathbf{I}_N\}$ covering arbitrary regions of the human body and being of different modalities $\mathcal{M} = \{1, \ldots, M\}$ so that $\mu(\mathbf{I}) \in \mathcal{M}$. The framework requires a set of templates $\mathcal{T} = \{\mathbf{T}_1, \ldots, \mathbf{T}_K\}$ where each modality is represented and $\mu(\mathbf{T}) \in \mathcal{M}$ to facilitate an unbiased normalization of a heterogeneous corpus of images. The templates are chosen in a way to cover natural variation such as size and sex. For each modality,

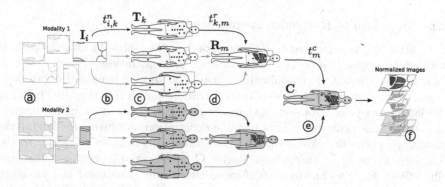

Fig. 2. Multi-modal multi-template normalization (a) The framework facilitates the processing of routine imaging data of different modalities and exhibiting different coverage of the human body. (b) During normalization, an image is aligned to multiple templates (c) of the same modality as the image. All templates are aligned with a modality specific atlas (d) supported by manually annotated landmarks. (e) Atlases of different modalities are carefully aligned to a central reference space using landmarks, body, bone and organmasks. Positions in an image are mapped to the reference atlas by concatenation of the three transformations ($t_{i,k}^n$, $t_{k,m}^r$ and t_m^c) that yield maximal registration quality. (f) After normalization, coordinates and label masks are mapped across the population.

a distinct atlas $\mathbf{R}^m$ is required as well as a central reference space $\mathbf{C}$. Each template is aligned with its modality specific atlas whereas each atlas is aligned with the central reference space. The alignment between template and atlas is unimodal but supported by some supervision to overcome a potential registration bias. The alignment of the atlases to the reference space are multi-modal and thus require an even higher degree of supervision.

2.1   Fragment to Template Registration

Registration of an image to a template is performed in two steps. (1) Estimation of the coverage and the position of the image in the human body and (2) a non-rigid transformation to the area estimated at step one. The estimation of position and coverage is performed by matching 3D scale invariant features (3DSIFT) according to [10] and performing affine registration by minimizing the Mean Squared Error (MSE) between the feature locations. Subsequently, a refinement step, by conducting an intensity based affine registration with high regularization on translation, optimizing the Normalized Cross Correlation (NCC) is performed. If no matching 3DSIFT features are identified, an intensity based affine registration with no regularization on translation is performed. After estimation of position and coverage, an intensity-based non-rigid registration is conducted [5]. In the following we define $t_{i,k}^a$ as the affine transformation between image $\mathbf{I}_i$ and template $\mathbf{T}_k$ so that $\mathbf{I}_i \approx \hat{\mathbf{I}}_{i,k}^a = t_{i,k}^a(\mathbf{T}_k)$. Further, we define $t_{i,k}^n$ as the concatenation of the affine and non-rigid transformation so that $\mathbf{I}_i \approx \hat{\mathbf{I}}_{i,k}^n = t_{i,k}^n(\mathbf{T}_k)$.

2.2   Template to Reference Registration

Each atlas $\mathbf{R}_m$ is registered to its corresponding templates $\mathbf{T}_k$ (if $m = \mu(\mathbf{T}_k)$) with a high degree of supervision. In our case, the registrations are supported by 58 manually annotated anatomical landmarks on specific bone and organ positions. We perform affine and b-spline registrations optimizing the MSE between the landmarks prior to a non-rigid image registration based on image intensities. As a refinement step a final b-spline registration optimizing for landmarks distance is performed. We define the transformations from an atlas $\mathbf{R}_m$ to the template $\mathbf{T}_k$ as $t_{k,m}^r$. The reference atlas $\mathbf{C}$ is registered to the modality specific atlases $\mathbf{R}_m$ and has to be performed either fully supervised and modality independent or multi-modal. We perform registrations using 58 landmarks, segmentations covering 20 organs and organ parts, a body mask and segmentations of all skeletal bones so that the registrations can be performed independently of the modality. We define the transformation from the reference space to an atlas $\mathbf{R}_m$ as t_m^c.

2.3   Template Selection

Template selection for an image $\mathbf{I}_i$ is performed in two steps: (1) calculation of $\hat{\mathbf{I}}_{i,k}^a$ for every k where $\mu(\mathbf{T}_k) = \mu(\mathbf{I}_i)$ and selection of the top C templates that

Table 1. Mean, median, standard deviation and number of misaligned (misalig.) landmarks. Results are given in mm of landmark distances to the ground truth for the template registration (T) and direct registration (D).

	CT		MR		ALL	
	T	D	T	D	T	D
Mean	**17.95**	32.42	**22.57**	62.91	**19.35**	41.64
Median	**13.39**	17.60	**17.13**	36.15	**14.13**	21.72
Std	**16.53**	63.65	**18.03**	63.65	**17.12**	63.65
misalig.	**3**	25	**0**	36	**3**	61
#cases	49		28		77	
#landm.	476		206		682	

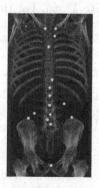

Fig. 3. Landmarks used to evaluate registration accuracy.

yield the highest $NCC(\mathbf{I}_i, \hat{\mathbf{I}}_{i,k}^a)$ so that $\mathcal{C}_i \subseteq \{1, \dots, K\}$ is the set of candidate templates for image $\mathbf{I}_i$. Subsequently, the non-rigid transformations between the image and templates $t_{i,k}^n$ are computed for $k \in \mathcal{C}$ only. In the second step (2) the transformations $t_{i,k}^n$ and $t_{k,m}^r$ are concatenated so that

$$\mathbf{I}_i \approx \hat{\mathbf{I}}_{i,k} = t_{i,k}^n(t_{k,m}^r(\mathbf{R_m})) \tag{1}$$

and we can select the template by

$$\underset{k}{\text{maximize}} NCC(\mathbf{I}_i, \hat{\mathbf{I}}_{i,k}) \qquad \text{subject to } \mu(\mathbf{I}_i) = \mu(\mathbf{T}_k), k \in \mathcal{C}_i \tag{2}$$

Given a template x_i that yields maximum NCC the final transformation from the reference space to an image is

$$t_i = t_{i,x_i}^n \circ t_{x_i, \mu(i)}^r \circ t_{\mu(i)}^c \tag{3}$$

3 Experiments

Test Data. We perform evaluation on a heterogeneous set of CT (#48) and MR-T1 (#28) images recorded in the daily clinical routine. The images cover different regions of the human body and vary in resolution, dose (CT) and sectioning (saggital and axial). We annotated 16 landmarks on specific bone and organ positions (aortic arch, trachea biforcation, cristia iliaca left, crista iliaca right, symphysis, aorta bifurcation, L5, L4, L3, L2, L1, xyphoideus, sternoclavicular left, sternoclavicular right, renalpelvis left and renalpelvis right) covering the chest and abdomen (Fig. 3). Depending on coverage, each image may only exhibit a subset of these landmarks. More formally, the evaluation set consists of tuples $\langle \mathbf{I}_i, \mathcal{V}_i \rangle$ where $\mathcal{V}_i \subseteq \{1, \dots, 16\}$.

Template Data. For the template set and the atlases we use 22 (11 CT and 11 MR) whole body volumes of the VISCERAL Anatomy 3 dataset [4], which provides manually annotated organ masks and landmarks. For each modality 10 volumes are used as templates and one is used as modality specific atlas. We defined the CT atlas to also represent the reference space.

Evaluation and Metric. For evaluation we assess robustness and accuracy of the approach and study the influence of number of templates used. We compare the proposed template framework to a direct registration approach where each image is registered to the modality specific atlases $\mathbf{R}_m$ directly. To allow for a fair comparison, the direct registration is performed according to the *image to template* registration method described in Sect. 2.1. To assess registration quality, we transform the landmark positions from the image space into the reference space. We then calculate the distance between the transformed landmarks and the corresponding landmarks in the reference. We report the registration accuracy for each landmark and compare the performance between MR and CT cases. To assess robustness we define landmarks that are off by more than 100 mm as misaligned.

Parameters and pre-processing. We perform bias-field correction [14] on the MR images and rescale all volumes (images and templates) to an isotropic voxel resolution of 2 mm prior to processing. If not stated differently, results reported are produced with parameters K (Number of templates) set to 10 and C (max. number of non-rigid registrations) set to 4.

4   Results

Figure 4 gives a comparison of registration errors to the ground truth when using the template approach compared to direct registration. Results are given for the full dataset(CT+MR) and each landmark. Distinct improvement in accuracy

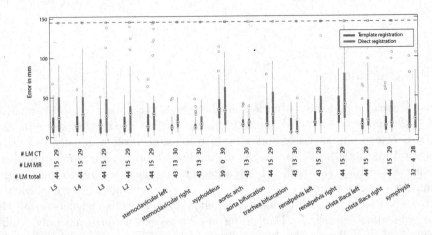

Fig. 4. Comparison of the template approach to direct registration for each landmark and the full dataset (CT + MR). Registration errors are given in mm distance to the ground truth landmarks.

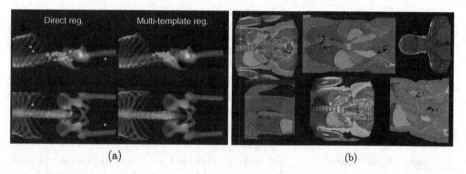

(a)   (b)

Fig. 5. (a) Mappings of the L5 vertebrae landmarks to the reference space comparing the direct registration to the robust template approach. (b) Organ labels mapped from the reference space to the images.

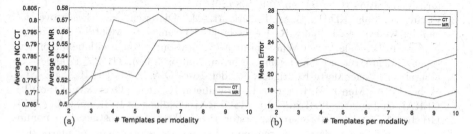

Fig. 6. Effect of varying number of templates on final Normalized Cross Correlation (NCC) (a) and registration error (b).

(lower median) but also stability (less outliers) can be seen. Table 1 shows averaged registration errors for the two modalities. For CT, median distance improves from 17.6 to 13.4 mm and for MR from 36.2 to 17.1 mm. The number of misaligned landmarks drops from 25/476 to 3 for CT and from 28/206 to 0 for MR. This indicates that especially MR benefits from the multiple templates in terms of stability. However, the 3 misaligned landmarks of the CT cases (xyphoideus and symphysis) are close to the image border and are therefore rather sensible for misalignment. Figure 5a illustrates an exemplary mapping of all landmarks on vertebra L5 to the reference space. Figure 5b shows labellings of organs mapped from the reference to the images. Figure 6b shows the effect of a varying number of templates on the mean landmark error and Fig. 6a shows the effect of a varying number of templates on the final NCC values (Eq. 2). Figure 1 illustrates localization, coverage and averaged intensity values of CT and MR images that were mapped from image to reference space.

5 Conclusion

This paper addresses the challenging task of mapping clinical routine images of different modalities to a common reference space. A multi-template approach for localization and deformable image registration is presented. Evaluation has been

performed on a representative dataset of CT and MR images using manually annotated landmarks. The results show a distinct improvement in registration accuracy
and stability compared to direct registration to an atlas. We believe, that spatial
normalization of routine images provides a useful tool to especially study systematic diseases such as multiple myeloma, metastatic cancer and others that exhibit
visual traits throughout the human body in different imaging modalities.

References

1. Crum, W.R., Griffin, L.D., Hill, D.L.G., Hawkes, D.J.: Zen and the art of medical image registration: correspondence, homology, and quality. NeuroImage **20**(3),
   1425–1437 (2003)
2. Degen, J., Heinrich, M.P.: Multi-atlas based pseudo-CT synthesis using multimodal
   image registration and local atlas fusion strategies. In: Computer Vision and Pattern Recognition (CVPR), pp. 160–168 (2016)
3. Fischl, B., Salat, D.H., Busa, E., Albert, M., Dieterich, M., Haselgrove, C.,
   van der Kouwe, A., Killiany, R., Kennedy, D., Klaveness, S., Montillo, A., Makris,
   N., Rosen, B., Dale, A.M.: Whole brain segmentation: automated labeling of neuroanatomical structures in the human brain. Neuron **33**(3), 341–355 (2002)
4. Goksel, O., Foncubierta-Rodriguez, A., del Toro, O.A.J., Müller, H., Langs, G.,
   Weber, M.A., Menze, B.H., Eggel, I., Gruenberg, K., et al.: Overview of the VIS-
   CERAL challenge at ISBI 2015. In: VISCERAL Challenge@ ISBI, pp. 6–11 (2015)
5. Gruslys, A., Acosta-Cabronero, J., Nestor, P.J.: Others: a new fast accurate nonlinear medical image registration program including surface preserving regularization.
   IEEE Trans. Med. Imaging **33**(11), 2118–2127 (2014)
6. Hamm, J., Ye, D.H., Verma, R., Davatzikos, C.: GRAM: a framework for geodesic
   registration on anatomical manifolds. Med. Image Anal. **14**(5), 633–642 (2010)
7. Heinrich, M.P., Jenkinson, M., Papież, B.W., Brady, S.M., Schnabel, J.A.: Towards
   realtime multimodal fusion for image-guided interventions using self-similarities.
   In: Mori, K., Sakuma, I., Sato, Y., Barillot, C., Navab, N. (eds.) MICCAI
   2013. LNCS, vol. 8149, pp. 187–194. Springer, Heidelberg (2013). doi:10.1007/
   978-3-642-40811-3_24
8. Iglesias, J.E., Sabuncu, M.R.: Multi-atlas segmentation of biomedical images: a
   survey. Med. Image Anal. **24**(1), 205–19 (2015)
9. Koch, L.M., Rajchl, M., Bai, W., Baumgartner, C.F., Tong, T., Passerat-Palmbach,
   J., et al.: Multi-Atlas Segmentation using Partially Annotated Data: Methods and
   Annotation Strategies, pp. 1–17. arXiv preprint, arxiv:1605.00029 (2016)
10. Toews, M., Wells, W.M.: Efficient and robust model-to-image alignment using 3D
    scale-invariant features. Med. Image Anal. **17**(3), 271–282 (2013)
11. Viola, P., Wells Iii, W.M.: Alignment by maximization of mutual information. Int.
    J. Comput. Vis. **9**(242), 22–137 (1997)
12. Wolz, R., Chu, C., Misawa, K., Fujiwara, M., Mori, K., Rueckert, D.: Automated
    abdominal multi-organ segmentation with subject-specific atlas generation. IEEE
    Trans. Med. Imaging **32**(9), 1723–1730 (2013)
13. Xie, L., Pluta, J.B., Das, S.R., Wisse, L.E., Wang, H., Mancuso, L., Kliot, D.,
    Avants, B.B., Ding, S.L., Manjón, J.V., Wolk, D.A., Yushkevich, P.A.: Multi-
    template analysis of human perirhinal cortex in brain MRI: explicitly accounting
    for anatomical variability. NeuroImage **144**, 183–202 (2017)
14. Zhang, Y., Brady, M., Smith, S.: Segmentation of brain MR images through a
    hidden Markov random field model and the expectation-maximization algorithm.
    IEEE Trans. Med. Imaging **20**(1), 45–57 (2001)

Automated Detection of Epileptogenic Cortical Malformations Using Multimodal MRI

Ravnoor S. Gill[(⊠)], Seok-Jun Hong, Fatemeh Fadaie,
Benoit Caldairou, Boris Bernhardt, Neda Bernasconi,
and Andrea Bernasconi

Neuroimaging of Epilepsy Laboratory, Montreal Neurological Institute
and Hospital, McGill University, Montreal, QC, Canada
ravnoor.gill@mail.mcgill.ca

Abstract. Focal cortical dysplasia (FCD), a malformation of cortical development, is a frequent cause of drug-resistant epilepsy. This surgically-amenable lesion is histologically characterized by cortical dyslamination, dysmorphic neurons, and balloon cells, which may extend into the immediate subcortical white matter. On MRI, FCD is typically associated with cortical thickening, blurring of the cortical boundary, and intensity anomalies. Notably, even histologically-verified FCD may not be clearly visible on preoperative MRI. We propose a novel FCD detection algorithm, which aggregates surface-based descriptors of morphology and intensity derived from T1-weighted (T1w) MRI, T2-weighted fluid attenuation inversion recovery (FLAIR) MRI, and FLAIR/T1w ratio images. Features were systematically sampled at multiple intracortical/subcortical levels and fed into a two-stage classifier for automated lesion detection based on ensemble learning. Using 5-fold cross-validation, we evaluated the approach in 41 patients with histologically-verified FCD and 38 age-and sex-matched healthy controls. Our approach showed excellent sensitivity (83%, 34/41 lesions detected) and specificity (92%, no findings in 35/38 controls), suggesting benefits for presurgical diagnostics.

Keywords: Magnetic resonance imaging · Epilepsy · Lesion detection · Multiparametric MRI · Presurgical diagnostics

1 Introduction

Focal cortical dysplasia (FCD), a malformation of cortical development, is a prevalent cause of drug-resistant epilepsy. Its surgical removal is currently the only treatment option to arrest seizures. Cardinal histopathological features of FCD include cortical dyslamination associated with various intra-cortical cytological anomalies, namely dysmorphic neurons (FCD Type-IIA) and balloon cells (FCD Type-IIB) [1].

On MRI, FCD is typically associated with varying degrees of cortical thickening, blurring of the interface between the grey and white matter and anomalous intensity profiles. Notably, even histologically-verified FCD may not be clearly visible on MRI [2]. Previous studies using voxel- [3, 4] or surface-based methods [5–7] modeled limited numbers of features derived from T1 or T2-weighted MRI, except for one [8]

© Springer International Publishing AG 2017
M.J. Cardoso et al. (Eds.): DLMIA/ML-CDS 2017, LNCS 10553, pp. 349–356, 2017.
DOI: 10.1007/978-3-319-67558-9_40

combining them in a small pediatric cohort. All these methods, however, were mainly applied to large- to medium-sized lesions visible on MRI and provided limited sensitivity [9]. Moreover, histological validation was present in 30% of cases only.

The current work proposes a novel *in vivo* surface-based automated detection algorithm modeling FCD at various depths within the cortex and the subcortical white matter. Our method exploits the diagnostic power of T1-weighted and T2-weighted FLAIR contrasts together with a synthetic FLAIR/T1 ratio map; the latter was specifically designed to increase the sensitivity for co-occuring FLAIR hyperintensity and T1w hypointensity present at the interface between the grey and white matter. Our framework was validated in a cohort of 41 patients with histopathologically proven FCD.

2 Methods

2.1 MRI Acquisition

Multimodal MRI was acquired on a 3T Siemens TimTrio using a 32-channel head coil, including a 3D T1-weighted MPRAGE (T1w; TR = 2300 ms, TE = 2.98 ms, flip angle = 9°, FOV = 256 mm^2, voxel size = 1 × 1 × 1 mm^3) and fluid-attenuated inversion recovery (FLAIR; TR = 5000 ms, TE = 389 ms, TI = 1.8 ms, flip angle = 120°, FOV = 230 mm^2, voxel size = 0.9 × 0.9 × 0.9 mm^3).

2.2 Multi-contrast MRI Pre-processing

T1w MRI underwent intensity inhomogeneity correction [10] followed by intensity standardization, linear registration to MNI152 space, and classification into white matter (WM), gray matter (GM), and cerebrospinal fluid (CSF) [11]. GM-WM and GM-CSF surface models were reconstructed using CLASP, an algorithm relying on intensity and geometric constraints [12]. Surface-based registration that aligned individual subjects based on cortical folding was used to increase across-subjects correspondence in measurement locations [13]. T1w images were linearly co-registered to FLAIR. After similar pre-processing, FLAIR images were divided by T1w images to generate a FLAIR/T1w ratio map; this ratio allows for additional correction of B1 intensity non-uniformity after N3 correction. Hyperintensities exceeding 1 SD from the mean ratio within the brain mask were excluded, generating the ratio image.

2.3 Multi-surface Generation

To examine intracortical GM, we positioned 3 surfaces between the inner and outer cortical surfaces at 25%, 50%, and 75% cortical thickness, guided by a straight line providing vertex-correspondence across surfaces [14]. Although these surfaces do not necessarily reflect cortical laminae, they capture relative differences along the axis perpendicular to the cortical mantle. To assess the WM immediately beneath the cortex, we generated 3 equidistant surfaces guided by a Laplacian field running between the GM-WM interface and the ventricles, with between-surface intervals adapted to the resolution of each modality.

2.4   Feature Extraction

The following features were extracted in the native space of a given contrast to minimize interpolation.

Intensity-Based Features. For each modality, *i.e.*, T1w, FLAIR, and FLAIR/T1w, we divided voxel-wise intensities by the mean GM-WM boundary intensity; this value was normalized with respect to the mode of the respective intensity histogram [15]. Normalized intensities were linearly interpolated at each surface-point (or vertex) of intra- and subcortical surface models. We did not sample intensity on the GM-CSF surface to avoid CSF contamination; values for all other surfaces were analytically corrected for partial volume effects [16]. We also computed intensity gradients in perpendicular and tangential direction relative to cortical surfaces, to model radial and tangential dyslamination [7].

Morphology. Cortical thickness was calculated as the Euclidean distance between vertices on the GM-WM and GM-CSF surfaces [12]. Small FCD lesions often occur at the bottom of a sulcus and display curvature changes [5]. We thus computed sulcal depth for each vertex as the shortest geodesic distance from a gyral crown, and measured absolute mean curvature along the 50% intracortical surface [17].

Feature Profiling. We assigned at each vertex a unique vector of intra and subcortical intensity and morphological features smoothed using a surface-based 5 mm full-width-at-half-maximum Gaussian kernel and z-normalized with respect to the distribution in healthy controls. For each intensity feature, we calculated an average across the 4 intracortical (25%, 50%, 75%, GM-WM interface) and 3 subcortical surfaces.

For each individual, we thus obtained 3 morphological maps (cortical thickness, sulcal depth, curvature), 6 intensity maps (T1w, FLAIR, FLAIR/T1w at intracortical and subcortical levels), 6 corresponding gradient maps (3 tangential, 3 perpendicular) together with their asymmetries yielding a total of 30 features.

3   Experiment and Results

3.1   Subjects

Our patient cohort consisted of 41 patients (20 males; mean ± SD age = 27 ± 9 years) admitted to our Hospital for the investigation of drug-resistant focal epilepsy. The presurgical workup included neurologic examination, assessment of seizure history, MRI, and video-EEG telemetry. In 33 (80%) patients, lesions were initially not seen on conventional radiological inspection of pre-operative MRI. Since the MRI was initially reported as unremarkable, the location of the seizure focus was established using intracerebral EEG; retrospective inspection revealed a subtle FCD in the seizure onset region in all. All patients underwent surgery and the diagnosis FCD was histopathologically verified.

The control group consisted of 38 age- and sex-matched healthy individuals (19 males; mean ± SD: age = 30 ± 7 years).

3.2 Manual Lesion Segmentation

Two experts, blinded to clinical information, independently segmented FCD lesions on co-registered T1 and FLAIR MRI. Inter-rater Dice agreement index ($D = 2|M_1 \cap M_2|/ [|M_1| + |M_2|]$; M_1, M_2: 1st, 2nd label; $M_1 \cap M_2$: intersection of M1 and M2) was 0.91 ± 0.11. Their consensus volume label was intersected with the surface models, thereby generating a surface-based lesion label.

3.3 Classification Paradigm (Fig. 1)

We built a two-stage classifier. A first vertex-wise classification was designed to maximize sensitivity (*i.e.*, detecting a maximum number of lesional clusters), whereas a subsequent cluster-wise classification aimed at improving specificity (*i.e.*, removing false positives while maintaining optimal sensitivity).

We used an ensemble of RUSBoosted decision trees to systematically test detection performance across both classification stages. RUSBoost [18] is a hybrid sampling/ boosting algorithm, which can learn while mitigating the class imbalance problem that occurs from the presence of imbalance between high number of non-lesional vertices in a given subject compared to lesional vertices.

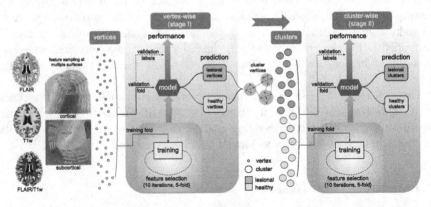

Fig. 1. Vertex-wise and cluster-wise classification schema. See text for details

Vertex-Level Classification. RUSBoost randomly undersampled from the pool of non-lesional vertices until a balanced distribution was achieved. *AdaBoost* [19], a common boosting algorithm, then iteratively built an ensemble of base learners (decision trees). During each iteration, higher penalty weights were assigned to misclassified vertices to improve classification accuracy for the next round. After the final iteration, all trained base learner models participate in a weighted vote to classify test vertices as lesional or non-lesional. On the resulting predictions, a cluster was defined as a collection of vertices that form 6-connected neighbors on the discrete cortical surface mesh. Lesional (true positives) and non-lesional (false positives) clusters were fed into subsequent cluster-level classification.

Cluster-Level Classification. For each cluster, we assessed the overall load of anomalies by computing the Mahalanobis distance (a multivariate z-transform between each patient's feature vector and the corresponding distribution in controls). For the set of vertices displaying the highest distance, we used the 15 original features (excluding their asymmetries) to compute the following higher-order features: statistical moments (mean, standard deviation, skewness, kurtosis, moment, and their asymmetries) representing the shape of the distribution of each feature, spatial location (as determined by anatomical parcellation and 3D coordinates, and lesion size. This process generated a total of 95 features per subject. Classification then proceeded using these cluster features in the same manner as vertex-wise classification.

Feature Selection. In both classification stages, we used an ensemble of extremely randomized trees in conjunction with RUSBoost to select features. This procedure introduced more randomness during feature selection and training, which has been shown to improve the bias/variance tradeoff and performance compared to conventional tree-based classifiers [20].

Partitioning of Training and Test Datasets. At each stage, classifiers were trained using 5-fold cross validation with 10 iterations: 4 folds of data were used for model training (200 RUSBoosted decision trees), and the remaining for testing (*i.e.*, lesion detection). For feature selection, nested 5-fold cross-validation was implemented. Optimal features were determined across 10 iterations and finally averaged. Selected features were used for classifier training and testing. This procedure permits a conservative assessment of performance and generalizability for previously-unseen FCD cases.

Evaluation of Classification Accuracy. Performance was assessed relative to manual labels. Sensitivity was the proportion of patients in whom a detected cluster co-localized with the lesion label. Specificity was determined with respect to controls (i.e., proportion of controls in whom no FCD lesion cluster was falsely identified). We also report the number of clusters detected in patients remote from the lesion label (*i.e.*, false positives).

3.4   Results

At the first vertex-wise stage, the classifier detected all but 4 lesions (37/41 = 90% sensitivity). However, it also detected false positives (mean $\pm$ SD clusters: 25 $\pm$ 23 in patients; 7 $\pm$ 5 in controls). Subsequent cluster-wise classification guaranteed a sensitivity of 83% (34/41 clusters co-localized with the label), while it dramatically reduced the number of false positives (mean $\pm$ SD clusters = 4 $\pm$ 5). We also obtained a high specificity of 92% with only a single cluster in 3 healthy subjects. The results for both stages are summarized in the Table 1. An example is shown in Fig. 2.

The highest normalized weights (*i.e.*, more than 10% of total weighting) were as follows: Vertex-wise classification was driven by perpendicular gradient in 69% of cases (derived from: FLAIR in 25%, T1w in 24%, FLAIR/T1w in 20%), followed by subcortical normalized intensity (derived from T1w in 31%; FLAIR in 16%, FLAIR/T1w in 15%). Cluster-wise classification was largely driven by cortical

Table 1. Summary of classification results.

Classifier	Sensitivity	FPs in patients (mean ± SD)	FPs in controls (mean ± SD)	Specificity (w.r.t controls)
Vertex-wise	90% (37/41)	25 ± 23	7 ± 5	N/A
Cluster-wise	83% (34/41)	4 ± 5	0.08 ± 0.27	92% (35/38)

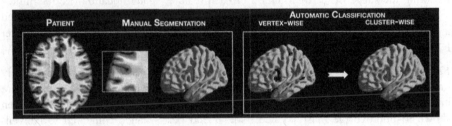

Fig. 2. Left: Axial T1w MRI showing the cortex harboring the FCD (dashed square), a close-up of the outline of the lesion label (solid green line), and its surface projection. Right: vertex-wise and subsequent cluster-wise automated classification results. (Color figure online)

thickness (34%), followed by FLAIR subcortical normalized intensity (33%), and lesion size (33%).

4 Discussion

The current study presents a novel machine intelligence system for the automated detection of FCD lesions. Our approach was developed and evaluated in a consecutive cohort of patients with histopathologically-validated lesions, with the majority of patients having lesions that were initially overlooked on standard radiological exam. Core to our approach was a surface-based integration of morphological markers as well as intensity and textural features derived from co-registered T1w and FLAIR data as well as their ratio. The latter was specifically chosen to enhance co-occuring FLAIR hyperintensity and T1w hypointensity occurring at the junction between the grey and white matter, an important FCD feature, so far not addressed [21].

To classify FCD lesions in a high-dimensional dataset, we chose a non-parametric boosted decision tree ensemble that can capture complex decision boundaries while avoiding overfitting. Circumventing class imbalance is another critical issue in lesion detection. Therefore, to reduce bias against minority class (i.e., lesional vertices), we implemented random-undersampling with adaptive boosting [18, 19], with robust 5-fold cross-validation; this, together with a comprehensive multi contrast modelling of FCD features, likely contributed to the highest performance to date compared to previous studies [3–8].

Operating on two sequential levels, our approach resulted in both high sensitivity and specificity. Notably, the number of false positive findings in controls was rather modest (only 1 cluster per subject in 3/38 healthy controls) as was the number of extra-lesional clusters in patients. Notably, as our patients were seizure free after surgery, we are confident that these extra-lesional clusters were indeed false positives. Our automated algorithm provided a 4 times higher detection rate than conventional radiological visual inspection, which allowed identifying only 20% of the lesions.

In conclusion, we designed our protocol to attain high specificity, which is critical in the management of patients who undergo presurgical evaluation for medically intractable seizures. This histologically-validated new FCD detection method, providing the highest performance to date compared to the literature, has the potential to become a useful clinical tool to assist in the diagnosis of subtle lesions that are frequently overlooked by conventional means of analysis.

References

1. Sisodiya, S.M., Fauser, S., Cross, J.H., Thom, M.: Focal cortical dysplasia type II: biological features and clinical perspectives. Lancet Neurol. **8**, 830–843 (2009)
2. Bernasconi, A., Bernasconi, N., Bernhardt, B.C., Schrader, D.: Advances in MRI for "cryptogenic" epilepsies. Nat. Rev. Neurol. **7**, 99–108 (2011)
3. Colliot, O., Bernasconi, N., Khalili, N., Antel, S.B., Naessens, V., Bernasconi, A.: Individual voxel-based analysis of gray matter in focal cortical dysplasia. NeuroImage **29**, 162–171 (2006)
4. Focke, N.K., Bonelli, S.B., Yogarajah, M., Scott, C., Symms, M.R., Duncan, J.S.: Automated normalized FLAIR imaging in MRI-negative patients with refractory focal epilepsy. Epilepsia **50**, 1484–1490 (2009)
5. Besson, P., Andermann, F., Dubeau, F., Bernasconi, A.: Small focal cortical dysplasia lesions are located at the bottom of a deep sulcus. Brain J. Neurol. **131**, 3246–3255 (2008)
6. Thesen, T., Quinn, B.T., Carlson, C., Devinsky, O., DuBois, J., McDonald, C.R., French, J., Leventer, R., Felsovalyi, O., Wang, X., Halgren, E., Kuzniecky, R.: Detection of epileptogenic cortical malformations with surface-based MRI morphometry. PLoS ONE **6**, e16430 (2011)
7. Hong, S.-J., Kim, H., Schrader, D., Bernasconi, N., Bernhardt, B.C., Bernasconi, A.: Automated detection of cortical dysplasia type II in MRI-negative epilepsy. Neurology **83**, 48–55 (2014)
8. Adler, S., Wagstyl, K., Gunny, R., Ronan, L., Carmichael, D., Cross, J.H., Fletcher, P.C., Baldeweg, T.: Novel surface features for automated detection of focal cortical dysplasias in paediatric epilepsy. NeuroImage Clin. **14**, 18–27 (2017)
9. Kini, L.G., Gee, J.C., Litt, B.: Computational analysis in epilepsy neuroimaging: a survey of features and methods. NeuroImage Clin. **11**, 515–529 (2016)
10. Sled, J.G., Zijdenbos, A.P., Evans, A.C.: A nonparametric method for automatic correction of intensity nonuniformity in MRI data. IEEE Trans. Med. Imaging **17**, 87–97 (1998)
11. Kim, H., Caldairou, B., Hwang, J.-W., Mansi, T., Hong, S.-J., Bernasconi, N., Bernasconi, A.: Accurate cortical tissue classification on MRI by modeling cortical folding patterns. Hum. Brain Mapp. **36**, 3563–3574 (2015)
12. Kim, J.S., Singh, V., Lee, J.K., Lerch, J., Ad-Dab'bagh, Y., MacDonald, D., Lee, J.M., Kim, S.I., Evans, A.C.: Automated 3-D extraction and evaluation of the inner and outer cortical surfaces using a Laplacian map and partial volume effect classification. NeuroImage **27**, 210–221 (2005)
13. Lyttelton, O., Boucher, M., Robbins, S., Evans, A.: An unbiased iterative group registration template for cortical surface analysis. NeuroImage **34**, 1535–1544 (2007)

14. Polimeni, J.R., Fischl, B., Greve, D.N., Wald, L.L.: Laminar analysis of 7T BOLD using an imposed spatial activation pattern in human V1. NeuroImage **52**, 1334–1346 (2010)
15. Hong, S.-J., Bernhardt, B.C., Schrader, D., Caldairou, B., Bernasconi, N., Bernasconi, A.: MRI-based lesion profiling of epileptogenic cortical malformations. In: Navab, N., Hornegger, J., Wells, W.M., Frangi, A.F. (eds.) Medical Image Computing and Computer-Assisted Intervention – MICCAI 2015, pp. 501–509. Springer International Publishing, Cham (2015)
16. Shafee, R., Buckner, R.L., Fischl, B.: Gray matter myelination of 1555 human brains using partial volume corrected MRI images. NeuroImage **105**, 473–485 (2015)
17. Im, K., Lee, J.M., Lyttelton, O., Kim, S.H., Evans, A.C., Kim, S.I.: Brain size and cortical structure in the adult human brain. Cereb. Cortex **18**, 2181–2191 (2008)
18. Seiffert, C., Khoshgoftaar, T.M., Van Hulse, J., Napolitano, A.: RUSBoost: a hybrid approach to alleviating class imbalance. IEEE Trans. Syst. Man Cybern. Part A Syst. Hum. **40**, 185–197 (2010)
19. Freund, Y., Schapire, R.E.: A decision-theoretic generalization of on-line learning and an application to boosting. J. Comput. Syst. Sci. **55**, 119–139 (1997)
20. Geurts, P., Ernst, D., Wehenkel, L.: Extremely randomized trees. Mach. Learn. **63**, 3–42 (2006)
21. Hong, S.-J., Bernhardt, B.C., Schrader, D.S., Bernasconi, N., Bernasconi, A.: Whole-brain MRI phenotyping in dysplasia-related frontal lobe epilepsy. Neurology **86**, 643–650 (2016)

Prediction of Amyloidosis
from Neuropsychological and MRI Data
for Cost Effective Inclusion of Pre-symptomatic
Subjects in Clinical Trials

Manon Ansart[1,2]([⊠]), Stéphane Epelbaum[1,2,3], Geoffroy Gagliardi[1,3],
Olivier Colliot[1,2,3,4], Didier Dormont[1,2,4], Bruno Dubois[1,3], Harald Hampel[1,3,5],
Stanley Durrleman[1,2], and for the ADNI, and the INSIGHT study group

[1] Sorbonne Universités, UPMC Univ Paris 06, Inserm, CNRS, Institut du cerveau
et de la moelle (ICM) - Pitié-Salpêtrière hospital, Bvd de l'hôpital, Paris, France
manon.ansart@inserm.fr
[2] Inria Paris, Aramis Project-team, 75013 Paris, France
[3] AP-HP, Pitié-Salpêtrière hospital, Department of Neurology, Institut de la
Mémoire et de la Maladie d'Alzheimer (IM2A), Paris, France
[4] AP-HP, Pitié-Salpêtrière Hospital, Department of Neuroradiology, Paris, France
[5] AXA Research Fund and UPMC Chair, Paris, France

Abstract. We propose a method for selecting pre-symptomatic subjects
likely to have amyloid plaques in the brain, based on the automatic
analysis of neuropsychological and MRI data and using a cross-validated
binary classifier. By avoiding systematic PET scan for selecting subjects,
it reduces the cost of forming cohorts of subjects with amyloid plaques
for clinical trials, by scanning fewer subjects but increasing the number
of recruitments. We validate our method on three cohorts of subjects at
different disease stages, and compare the performance of six classifiers,
showing that the random forest yields good results more consistently,
and that the method generalizes well when tested on an unseen data set.

1 Introduction

One of the lesions defining Alzheimer's disease (AD) is the formation of amyloid
plaques in the brain. A commonly accepted hypothesis is that this plaque formation is the starting point that triggers a cascade of events leading to neuronal
loss, cognitive decline and then dementia [9]. Those plaques appear very early
in the disease course, often way before any signs of cognitive decline and diagnosis [5,12]. They are the consequence of the aggregation of beta-amyloid (Aβ)
peptides, and together with neurofibrillary tangles, they are thought to cause
the death of neurons, hence being the potential cause of cognitive decline.

Consequently, amyloid plaques are targeted by several molecules at different stages of their formation, with the aim that preventing their formation or
clearing them would stop the process resulting in AD. Several of those potential drugs, such as solanezumab and bapineuzumab have already been tested on

© Springer International Publishing AG 2017
M.J. Cardoso et al. (Eds.): DLMIA/ML-CDS 2017, LNCS 10553, pp. 357–364, 2017.
DOI: 10.1007/978-3-319-67558-9_41

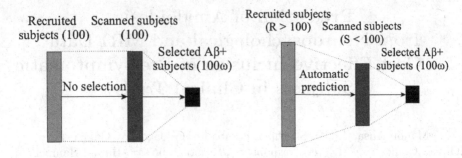

Fig. 1. Current (left) and proposed (right) processes for Aβ+ subjects selection

mild-to-moderate AD subjects and did not prove to slow down the progression of the symptoms of AD [4]. A possible explanation for these failures is that the treatments have been tested on subjects too late in the disease course, and for some of the trials on subjects without confirmation of amyloidosis. Drugs may stop the formation of amyloid plaques or clear them effectively, but they cannot repair the damage that has already been caused by the plaques. An hypothesis is that applying those treatments specifically on pre-symptomatic subjects with amyloidosis would make them more effective.

Testing these molecules at the preclinical stage raises the problem of recruiting pre-symptomatic subjects with amyloid plaques [15]. Positron emission tomography (PET) imaging with amyloid ligands is, together with lumbar puncture, the most widely used techniques to assess amyloid plaques presence *in vivo*. However, the prevalence of Aβ positive (Aβ+) subjects among asymptomatic elderly people is rather low: about 30% [2], resulting in 333 subjects to recruit and scan to get 100 Aβ+ subjects. A PET scan is however quite costly, about 1,000€ in Europe, and 5,000$ in the USA, so creating cohorts of pre-symptomatic Aβ+ subjects amounts to be very expensive. To ease the economic burden, we propose here to introduce a pre-screening phase to select subjects with higher risk of being Aβ+ than in the general population, and perform a confirmatory PET scan to those subjects only. We propose to predict the presence of amyloidosis in subjects by the automatic analysis of their neuropsychological assessments and structural imaging data, which are exams that are less expensive. We propose to use machine learning algorithms to find the patterns in these data that best predict the presence of amyloid plaques in the brain.

Methods to automatically predict amyloidosis from socio-demographic, genetic and cognitive variables have been proposed in [11,13]. In particular, they studied how univariate methods perform compared with multivariate ones. The threshold of the logistic regression in [13] was set a priori, and not optimized. The approach in [11] aimed to maximize the Positive Predictive Value (PPV). This value might be arbitrarily good by using a more and more stringent detection threshold, but that implies that more and more subjects need to be recruited for a given target number of Aβ+ subjects, as many positive subjects are discarded as false negatives. To better reflect this trade-off, we propose to translate the specificity and sensitivity of a classifier into a number of subjects to be recruited (R) and a number of subjects to be scanned (S), as shown Fig. 1. Each value of R and S corresponds to a given cost for the constitution of the cohort. The aim of our approach is to find the threshold minimizing this cost.

. In this paper, we will benchmark an array of cross-validated machine learning algorithms for the prediction of amyloidosis from several feature sets extracted from clinical and structural imaging data. We will validate these algorithms on three different cohorts with subjects at various disease stages. The hyper-parameters will be tuned by maximizing the area under the ROC curve (AUC). The score threshold will be chosen so as to minimize the cost.

2   Materials and Methods

2.1   Validation Cohorts

The method is validated on 3 cohorts: INSIGHT, ADNI-CN and ADNI-MCI. INSIGHT is a monocentric French study including asymptomatic subjects with a subjective memory complaint (SMC). 318 subjects have an AV45 PET scan and hence an $A\beta$ standardized uptake value ratio (SUVr) for baseline, among which 88 (27.7%) are $A\beta+$.

The Alzheimer's Disease Neuroimaging Initiative (ADNI) is a multicentric longitudinal study. We use the cognitively normal subjects (ADNI-CN) and the subjects with mild cognitive impairments (ADNI-CN) that have an $A\beta$ status assessed by AV45 PET scan or CSF biomarkers in the absence of PET scan. The baseline visit of the subjects who stay $A\beta+$ or $A\beta-$ for all visits is used. 431 CN subjects (37.6% of $A\beta+$) and 596 MCI subjects (62.9% of $A\beta+$) are available.

2.2   Input Features

Socio-demographic (age, gender, education), genetic (APOE) and cognitive features are used as inputs. For ADNI, the Alzheimer's Disease Assessment Scale cognitive sub-scale (ADAScog) is divided into memory, language, concentration and praxis, and for INSIGHT SMC questionnaires and cognitive tests (targeting memory, executive functions, behavior or overall cognitive skills) are used. MRI features are also used and compared with cognitive assessments in terms of prediction power. Cortical thicknesses averaged on 72 regions are extracted using FreeSurfer and divided by the total cortical thickness. The hippocampal volume is computed using FreeSurfer for ADNI and SACHA [3] for INSIGHT.

2.3   Algorithms

The classification is made using different algorithms in order to compare their performance. Hyper-parameters are tuned using cross-validation to maximize the AUC. The used algorithms are: random forest [1] (validation of the number and depth of the trees), regularized logistic regression [8] (validation of the regularization parameter), linear support vector machine [14] (SVM) (validation of

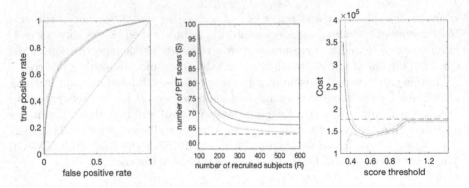

Fig. 2. Example of ROC curve (left), S vs. R curve (middle) and corresponding cost curve (right)

the penalty parameter), additive logistic regression [7] (AdaLogReg) (validation of the number and depth of the learners and of the learning rate for shrinkage), and adaptive boosting [6] (AdaBoost) (same hyper-parameters as AdaLogReg).

The data set is randomly split into a training (70%) and a test set (30%) 50 times, and a 5-fold validation is performed on the training set to automatically choose the algorithm hyper-parameters. All algorithms are trained on the whole training set, and their performance is evaluated on the test set. The performance mean and standard deviation (std) are computed and used to perform t-tests.

2.4   Performance Measures

The AUC is used to evaluate the overall performance of the methods and to tune the hyper-parameters. The maximum balanced accuracy (average of sensitivity and specificity, noted BAcc), which corresponds to a specific point on the ROC curve, is also used. The last measure is the minimal cost for recruiting $100 * \omega$ Aβ+ subjects, where ω is the proportion of positive subjects in the data set. In order to compute this cost, the ROC curve is computed (Fig. 2, left), then the number of subjects that have to be recruited (R) and the number of subjects that have to be scanned (S) are computed for each point on the ROC curve (Fig. 2, middle):

$$S = 100 * \omega * \frac{TP + FP}{TP} \quad (1) \qquad\qquad R = 100 * \omega * \frac{N}{TP} \quad (2)$$

where TP stands for number of True Positives, FP for number of False Positives and N for number of tested subjects. The corresponding cost is computed at each point (Fig. 2, right). The point with the minimal cost is kept, and the corresponding cost is used as a performance measure. We made the hypothesis that recruiting a subject (with cognitive scores and genetic information) costs 100€, doing an MRI costs 400€ and a PET scan 1,000€. As a comparison, recruiting $100 * \omega$ Aβ+ subjects doing a confirmatory PET scan for all subjects would correspond to a cost of 110,000€ (100 recruitments and PET scans).

Table 1. Benchmark of algorithms, given in the form: average performance (std)

	Data set	Random forest	Logistic regression	SVM	AdaLogReg	AdaBoost
AUC	INSIGHT	67.5 (5.5)	62.7 (6.1)	62.0 (5.8)	67.5 (5.7)	67.2 (6.9)
	ADNI-CN	69.1 (4.0)	69.5 (4.1)	67.3 (5.0)	66.4 (4.6)	66.5 (5.1)
	ADNI-MCI	83.8 (2.8)	82.5 (2.6)	82.4 (2.7)	82.6 (2.8)	83.1 (3.3)
BAcc	INSIGHT	63.9 (1.5)	60.1 (1.6)	59.6 (1.4)	62.3 (1.5)	62.3 (1.3)
	ADNI-CN	63.3 (1.0)	63.8 (1.1)	62.3 (0.9)	61.6 (1.3)	61.4 (1.3)
	ADNI-MCI	74.5 (0.9)	74.2 (0.8)	74.5 (0.8)	73.6 (0.8)	73.4 (1.0)
Cost (€)	INSIGHT	80,697 (15,900)	91,866 (15,811)	96,813 (13,147)	85,134 (16,137)	85,118 (18,944)
	ADNI-CN	88,206 (86,88)	84,833 (7,694)	88,404 (9,049)	93,231 (8,216)	92,921 (9,192)
	ADNI-MCI	85,673 (30,56)	86,460 (2,522)	86,436 (2,642)	86,056 (2,566)	86,269 (3,485)

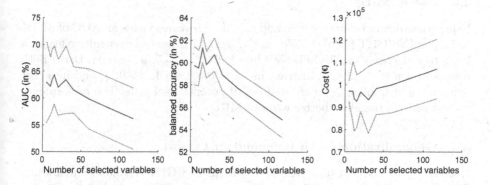

Fig. 3. Performance variations depending on the number of kept lasso variables

3 Experiments and Results

3.1 Algorithm Benchmark

An algorithm benchmark (using socio-demographic, genetic and cognitive features) is presented in Table 1. There is no algorithm that consistently outperforms the others for all criterion. However, if a choice has to be made, the random forest is consistently among the best algorithms for all measures and all data sets. Its performances are the best for INSIGHT and ADNI-MCI, and are slightly below the ones of the logistic regression for the ADNI-CN. Using a random forest leads to a significant decrease in the cost of recruiting $100 * \omega$ Aβ+ compared to the initial cost of 110,000€ (p < 0.001).

3.2 Feature Selection

Using all the INSIGHT available features (117 features including 112 cognitive ones) for prediction gives an AUC of 56.2% (±7.5). Dimension reduction is therefore considered, comparing several methods. Principal Component Analysis (PCA) and Independent Component Analysis (ICA) using fastICA [10] are first considered with a variable number of selected dimensions, but both give less than 52% of AUC. Alternatively, Lasso feature selection is performed, using a linear regression followed by a random forest and keeping from 5 up to 60 features (Fig. 3). The best results, obtained on 15 features, correspond to an AUC of 64.3% (±5.2), which is significantly better than using all features (p < 0.001). Another strategy is forming aggregates for each cognitive test using expert knowledge on which test variables are most likely to be a marker of AD. 26 cognitive summary variables are constructed this way, and using them in place of the 112 original cognitive features gives an AUC of 67.5% (±5.5), which is significantly better (p < 0.005) than the performance reached using automatic methods.

3.3   Use of MRI

Using socio-demographic, genetic and cognitive features yields an AUC of 67.5% (±5.5) on INSIGHT (Table 1, column 1). Using MRI instead of cognitive features leads to a decrease in AUC (Table 2 line 1: 61.9% ± 6.5, p < 0.001). Using both results in a non-significant increase in AUC (68.8 ± 4.4, p > 0.1), and in a cost increase, as it implies to do an MRI on all potential subjects. The performance vs cost ratio is therefore better without MRI.

3.4   Generalization on an Independent Cohort

INSIGHT and ADNI are different databases, as INSIGHT is a monocentric study focused on SMC, and ADNI is multicentric with different inclusion criterion and goals. The hippocampal volumes have also been extracted using different softwares. In order to see if the proposed method could generalize well to other data sets, it is trained on ADNI-CN and tested on INSIGHT, as they correspond to the most similar subject profiles. The socio-demographic, genetic and MRI variables are used, and a lasso selection of 12 features is performed on the MRI variables. In order to have a fair comparison, training and test set are created with the same size as the training and test data sets coming from INSIGHT, by randomly selecting $318 * 0.7 = 223$ subjects from ADNI-CN for the training set and $318 * 0.3 = 95$ from INSIGHT for the test set. This sampling and the

Table 2. Results using MRI variables, socio-demographic and genetic information on different data sets

Data set	AUC in % (std)	BAcc in % (std)	Cost in € (std)
Trained and tested on INSIGHT	61.9 (6.5)	59.3 (1.5)	14,6147 (4,975)
Trained on ADNI-CN, tested on INSIGHT	62 (6.6)	58 (1.7)	14,5989 (5,112)
Trained on ADNI-CN, tested on INSIGHT (all samples)	66.1 (3.6)	62.5 (1.1)	14,5896 (2,663)
Trained and tested on [INSIGHT ADNI-CN]	61.3 (6.6)	58.5 (1.5)	14,5642 (3,897)
Trained and tested on [INSIGHT ADNI-CN] (all samples)	66.7 (3.7)	62.3 (1.0)	14,6613 (5,859)

classification are performed 50 times in order to get an average performance. The performances obtained by learning on either INSIGHT alone (Table 2 line 1) or ADNI-CN (Table 2 line 2) are very similar, which means the proposed method is likely to give similar results if applied on a new data set of CN elderly subjects.

3.5 Pooling Data Sets

A new data set is created by pooling subjects from ADNI-CN and INSIGHT, while keeping the same total cohort size as in INSIGHT. The method gives similar performances when validated on this pooled data set or on INSIGHT (Table 2, lines 1 and 4), which shows that the heterogeneity of pooled data sets does not alter the classification performances.

3.6 Effect of Sample Size

When the classifier is trained and tested on INSIGHT, $318 * 0.7 = 223$ subjects are used for training. The training set can contain up to 431 subjects when training on ADNI and testing on INSIGHT, and 524 subjects when using the pooled data set, which is respectively 2.30 and 1.86 times larger. We can therefore train the method on larger and larger data sets, keeping the same proportion between the training and the test set (70%–30%) for comparison. The results, reported in Table 2 lines 3 and 5, show a significant increase in the AUC ($p < 0.001$) when the size of the data set increases, which comforts the need to create large databases, or pool existing databases, to create more accurate medical models.

4   Conclusion

We proposed a method for creating cohorts of $A\beta+$ pre-symptomatic subjects, by building a classifier optimized to minimize cohort creation costs. The proposed method identifies in a pre-screening phase a sub-set of subjects with a much higher prevalence of $A\beta+$ cases. We benchmarked cross-validated algorithms and showed that the random forest consistently yields good results. We tested our method on 3 data sets and showed that it always results in a significant cost decrease for creating such cohorts. We showed that the method generalizes well when trained on a cohort and tested on an independent one, therefore showing its potential for being used in real clinical environment with heterogeneous procedures for subject selection, data acquisition and processing. The best costs are achieved by using socio-demographic, genetic and cognitive features chosen using expert knowledge. Using MRI features increases the overall costs, but the performances could be increased by extracting more complex features, or by using a priori knowledge for selecting relevant variables.

Acknowledgement. This work was partly funded by ERC grant N°678304, H2020 EU grant N°666992 and ANR grant ANR-10-IAIHU-06. HH is supported by the AXA Research Fund, the Fondation UPMC and the Fondation pour la Recherche sur Alzheimer, Paris, France. OC is supported by a "contrat d'interface local" from AP-HP.

References

1. Breiman, L.: Random forests. Mach. Learn. **45**(1), 5–32 (2001)
2. Chetelat, G., La Joie, R., Villain, N., Perrotin, A., de La Sayette, V., Eustache, F., Vandenberghe, R.: Amyloid imaging in cognitively normal individuals, at-risk populations and preclinical alzheimer's disease. Neuroimage Clin. **2**, 356–365 (2013)
3. Chupin, M., Hammers, A., Liu, R.S.N., Colliot, O., Burdett, J., Bardinet, E., Duncan, J.S., Garnero, L., Lemieux, L.: Automatic segmentation of the hippocampus and the amygdala driven by hybrid constraints: method and validation. NeuroImage **46**(3), 749–761 (2009)
4. Doody, R.S., Thomas, R.G., Farlow, M., Iwatsubo, T., Vellas, B., Joffe, S., Kieburtz, K., Raman, R., Sun, X., Aisen, P.S., Siemers, E., Liu-Seifert, H., Mohs, R.: Phase 3 trials of solanezumab for mild-to-moderate alzheimer's disease. N. Engl. J. Med. **370**(4), 311–321 (2014)
5. Dubois, B., Hampel, H., Feldman, H.H., Scheltens, P., Aisen, P., Andrieu, S., Bakardjian, H., Benali, H., Bertram, L., Blennow, K., Broich, K., Cavedo, E., Crutch, S., Dartigues, J.F., Duyckaerts, C., Epelbaum, S., Frisoni, G.B., Gauthier, S., Genthon, R., Gouw, A.A., et al.: Preclinical alzheimer's disease: definition, natural history, and diagnostic criteria. Alzheimer's Dement. **12**(3), 292–323 (2016)
6. Friedman, J.: Greedy function approximation: A gradient boosting machine. Ann. Stat. **29**(5), 1189–1232 (2001)
7. Friedman, J., Hastie, T., Tibshirani, R.: Additive logistic regression: a statistical view of boosting. Ann. Stat. **28**(2), 337–407 (2000)
8. Friedman, J., Hastie, T., Tibshirani, R.: Regularization paths for generalized linear models via coordinate descent. J. Stat. Softw. **33**(1), 1 (2010)
9. Hardy, J.A., Higgins, G.A.: Alzheimer's disease: the amyloid cascade hypothesis. Science **256**(5054), 184–185 (1992)
10. Hyvarinen, A.: Fast and robust fixed-point algorithms for independent component analysis. IEEE Trans. Neural Netw. **10**(3), 626–634 (1999)
11. Insel, P.S., Palmqvist, S., Mackin, R.S., Nosheny, R.L., Hansson, O., Weiner, M.W., Mattsson, N.: Assessing risk for preclinical β-amyloid pathology with APOE, cognitive, and demographic information. Alzheimer's Dement. Diagn. Assess. Dis. Monit. **4**, 76–84 (2016)
12. Jack, C.R., Knopman, D.S., Jagust, W.J., Shaw, L.M., Aisen, P.S., Weiner, M.W., Petersen, R.C., Trojanowski, J.Q.: Hypothetical model of dynamic biomarkers of the alzheimer's pathological cascade. Lancet Neurol. **9**(1), 119 (2010)
13. Mielke, M.M., Wiste, H.J., Weigand, S.D., Knopman, D.S., Lowe, V.J., Roberts, R.O., Geda, Y.E., Swenson-Dravis, D.M., Boeve, B.F., Senjem, M.L., Vemuri, P., Petersen, R.C., Jack, C.R.: Indicators of amyloid burden in a population-based study of cognitively normal elderly. Neurology **79**(15), 1570–1577 (2012)
14. Muller, K.R., Mika, S., Ratsch, G., Tsuda, K., Scholkopf, B.: An introduction to kernel-based learning algorithms. IEEE Trans. Neural Netw. **12**(2), 181–201 (2001)
15. O'Brien, J.T., Herholz, K.: Amyloid imaging for dementia in clinical practice. BMC Medicine **13**, 163 (2015)

Automated Multimodal Breast CAD Based on Registration of MRI and Two View Mammography

T. Hopp$^{(\boxtimes)}$, P. Cotic Smole, and N.V. Ruiter

Institute for Data Processing and Electronics, Karlsruhe Institute of Technology,
Karlsruhe, Germany
torsten.hopp@kit.edu

Abstract. Computer aided diagnosis (CAD) of breast cancer is mainly focused on monomodal applications. Here we present a fully automated multimodal CAD, which uses patient-specific image registration of MRI and two-view X-ray mammography. The image registration estimates the spatial correspondence between each voxel in the MRI and each pixel in cranio-caudal and mediolateral-oblique mammograms. Thereby we can combine features from both modalities. As a proof of concept we classify fixed regions of interest (ROI) into *normal* and *suspect* tissue. We investigate the classification performance of the multimodal classification in several setups against a classification with MRI features only. The average sensitivity of detecting suspect ROIs improves by approximately 2% when combining MRI with both mammographic views compared to MRI-only detection, while the specificity stays at a constant level. We conclude that automatically combining MRI and X-ray can enhance the result of a breast CAD system.

Keywords: Computer aided diagnosis · Multimodal image registration · X-ray mammography · MRI

1 Introduction

Computer aided diagnosis (CAD) for breast cancer detection has been widely studied in the last years. Most applications of breast CAD have been developed for X-ray mammography [1]. Furthermore, CAD has often been applied for breast magnetic resonance imaging (MRI) [2] and breast sonography [3]. While breast CAD is thereby mostly limited to monomodal imaging, several studies, e.g. [4], have shown that the combination of modalities can lead to better detection rates. There are only few approaches combining the diagnostic information of two imaging modalities for breast CAD, e.g. [5]. One reason is the challenging spatial correlation of tissue structures as in X-ray mammography, MRI and sonography the patient positioning and compression state of the breast is considerably different. In order to apply a multimodal CAD this leads to manual selection of corresponding tissue structures in multiple modalities [5], which requires experienced radiologists and is time consuming.

© Springer International Publishing AG 2017
M.J. Cardoso et al. (Eds.): DLMIA/ML-CDS 2017, LNCS 10553, pp. 365–372, 2017.
DOI: 10.1007/978-3-319-67558-9_42

In our previous work we developed and evaluated an automated method for MRI to X-ray mammography image registration based on a biomechanical model [6]. It allows estimating the position of a tissue structure in the X-ray mammogram given its location in the MRI, i.e. for each voxel in the MRI the corresponding pixel in the mammogram is computed. Based on this registration we proposed an automated multimodal CAD approach using the combination of X-ray mammograms and MRI [7]. While this first study was limited to cranio-caudal mammograms, we now extended the method to two-view mammograms allowing the automated combination of cranio-caudal mammograms, mediolateral-oblique mammograms and MRI in a CAD system. In this paper we present a proof of concept and give a first estimate of the gain by automatically integrating information of MRI and two-view mammography in breast CAD.

2   Methods

2.1   Image Registration

For automated combination of modalities, the spatial correspondence between MRI and both mammographic views needs to be estimated. We apply an image registration which uses a biomechanical model of the breast to simulate the mammographic compression. The patient-specific biomechanical model is generated from the segmented MR volume. For the segmentation, a fuzzy C-means clustering similar to [8] and edge detection is applied. The model geometry is assembled by a tetrahedral mesh differentiating fatty and glandular tissue. Both tissue types are modeled as hyperelastic neo-hookean material with individual material parameters for fatty and glandular tissue. Mammographic compression is simulated by adding compression plates into the simulation and formulating a contact problem which is solved by the Finite Element method and solved using the dynamic solver of the software package ABAQUS. Based on the estimated deformation field, each three-dimensional point in the MR volume can be mapped to a two-dimensional point in the X-ray mammogram. This is achieved by simulating a perspective X-ray casting on the deformed MRI. The registration was carried out for both mammographic views, i.e. cranio-caudal (CC) and mediolateral-oblique (MLO) mammograms. For more details refer to our earlier publications [6,9].

2.2   CAD System

As a proof of concept, we designed a CAD system which aims to classify distinct cubic regions of interest (ROIs) into one of the categories *normal* or *suspect*. For this purpose, the breast in the MR volume is quantized into ROIs of size $(10 \times 10 \times 10)\,mm^3$. To extract multimodal information, the eight vertices of the MRI ROIs are mapped to the CC mammogram as well as the MLO mammogram based on the deformation fields computed during the image registration. The mammography ROIs are then formed by the convex hull of these eight automatically mapped points (Fig. 1).

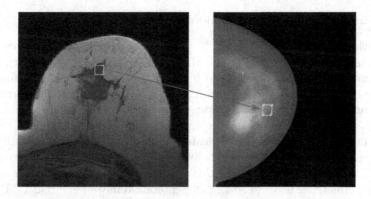

Fig. 1. Mapping from the MRI (left) to the CC mammogram (right) based on the estimated deformation field. The vertices of the MRI ROI (yellow rectangle) are mapped to the X-ray mammogram (yellow dots). The mammography ROI is formed by the convex hull of these points (green). (Color figure online)

For each 3D MRI ROI, 64 features are extracted. Intensity based features include e.g. the mean intensity and its variance at three time points (pre-contrast, 1 min and 6–7 min post-contrast). Texture features are based on 3D gray-level co-occurence matrices [10]. Temporal features analyze the contrast enhancement from pre-contrast to 1 min post-contrast and from 1 min post-contrast to the last time point similar to the three time points method [11].

For each mapped mammography ROI, 54 features are extracted. Similar to the MRI intensity features, e.g. the mean intensity and its median and variance in the ROI are computed. Texture features are based on gray-level co-occurence matrices [12] and gray-level runlength matrices [13]. Furthermore a multilevel Otsu thresholding [14] is applied and features based on morphological enhancement [15] are added. All features from MRI and mammography are gathered in a combined feature vector. Combining the MRI features with CC and MLO mammography features, a total of up to 172 features are used to classify each ROI.

The classification problem is addressed by the WEKA pattern recognition toolbox [16]. A correlation based feature subset selection is performed with a best-first search method [17] using a ten-fold cross-validation. Features selected in at least one of ten folds were then selected for classification. We repeated the cross-validation for the feature selection three times with varying random seed in order to reduce the sensibility of our analysis due to the dataset splitting.

To demonstrate the feasibility of the approach, we applied a random forest classifier as it proved to provide robust results in our earlier study [7]. The classifier performance was again evaluated by a ten-fold cross-validation and repeated three times with a random seed.

2.3   Clinical Datasets and Evaluation Methods

The method was evaluated using 43 patient datasets from a previous study for which an image registration of MRI with CC as well as with the MLO oblique mammogram was carried out [6,18]. Each dataset included a time series of T1-weighted dynamic contrast enhanced MR volumes and the corresponding CC and MLO mammogram of the same patient. MR images were acquired on 1.5T scanners (Siemens Magnetom Symphony, Sonata, Avanto) with the patient in prone position using dedicated bilateral breast coils. The MRI parameters were as follows: matrix size = 384×384, slices = 33, spatial resolution = $0.9 \times 0.9 \times 3.0\,\mathrm{mm}^3$, time of acquisition = 1 min. per measurement. As a contrast agent, $0.1\,\mathrm{mmol/(kg}$ body weight) gadopentetate-dimeglumine (Gd-DTPA) was administered at $3\,\mathrm{ml/s}$ intravenously using a power injector for standardized injection. The contrast agent injection bolus was followed by $20\,\mathrm{ml}$ of physiological saline solution. Full field digital mammograms were acquired on GE Senograph 2000D units.

The ground truth for the classifier training and evaluation is given based on expert annotations. Each annotation circumscribes the lesion in the 3D MR volume with a freehand tool. The annotated lesions included a mixture of malign and benign lesions verified by histology or follow-up diagnosis. A ROI of our CAD approach was labeled as *suspicious* if at least 50% of its volume was covered by the expert's lesion annotation.

As a proof of principle, a ROI size in the MRI of $10 \times 10 \times 10\,\mathrm{mm}^3$ was chosen as a tradeoff between detectable lesion size and the time consumption for mapping the ROI from MRI to both mammographic views. This resulted in a total of 31,239 ROIs of which 634 were labeled as *suspicious*.

The image registration of these datasets was performed in previous studies [6,9]. The target registration error (TRE) was estimated based on the Euclidean distance between the annotated lesion center in the mammogram and the annotated lesion center projected from the 3D MRI into the 2D mammogram. The average TRE was approximately $13.6\,\mathrm{mm}$ (Standard deviation (SD) $9.6\,\mathrm{mm}$) [6] for the CC view mammograms and $16.3\,\mathrm{mm}$ (SD $8.7\,\mathrm{mm}$) for the MLO view mammograms.

For the evaluation we tested our CAD system with four different setups: (1) using only features from MRI, (2) using features from MRI and CC-view mammograms, (3) using features from MRI and MLO-view mammograms, (4) using features from MRI and CC- and MLO-view mammograms. For each setup the previously described feature selection and classifier evaluation by tenfold cross-validation was carried out. Hence, in total we obtained nine classification evaluation runs per setup with all combinations of three subsets of the features and three random seeds for the classifier cross-validation. To analyze the classifier performance, the error rate of incorrectly classified ROIs as well as the ROI related sensitivity and specificity were calculated from the true and false positive respectively true and false negative rates, where *suspicious* ROIs are positive.

3   Results

Figures 2 and 3 show box plots of the evaluation of the four setups. The average sensitivity increases from an average of 76.1% (± standard deviation 0.8%) if only MRI features are considered (setup 1) to 76.9% (±0.4%), 77.1% (±1.0%) and 77.9% (±1.2%) for the multimodal setups 2 (MRI + CC), 3 (MRI + MLO) and 4 (MRI + CC + MLO) respectively. At the same time the ROI related specificity stays at a constant level of 99.8% for all setups. This trend is also present in the error rate of wrongly classified ROIs: it decreases from 0.68% (MRI) to 0.64% (MRI + CC), 0.63% (MRI + MLO) and 0.62% (MRI + CC + MLO). Hence the best result is obtained by combining the multimodal information from MRI and both mammographic views.

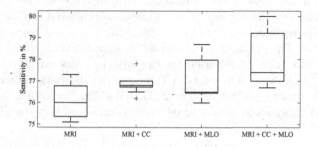

Fig. 2. Classification result: sensitivity of detecting *suspicious* ROIs with the evaluated four setups. The tops and bottoms of each "box" are the 25th and 75th percentiles of the samples respectively, the line within the box gives the median value. The vertical dashed lines indicate the whiskers which include all datasets not considered as outliers, i.e. are below 1.5 times the interquartile range.

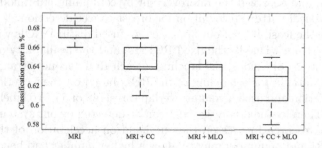

Fig. 3. Classification result: percentage of wrongly classified ROIs with the evaluated four setups. The tops and bottoms of each "box" are the 25th and 75th percentiles of the samples respectively, the line within the box gives the median value. The vertical dashed lines indicate the whiskers which include all datasets not considered as outliers, i.e. are below 1.5 times the interquartile range.

In order to analyze the influence of the TRE on the classification performance, we selected a subset of all datasets with the TRE below 10 mm in both the

CC-view registration and the MLO-view registration. Thereby the TRE equals the size of a ROI in our evaluated scenario, which ensures that the true lesion positions in both mammographic views overlap with the mapped ROIs in all cases.

For this data subset the same evaluation as before was carried out. The sensitivity for setup 1, 2, 3 and 4 was 76.2% ($\pm$4.2%), 79.6% ($\pm$3.2%), 75.7% ($\pm$4.2%) and 78.3% ($\pm$3.5%). At the same time the specificity changes only slightly and increases by adding multimodal information: for setup 1, 2, 3 and 4, the average specificities are 99.6% ($\pm$0.1%), 99.7% ($\pm$0.1%), 99.7% ($\pm$0.1%) and 99.7% ($\pm$0.1%). The according classification errors are 0.81% ($\pm$0.11%), 0.69% ($\pm$0.12%), 0.83% ($\pm$0.05%) and 0.62% ($\pm$0.02%).

Adding information from CC view mammograms increases the sensitivity considerably more than in the evaluation with all datasets. Yet, in the MLO case the sensitivity slightly decreases, while in combined CC-view and MLO-view case, intermediate results can be observed. The same holds for the classification error. The specificity remains at an approximately constant level for all multimodal setups. One reason for the mixed results might be the limited size of the data subset: it consists of three patient datasets only with a TRE below 10 mm which equals a total of 2195 ROIs. This hypothesis is supported by the considerably larger standard deviations compared to the evaluation with all datasets.

4   Discussion and Conclusion

In this paper we presented a method for an automated multimodal CAD system based on an image registration of 3D MRI and 2D two-view X-ray mammography. We extend our earlier approach by adding MLO mammograms into the CAD system and analyzed the improvement by combining multimodal information in a classical pattern recognition for breast cancer detection.

To check the feasibility of our approach, we used distinct ROIs with a size of $10 \times 10 \times 10\,\text{mm}^3$. Due to the fixed ROI size and the availability of only free-hand annotations of lesions, the labeling of each ROI was not straight forward. We decided initially for a labeling of the ROIs based on their overlap with the lesion annotation and used a volume overlap threshold of 50% to label the ROI as *suspicious*. Thereby the lesion size that can be detected by our system is limited. Our current research focuses on the one hand on acceleration of the mapping between MRI and mammography to allow a higher number and hence a smaller size of ROIs or even a voxel based classification using a sliding window approach. On the other hand the influence of the labeling threshold will be investigated further and we are planning an additional review of the labeling with experts to clean the dataset from outliers.

Similar to our earlier study, this work is limited to datasets selected retrospectively from clinical routine where all MRI examinations were carried out with the same protocol. Mammograms were all acquired with the same mammography system leading to homogeneous image characteristics. Datasets were furthermore selected such that a lesion could be delineated in both the MRI and X-ray mammogram by an expert in order to evaluate the TRE.

For evaluation of our CAD approach, a basic set of commonly used features was extracted for MRI and both mammographic views. This might not yet tap the full potential of a CADe system, yet we allow easy extension of the system in future by a plugin-like feature extraction software architecture.

Despite the limitations of this study, the initial results are promising: Approximately 78% of the *suspect* labeled ROIs could be identified by our proposed method. Though the average registration error was larger than the ROI size, the results improved when combining multimodal information. Due to nonlinear deformations, complex tissue structures and manual interactions during the patient positioning, the accurate registration of X-ray mammograms and MRI is challenging and a lively field of research. Yet a TRE in the range of current MRI to mammography registration approaches [6,19,20] already leads to improvements of the CADe performance. We showed in this study that including a second mammographic view can further enhance our multimodal CAD system for breast cancer detection.

References

1. Cheng, H., Shi, X., Min, R., Hu, L., Cai, X., Du, H.: Approaches for automated detection and classification of masses in mammograms. Pattern Recogn. **39**(4), 646–668 (2006)
2. Dorrius, M., van der Weide, M., van Ooijen, P., Pijnappel, R., Oudkerk, M.: Computer-aided detection in breast MRI: a systematic review and meta-analysis. Eur. Radiol. **21**(8), 1600–1608 (2011)
3. Cheng, H., Shan, J., Ju, W., Guo, Y., Zhang, L.: Automated breast cancer detection and classification using ultrasound images: a survey. Pattern Recogn. **43**(1), 299–317 (2010)
4. Lord, S., Lei, W., Craft, P., Cawson, J., Morris, I., Walleser, S., Griffiths, A., Parker, S., Houssami, N.: A systematic review of the effectiveness of magnetic resonance imaging (MRI) as an addition to mammography and ultrasound in screening young women at high risk of breast cancer. Eur. J. Cancer **43**(13), 1905–1917 (2007)
5. Yuan, Y., Giger, M.L., Li, H., Bhooshan, N., Sennett, C.A.: Multimodality computer-aided breast cancer diagnosis with FFDM and DCE-MRI. Acad. Radiol. **17**(9), 1158–1167 (2010)
6. Hopp, T., Dietzel, M., Baltzer, P., Kreisel, P., Kaiser, W., Gemmeke, H., Ruiter, N.: Automatic multimodal 2D/3D breast image registration using biomechanical FEM models and intensity-based optimization. Med. Image Anal. **17**(2), 209–218 (2013)
7. Hopp, T., Neupane, B., Ruiter, N.V.: Automated multimodal computer aided detection based on a 3d–2d image registration. In: Proceedings of 13th International Workshop on Breast Imaging, IWDM 2016, Malmö, Sweden, pp. 400–407 (2016)

8. Wu, S., Weinstein, S., Keller, B.M., Conant, E.F., Kontos, D.: Fully-automated fibroglandular tissue segmentation in breast MRI. In: Maidment, A.D.A., Bakic, P.R., Gavenonis, S. (eds.) IWDM 2012. LNCS, vol. 7361, pp. 244–251. Springer, Heidelberg (2012). doi:10.1007/978-3-642-31271-7_32

9. Hopp, T., de Barros Rupp Simioni, W., Perez, J.E., Ruiter, N.: Comparison of bio-mechanical models for MRI to X-ray mammography registration. In: Proceedings 3rd MICCAI Workshop on Breast Image Analysis, pp. 81–88 (2015)

10. Chen, W., Giger, M.L., Li, H., Bick, U., Newstead, G.M.: Volumetric texture analysis of breast lesions on contrast-enhanced magnetic resonance images. Magn. Reson. Med. **58**(3), 562–571 (2007)

11. Degani, H., Gusis, V., Weinstein, D., Fields, S., Strano, S.: Mapping pathophys-iological features of breast tumors by MRI at high spatial resolution. Nat. Med. **3**(7), 780–782 (1997)

12. Haralick, R.M., Shanmugam, K., Dinstein, I.: Textural features for image classifi-cation. IEEE Trans. Syst. Man Cybern. **SMC-3**(6), 610–621 (1973)

13. Galloway, M.M.: Texture analysis using gray level run lengths. Comput. Graph. Image Process. **4**(2), 172–179 (1975)

14. Otsu, N.: A threshold selection method from gray-level histograms. IEEE Trans. Syst. Man Cybern. **9**(1), 62–66 (1979)

15. Li, H., Wang, Y.J., Liu, K.J.R., Lo, S.C.B., Freedman, M.T.: Computerized radi-ographic mass detection - part i: lesion site selection by morphological enhancement and contextual segmentation. IEEE Trans. Med. Imag. **20**, 289–301 (2001)

16. Hall, M., Frank, E., Holmes, G., Pfahringer, B., Reutemann, P., Witten, I.H.: The WEKA data mining software: an update. SIGKDD Expl. **11**(1), 10–18 (2009)

17. Hall, M.A.: Correlation-based feature subset selection for machine learning. Ph.D. thesis, The University of Waikato, Hamilton, New Zealand (1999)

18. Hopp, T., Ruiter, N.V.: 2D/3D registration for localization of mammographically depicted lesions in breast MRI. In: Maidment, A.D.A., Bakic, P.R., Gavenonis, S. (eds.) IWDM 2012. LNCS, vol. 7361, pp. 627–634. Springer, Heidelberg (2012). doi:10.1007/978-3-642-31271-7_81

19. Mertzanidou, T., Hipwell, J., Johnsen, S., Han, L., Eiben, B., Taylor, Z., Ourselin, S., Huisman, H., Mann, R., Bick, U., Karssemeijer, N., Hawkes, D.: MRI to x-ray mammography intensity-based registration with simultaneous optimisation of pose and biomechanical transformation parameters. Med. Image Anal. **18**(4), 674–683 (2014)

20. Lee, A., Rajagopal, V., Gamage, T.P.B., Doyle, A.J., Nielsen, P., Nash, M.: Breast lesion co-localisation between X-ray and MR images using finite element modelling. Med. Image Anal. **17**(8), 1256–1264 (2013)

EMR-Radiological Phenotypes in Diseases of the Optic Nerve and Their Association with Visual Function

Shikha Chaganti[1](✉), Jamie R. Robinson[2],
Camilo Bermudez[3], Thomas Lasko[4], Louise A. Mawn[5],
and Bennett A. Landman[6]

[1] Department of Computer Science, Vanderbilt University,
Nashville, TN 37235, USA
shikha.chaganti@vanderbilt.edu

[2] Vanderbilt University Medical Center, 1211 Medical Center Dr,
Nashville, TN 37232, USA

[3] Department of Biomedical Engineering, Vanderbilt University,
Nashville, TN 37235, USA

[4] Department of Biomedical Informatics, Vanderbilt University,
Nashville, TN 37235, USA

[5] Vanderbilt Eye Institute, 2311 Pierce Avenue, Nashville, TN 37232, USA

[6] Department of Electrical Engineering, Vanderbilt University,
Nashville, TN 37235, USA

Abstract. Multi-modal analyses of diseases of the optic nerve, that combine radiological imaging with other electronic medical records (EMR), improve understanding of visual function. We conducted a study of 55 patients with glaucoma and 32 patients with thyroid eye disease (TED). We collected their visual assessments, orbital CT imaging, and EMR data. We developed an image-processing pipeline that segmented and extracted structural metrics from CT images. We derived EMR phenotype vectors with the help of PheWAS (from diagnostic codes) and ProWAS (from treatment codes). Next, we performed a principal component analysis and multiple-correspondence analysis to identify their association with visual function scores. We found that structural metrics derived from CT imaging are significantly associated with functional visual score for both glaucoma ($R^2 = 0.32$) and TED ($R^2 = 0.4$). Addition of EMR phenotype vectors to the model significantly improved ($p < 1E{-}04$) the R^2 to 0.4 for glaucoma and 0.54 for TED.

Keywords: CT imaging · EMR · Regression · Optic nerve · MCA · PCA

1 Introduction

Pathologies of the optic nerve affect millions of Americans each year and can severely affect an individual's quality of life due to loss of visual function [1]. Accurate characterization of these diseases and timely intervention can preserve visual function. 3D computed tomography (CT) imaging of the eye orbit can captures structural changes in

© Springer International Publishing AG 2017
M.J. Cardoso et al. (Eds.): DLMIA/ML-CDS 2017, LNCS 10553, pp. 373–381, 2017.
DOI: 10.1007/978-3-319-67558-9_43

the eye orbit, which indicate the extent of disease progression and characterizes pathology. In prior studies [2, 3], a quantitative relationship between 3D structural metrics of the eye orbit was shown to be associated with visual outcomes such as visual acuity and field vision in patients with optic nerve disorders. However, the percentage of explained variance due to structural data was low ($R^2 \sim 0.1$–0.2). Several factors influence a model's ability to explain outcomes, particularly the selection of predictive features. Also, while information is available in radiological imaging, evaluation of radiology within the context of an individual's health history is important in determining functional changes, progression of disease, and prognosis. With the rise in adoption of digital electronic medical record (EMR) systems in the US health care system [4, 5], these records are available to medical research scientists with increasing ease.

In this study we develop an automated pipeline for segmentation and metric calculation of CT eye orbits for glaucoma and thyroid eye disease (TED). Further, we show that integrating EMR data, such as ICD-9 (International Classification of Diseases - 9) codes, and CPT (Current Procedural Terminology) codes, with imaging biomarkers improves the explained variance of disease outcomes.

2   Methods

2.1   Data

The study was conducted on a retrospective cohort of patients at Vanderbilt University Medical Center. Subjects were retrieved under Institutional Review Board (IRB) approval based on both having met clinical criteria for eye disease and undergoing CT imaging as part of their regular clinical care. The data collected include imaging records, visual testing, demographic data, complete ICD-9 codes and CPT codes. The disease groups included in this study are glaucoma (n = 55) and TED (n = 32).

2.2   Outcomes: Visual Function Scores

The outcomes in this study were calculated based on clinical visual acuity and visual field testing. Nine different outcome measures are calculated for a complete visual function evaluation as defined by the American Medical Association [6]. Right and left visual acuity scores are calculated as VAS_{od} and VAS_{os} respectively. The visual acuity for both eyes, VAS_{ou} is calculated as the best of VAS_{od} and VAS_{os}. The functional acuity score, FAS is a weighted score of VAS_{od}, VAS_{os}, and VAS_{ou} with weights 1:1:3. The scores from visual field testing, VFS_{od}, VFS_{os}, VFS_{ou}, and FFS are calculated similarly. A final score of visual function called functional visual score (FVS), is calculated as the average of FAS and FFS.

2.3   Image Processing

Figure 1 shows the image segmentation pipeline. First, multi-atlas segmentation was employed to identify four labels: the globe, the optic nerve, the extraocular muscles and

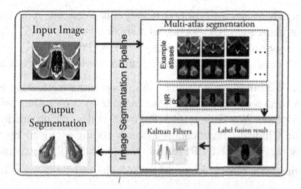

Fig. 1. Overview of image segmentation. Multi-atlas label fusion is used to segment the optic nerve, globe, muscle, and orbital fat. Kalman filters are used to segment the four individual extraocular muscles based on the result to achieve the final 3D segmentation result.

the periorbital fat. A set of twenty-five expertly labeled example 3D CT atlases is used as training examples to obtain the segmentation from a new input scan. Each of the example atlases is non-rigidly registered to the cropped input image space [7]. The corresponding labels of the example atlases are propagated to the input image space using the non-rigid deformations. Next, non-local statistical label fusion is used to obtain a segmented result with the four labels [8]. Segmenting the individual extraocular rectus muscles is challenging in diseased eyes, since obtaining true labels is difficult at the back of the orbit due to inflammation. So, we employ Kalman filters to segment muscle labels obtained from the multi-atlas algorithm [3] to identify the superior rectus muscle, the inferior rectus muscle, the lateral rectus muscle, the medial rectus muscle. Once the final segmentation is obtained twenty-five structural metrics are computed bilaterally [2]. For each structure, the volume, cross-sectional area, and diameter/length are measured. Indices of orbital crowding, i.e., Barrett's muscle index and volumetric crowding index are computed. In addition, degree of proptosis and orbital angle are computed. For each patient, i, a vector with 50 elements is constructed for 25 structural metrics computed bilaterally,

$$x_{CT}^{\{i\}} = [sm_{1\_os}\ sm_{2\_os} \cdots sm_{25\_os}\ sm_{1\_od}\ sm_{2\_od} \cdots sm_{25\_od}]$$

where, sm_{k_os} indicates k^{th} structural metric of the left eye and sm_{k_od} indicates k^{th} structural metric of the left eye.

2.4 EMR Features

From the EMR, complete ICD-9 codes and CPT codes were extracted for diagnostic and treatment information for each patient. However, only the ICD-9 and CPT codes available one month or more before the diagnosis are considered, since we are interested in understanding how a patient's history provides a context for imaging information.

PheWAS Codes. There are over 14,000 ICD-9 codes defined. A hierarchical system was defined that maps each ICD-9 code to a smaller group of 1865 phenotype codes originally used in phenome-wide association studies (PheWAS) [9]. Each phenotype, called a PheWAS code, indicates a related group of medical diagnoses and conditions.

ProWAS Codes. We introduce a similar hierarchical grouping to map each CPT code to a group of related procedures, which we indicate by a procedure wide association study (ProWAS) code. We define 1682 ProWAS codes, which are finer granularity subgroups of the Clinical Classification Systems coding provided by the Healthcare Cost and Utilization Project (HCUP) Agency for Healthcare Research and Quality [10].

For each patient, i, a binary vector with 1865 elements, $x_{PheWAS}^{\{i\}}$ is defined,

$$x_{PheWAS}^{\{i\}} = [d_1 \ d_2 \ldots d_{1865}]$$

where, d_k is 1 if the patient i has had the diagnosis phenotype d_k in the past and 0 otherwise. Similarly, a binary vector, $x_{ProWAS}^{\{i\}}$ is defined with 1682 elements,

$$x_{ProWAS}^{\{i\}} = [t_1 \ t_2 \ldots t_{1682}]$$

where, t_k is 1 if the patient i has had the treatment phenotype t_k in the past and 0 otherwise.

2.5 Dimensionality Reduction: PCA and MCA

A large amount of data is available for each patient; the final data vector for a patient i has 3597 elements in it. However, the data are correlated with each other, and it is possible to find underlying principal variables in the data. For the structural metrics, a principal component analysis (PCA) [11] is performed to reduce the dimensionality of the dataset. The first five principal components explaining about three fourths of the variance are extracted to give, for subject i,

$$x_{CT\_pca}^{\{i\}} = \left[sm_1' \ sm_2' \ldots sm_5' \right] \tag{1}$$

For the PheWAS and ProWAS binary vectors, multiple correspondence analysis (MCA) [12] is used to extract orthogonal components that are decomposed using the χ^2-statistic. The first five components are considered for both PheWAS and ProWAS vectors. As a result of MCA, we get two vectors of smaller dimensionality for each patient,

$$x_{PheWAS\_mca}^{\{i\}} = \left[d_1' \ d_2' \ldots d_5' \right] \tag{2}$$

$$x_{ProWAS\_mca}^{\{i\}} = \left[t_1' \ t_2' \ldots t_5' \right] \tag{3}$$

2.6   Stepwise Generalized Linear Model

The visual acuity scores are between 0 and 100 with most patients having scores close
to 100 and values closer to 0 being extremely rare. This makes the distribution of the
visual outcomes left skewed. Therefore, a generalized regression model (GLM) with a
Poisson distribution [13] is used to find the explanatory value of each set of datasets,
given by Eqs. (1), (2) and (3), and all the data together. These datasets are regressed
over the visual outcome scores s_v, where $s_v \in \{VAS_{ou}, VAS_{od}, VAS_{os}, VAS, FAS,$
$VFS_{ou}, VFS_{od}, VFS_{os}, FFS, FVS\}$. Four models are defined for each v,

$$M1 : s_v = \beta_0 + \beta_1 sm'_1 + \ldots + \beta_k sm'_k + \beta_{k+1} d'_k + \ldots + \beta_{k+l} d'_l + \beta_{k+l+1} t'_1 + \ldots$$
$$+ \beta_{k+l+m} t'_m + \beta_{k+l+m+1} age + \beta_{k+l+m+2} sex + \in$$
$$M2 : s_v = \beta_0 + \beta_1 sm'_1 + \ldots + \beta_k sm'_k + \beta_{k+1} age + \beta_{k+2} sex + \in$$
$$M3 : s_v = \beta_0 + \beta_1 d'_1 + \ldots + \beta_l d'_l + \beta_{l+1} age + \beta_{l+2} sex + \in$$
$$M4 : s_v = \beta_0 + \beta_1 t'_1 + \ldots + \beta_m t'_m + \beta_{m+1} age + \beta_{m+2} sex + \in$$

The four models are built using stepwise regression [14], with forward selection of
variables. At each step, the variable that most significantly improves the model
deviance is added until there is no more improvement. The explained variance of each
model, R^2 is noted.

2.7   Test of Deviance

The deviance of a model M, with fitted parameters $\hat{\theta}$ is given by,

$$D(M) = -2\left(\log\left(p\left(y|\hat{\theta}\right)\right) - \log(p(y|\theta_s))\right)$$

where, θ_s are the parameters of the saturated model, i.e., a model with parameters for
each data point such that it is fitted exactly. The deviance can be used to test signifi-
cance between two nested models $M_p(\hat{\theta}_p|X)$ and $M_q(\hat{\theta}_q|X)$, where $\hat{\theta}_p \subset \hat{\theta}_q$ and the
difference in the parameters between the two models is given by δ. The difference of
the deviance between the two models follows a χ^2- squared distribution with degree of
freedom δ. The null hypothesis, H_0 for the test of deviance is that adding δ parameters
to model M_p to get M_q does not improve the model. This test is used to compare models
M2-4 with M1.

3   Results

The average age group for glaucoma cohort is 65.4 ± 19.5 years and 72% of the
subjects were female. 91% of TED subjects were female, and the average age for this
group is 57.8 ± 16.2 years. On an average, each patient had 410 ICD-9 codes, and 660
CPT codes recorded. Figure 2 shows the individual distribution by sex along the first
two components of the three datasets in models M2, M3, and M4. For glaucoma, the
first component of the PCA on structural metrics corresponded to muscle and optic

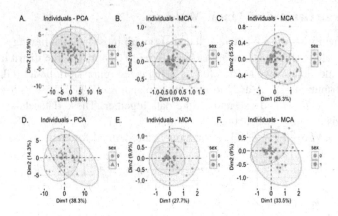

Fig. 2. Distribution of individuals by sex along the first two components from Eqs. (1), (2), and (3). Red and blue indicate 95% confidence ellipses for females and males respectively. (A) x_{CT_pca} for glaucoma. (B) x_{PheWAS_mca} for glaucoma. (C) x_{ProWAS_mca} for glaucoma. (D) x_{CT_pca} for TED. (E) x_{PheWAS_mca} for TED. (F) x_{ProWAS_mca} for TED. (Color figure online)

nerve measurements, and the second component corresponded to orbital and globe measurements. For TED, the first component corresponded to mostly muscle measurements, and the second component corresponded to measurements of the optic nerve. Some of the conditions associated with the first MCA component of the ICD-9 vector for Glaucoma are malaise, osteoarthrosis, and hypovolemia, and conditions associated with the second component included female genitourinary symptoms and symptoms associated with the eye such as pain and swelling. The first MCA component for TED's ICD-9 vector was associated with conditions including hyperlipidemia, diabetes, and circulatory problems, and some of the conditions most associated with the second MCA component were myalgia and abnormal blood chemistry. For the CPT vector for glaucoma, the first dimension was associated with a wide range of procedures such as CT scans, and pathology labs, and the second component was associated with cardiac testing. For TED, the first component was associated with procedures such as urinalysis and blood work, the second component was associated with physical therapy related procedures.

Tables 1 and 3 show the R^2 values of models M1, M2, M3, and M4 regressed over functional visual scores for glaucoma and TED respectively. The behavior of the models is the same for both the diseases. Addition of treatment and diagnostic

Table 1. Explained variance in Glaucoma. **indicates that model M1 is significantly better than model M2, i.e. using structural metrics alone, based on the p-values in Table 2.

R^2	VAS_{od}	VAS_{os}	VFS_{od}	VFS_{od}	VAS_{ou}	VFS_{ou}	FFS	FVS	FAS
M1	0.48^{**}	0.33^{**}	0.47^{**}	0.39^{**}	0.24	0.27^{**}	0.33^{**}	0.40^{**}	0.35
M2	0.45	0.13	0.37	0.16	0.24	0.18	0.23	0.33	0.35
M3	0.07	0.07	0.16	0.08	0.08	0.05	0.08	0.07	0.00
M4	0.04	0.00	0.20	0.06	0.00	0.05	0.09	0.06	0.00

Table 2. Test of deviance. Na indicates that the model is the same in both M1 and M2, as the same features were selected.

p-value	VAS_{od}	VAS_{os}	VFS_{od}	VFS_{od}	VAS_{ou}	VFS_{ou}	FFS	FVS	FAS
M1 vs. M2	4.30E−05	5.89E−09	2.92E−26	3.31E−38	na	5.38E−07	1.34E−07	1.18E−07	na
M1 vs. M3	1.66E−28	1.55E−10	3.17E−62	9.70E−46	0.001	2.65E−11	8.46E−15	2.39E−26	2.40E−06
M1 vs. M4	1.42E−29	3.59E−12	3.22E−58	8.32E−47	0.0005	8.48E−12	2.35E−14	3.15E−26	2.40E−06

Table 3. Explained variance in TED. **indicates that model M1 is significantly better than model M2, i.e. using structural metrics alone, based on the p-values in Table 4.

R^2	VAS_{od}	VAS_{os}	VFS_{od}	VFS_{od}	VAS_{ou}	VFS_{ou}	FFS	FVS	FAS
M1	0.61^{**}	0.30^{**}	0.59^{**}	0.37^{**}	0.28	0.36^{**}	0.42^{**}	0.54^{**}	0.44
M2	0.49	0.23	0.45	0.26	0.28	0.23	0.28	0.40	0.44
M3	0.30	0.22	0.33	0.24	0.16	0.20	0.24	0.28	0.20
M4	0.28	0.18	0.29	0.18	0.16	0.17	0.19	0.29	0.20

Table 4. Test of deviance. Na indicates that the model is the same in both M1 and M2, as the same features were selected.

p-value	VAS_{od}	VAS_{os}	VFS_{od}	VFS_{od}	VAS_{ou}	VFS_{ou}	FFS	FVS	FAS
M1 vs. M2	4.56E−06	0.00027	5.15E−24	6.75E−07	na	4.38E−08	7.50E−09	9.70E−10	na
M1 vs. M3	1.09E−17	0.00049	7.27E−45	1.23E−08	na	7.02E−10	1.88E−12	3.53E−18	1.12E−05
M1 vs. M4	1.60E−18	6.84E−05	4.61E−52	1.30E−10	na	9.51E−12	6.77E−15	2.36E−19	1.12E−05

phenotypes to model M2 to get model M1 results in significant improvement of explained variance in most of the visual outcomes: FVS, FFS, VFS_{ou}, VFS_{od}, VFS_{os}, VAS_{od} and VAS_{os}. The R^2 values that improve between model M2 to M1 are indicated by ** in Tables 1 and 3. The statistical significance of this improvement is tested using the test of deviance as described in Sect. 2.7. Tables 2 and 4 show the p-values of the tests of deviance performed between M1 and its nested models M2, M3, and M4.

However, it is interesting to note that composite visual acuity scores VAS_{ou} and FAS do not show an improvement between models M2 and M1, even though the right and left acuity scores VAS_{od} and VAS_{os} do. Note from the definition of these scores that they weight the best performing eye higher. This might indicate that changes in visual acuity might not be bilateral in these conditions. Whereas, for visual field scores the behavior of the individual eye scores is reflected in the composite scores, indicating that visual field changes might be bilateral in glaucoma and TED.

4 Discussion

To identify imaging biomarkers associated with diseases of the optic nerve such as glaucoma and thyroid eye disease, their relationship with visual function scores must be established. This study shows that addition of treatment and diagnostic phenotypes derived through MCA on ProWAS and PheWAS data can improve traditional imaging biomarker studies by providing the context of an individual's health history from clinical data. This is the first known study with the application of ProWAS mapping to identify treatment phenotypes for eye disease. We show that structural metrics of the eye orbit derived from CT imaging, treatment, and diagnostic phenotypes show a significant association with visual function scores and explain about 40%–60% of the variance for visual outcomes in glaucoma and thyroid eye disease.

Acknowledgements. This research was supported by NSF CAREER 1452485 and NIH grants 5R21EY024036. This research was conducted with the support from Intramural Research Program, National Institute on Aging, NIH. This study was in part using the resources of the Advanced Computing Center for Research and Education (ACCRE) at Vanderbilt University, Nashville, TN. This project was supported in part by ViSE/VICTR VR3029 and the National Center for Research Resources, Grant UL1 RR024975-01, and is now at the National Center for Advancing Translational Sciences, Grant 2 UL1 TR000445-06.

References

1. Rein, D.B., Zhang, P., Wirth, K.E., Lee, P.P., Hoerger, T.J., McCall, N., Klein, R., Tielsch, J.M., Vijan, S., Saaddine, J.: The economic burden of major adult visual disorders in the United States. Arch. Ophthalmol. **124**, 1754–1760 (2006)
2. Xiuya Yao, S.C., Nabar, K.P., Nelson, K., Plassard, A., Harrigan, R.L., Mawn, L.A., Landman, B.A.: Structural-functional relationships between eye orbital imaging biomarkers and clinical visual assessments. In: Proceedings of the SPIE Medical Imaging Conference
3. Chaganti, S., Nelson, K., Mundy, K., Luo, Y., Harrigan, R.L., Damon, S., Fabbri, D., Mawn, L., Landman, B.: Structural functional associations of the orbit in thyroid eye disease: Kalman filters to track extraocular rectal muscles. In: SPIE Medical Imaging, vol. 97847, p. 97841G. International Society for Optics and Photonics
4. Xierali, I.M., Hsiao, C.-J., Puffer, J.C., Green, L.A., Rinaldo, J.C., Bazemore, A.W., Burke, M.T., Phillips, R.L.: The rise of electronic health record adoption among family physicians. Ann. Fam. Med. **11**, 14–19 (2013)
5. Patel, V., Jamoom, E., Hsiao, C.-J., Furukawa, M.F., Buntin, M.: Variation in electronic health record adoption and readiness for meaningful use: 2008–2011. J. Gen. Intern. Med. **28**, 957–964 (2013)
6. Rondinelli, R.D., Genovese, E., Brigham, C.R.: Guides to the Evaluation of Permanent Impairment. American Medical Association, Chicago (2008)
7. Avants, B.B., Epstein, C.L., Grossman, M., Gee, J.C.: Symmetric diffeomorphic image registration with cross-correlation: evaluating automated labeling of elderly and neurodegenerative brain. Med. Image Anal. **12**, 26–41 (2008)
8. Asman, A.J., Landman, B.A.: Non-local statistical label fusion for multi-atlas segmentation. Med. Image Anal. **17**, 194–208 (2013)
9. Denny, J.C., Bastarache, L., Ritchie, M.D., Carroll, R.J., Zink, R., Mosley, J.D., Field, J.R., Pulley, J.M., Ramirez, A.H., Bowton, E.: Systematic comparison of phenome-wide association study of electronic medical record data and genome-wide association study data. Nat. Biotechnol. **31**, 1102–1111 (2013)
10. https://www.hcup-us.ahrq.gov/toolssoftware/ccs_svcsproc/ccssvcproc.jsp
11. Shlens, J.: A tutorial on principal component analysis. arXiv preprint arXiv:1404.1100 (2014)

12. Abdi, H., Valentin, D.: Multiple correspondence analysis. In: Encyclopedia of Measurement and Statistics, pp. 651–657 (2007)
13. McCullagh, P.: Generalized linear models. Eur. J. Oper. Res. **16**, 285–292 (1984)
14. Draper, N.R., Smith, H., Pownell, E.: Applied Regression Analysis. Wiley, New York (1966)

Erratum to: Fast Predictive Simple Geodesic Regression

Zhipeng Ding[1]([⊠]), Greg Fleishman[3,4], Xiao Yang[1], Paul Thompson[3], Roland Kwitt[5], Marc Niethammer[1,2], and The Alzheimer's Disease Neuroimaging Initiative

[1] Department of Computer Science, UNC Chapel Hill, Chapel Hill, USA
zp-ding@cs.unc.edu
[2] Biomedical Research Imaging Center, UNC Chapel Hill, Chapel Hill, USA
[3] Imaging Genetics Center, USC, Los Angeles, USA
[4] Department of Radiology, University of Pennsylvania, Philadelphia, USA
[5] Department of Computer Science, University of Salzburg, Salzburg, Austria

Erratum to:
Chapter "Fast Predictive Simple Geodesic Regression" in: M.J. Cardoso et al. (Eds.), Deep Learning in Medical Image Analysis and Multimodal Learning for Clinical Decision Support, LNCS 10553, https://doi.org/10.1007/978-3-319-67558-9_31

The attribution of the affiliations to the authors was not correct. By mistake, affiliation 5 (Department of Computer Science, University of Salzburg, Austria) was attributed to all authors. However, only the 5th author, Roland Kwitt, has this affiliation.

The updated online version of this chapter can be found at
https://doi.org/10.1007/978-3-319-67558-9_31

© Springer International Publishing AG 2017
M.J. Cardoso et al. (Eds.): DLMIA/ML-CDS 2017, LNCS 10553, p. E1, 2017.
https://doi.org/10.1007/978-3-319-67558-9_44

Author Index

Printed in the United States
By Bookmasters